MODELING COMPLEX TURBULENT FLOWS

ICASE/LaRC Interdisciplinary Series in Science and Engineering

Volume 7

Modeling Complex Turbulent Flows

Edited by

Manuel D. Salas

Institute for Computer Applications in Science and Engineering,
NASA Langley Research Center,
Hampton, Virginia,
U.S.A.

Jerry N. Hefner

NASA Langley Research Center,
Hampton, Virginia,
U.S.A.

and

Leonidas Sakell

Air Force Office of Scientific Research,
Bolling Air Force Base, DC,
U.S.A.

SPRINGER-SCIENCE+BUSINESS MEDIA, B.V.

A C.I.P. Catalogue record for this book is available from the Library of Congress.

ISBN 978-94-010-5986-2 ISBN 978-94-011-4724-8 (eBook)

DOI 10.1007/978-94-011-4724-8

ACKNOWLEDGEMENT FOR COVER PHOTO:

Cover Photo: "Leonardo da Vinci Nature Studies from the Royal Library at Windsor Castle", Page 43, Plate 29A, catalogue by Carlo Pedretti, 1980. Sheet of studies of water passing obstacles and falling into a pool (1507-9). Reproduced with permission from the Royal Collection Enterprises, Ltd.

Printed on acid-free paper

TABLE OF CONTENTS

Preface ... ix

Current and Future Needs in Turbulence Modeling
Jerry N. Hefner ... 1

Army Turbulence Modeling Needs
Thomas L. Doligalski ... 5

The Best Turbulence Models for Engineers
Peter Bradshaw ... 9

The Modeling of Turbulent Flows with Significant Curvature or Rotation
B.E Launder ... 29

A Perspective on Turbulence Modeling
Stephen B. Pope ... 53

Developments in Structure-based Turbulence Modeling
S.C. Kassinos and W.C. Reynolds ... 69

The Low Dimensional Approach to Turbulence
John L. Lumley and Peter Blossey ... 89

Modeling Non-equilibrium Turbulent Flows
Charles G. Speziale ... 107

Development of Algebraic Reynolds Stress Model for Non-equilibrium Turbulence
Sharath S. Girimaji ... 139

Toward a Vortex Method Simulation of Non-equilibrium Turbulent Flows
Peter S. Bernard 161

Two-point Closures and Statistical Equilibrium
Timothy T. Clark 183

Modeling the Turbulent Wall Flows Subjected to Strong Pressure Variations
K. Hanjalić, I. Hadžić, S. Jakirlić, and B. Basara 203

Some Structural Features of Pressure-driven Three-dimensional Turbulent Boundary Layers from Experiments
Roger L. Simpson, M. Semih Ölcmen, J.L. Fleming, and D.S. Ciochetto 223

Physics and Computations of Flows with Adverse Pressure Gradients
P.G. Huang 245

Computations of Complex Turbulent Flows Using the Commercial Code FLUENT
S.-E. Kim, D. Choudhury, and B. Patel 259

Simulation of Shock Wave-turbulent Boundary Layer Interactions Using the Reynolds-averaged Navier-Stokes Equations
Doyle D. Knight 277

Some Results Relevant to Statistical Closures for Compressible Turbulence
J.R. Ristorcelli 297

Transport Coefficients in Rotating Weakly Compressible Turbulence
Robert Rubinstein, Ye Zhou, and Gordon Erlebacher 329

Development of a Turbulence Model for Flows with Rotation and Curvature
S. Thangam, X. Wang, and Y. Zhou 349

Modeling and Analysis of Turbulent Flows in the Presence of Rotation and Curvature
X. Wang, S. Thangam, Y. Zhou, and A. Zigh 361

List of Attendees 373

PREFACE

Turbulence modeling both addresses a fundamental problem in physics, 'the last great unsolved problem of classical physics,' and has far-reaching importance in the solution of difficult practical problems from aeronautical engineering to dynamic meteorology. However, the growth of supercomputer facilities has recently caused an apparent shift in the focus of turbulence research from modeling to direct numerical simulation (DNS) and large eddy simulation (LES).

This shift in emphasis comes at a time when claims are being made in the world around us that scientific analysis itself will shortly be transformed or replaced by a more powerful 'paradigm' based on massive computations and sophisticated visualization. Although this viewpoint has not lacked articulate and influential advocates, these claims can at best only be judged premature. After all, as one computational researcher lamented, 'the computer only does what I tell it to do, and not what I want it to do.'

In turbulence research, the initial speculation that computational methods would replace not only model-based computations but even experimental measurements, have not come close to fulfillment. It is becoming clear that computational methods and model development are equal partners in turbulence research: DNS and LES remain valuable tools for suggesting and validating models, while turbulence models continue to be the preferred tool for practical computations.

We believed that a symposium which would reaffirm the practical and scientific importance of turbulence modeling was both necessary and timely. This belief led to the ICASE/LaRC/AFOSR Symposium on Modeling Complex Turbulent Flows, organized by the Institute for Computer Applications in Science and Engineering, NASA Langley Research Center, and the Air Force Office of Scientific Research. The symposium was held August 11-13, 1997 at the Radisson Hotel in Hampton, Virginia.

The symposium focused on complex turbulent flows, complexity being understood to indicate the presence of agencies which drive turbulence away from the Kolmogorov steady-state which underlies both elementary mixing length models and the simplest two-equation models. Sound modeling will remain the only practical way to compute such flows for the foreseeable future. The purposes of the symposium were:

– to evaluate recent progress in turbulence modeling
– to anticipate future modeling requirements
– to preview future directions for research.

The choice of particular topics for the symposium relied heavily on the outcome of the two Industry Roundtables co-sponsored by ICASE and LaRC. The symposium topics: compressible turbulence, curved and rotating flows, adverse pressure gradient flows, and nonequilibrium turbulence are all pacing issues in a wide range of industrial applications. For example, the turbulent flow over high-lift devices currently being investigated by Boeing and NASA, exhibits both strong adverse pressure gradients and substantial streamline curvature. Plans for a high-speed civil transport have brought renewed attention to compressible turbulent flows. Finally, aircraft maneuvering and control take place in a time-dependent, nonequilibrium turbulence environment. Lack of time ruled out consideration of the problem of predicting transition, which remains a difficult and crucial problem in aerodynamics. The range and complexity of the transition problem would demand a separate symposium to do it justice.

The editors would like to thank the participants for their contributions to the symposium and cooperation in making the symposium a success, and for their timely submission of the articles in this volume. The contribution of Ms. Emily Todd to organizing the symposium and the editorial assistance of Mrs. Shannon Verstynen are gratefully acknowledged.

CURRENT AND FUTURE NEEDS IN TURBULENCE MODELING

JERRY N. HEFNER
NASA Langley Research Center
Hampton, Virginia

The environment for conducting definitive turbulence modeling research has changed drastically over the past several years. With downsizing and reduced budgets in both industry and government, there is obviously reduced funding available for turbulence modeling research even though better turbulence modeling is still critical to computational fluid dynamics becoming more efficient, accurate, and useful. Within NASA, the funding reductions are compounded by the transition processes resulting from a restructuring and reorganization of the research and technology base program. The NASA R&T base program under the Aeronautics and Space Transportation Technology Enterprise is now outcome oriented and this is clearly evident from the goals of the Enterprise's Three Pillars for Success for Aviation and Space Transportation in the 21st Century. The Three Pillars' goals focus on three areas: global civil aviation, revolutionary technology leaps, and access to space. Under global civil aviation there are goals in aviation safety, environment (emissions and noise), and affordability (capacity and cost). Under revolutionary technology leaps, there are goals in barriers to high-speed travel, general aviation revitalization, and next-generation design tools and experimental aircraft. In the third area, there are goals aimed at reduced payload cost to low earth orbit. These goals are providing the framework and focus for NASA's Aeronautics and Space Transportation R&T program for the future; therefore, fundamental research to develop the needed turbulence modeling will have to be advocated and conducted in a manner to explicitly support these goals. There will be no single funding source for turbulence modeling research; instead, funding support will have to be derived from programs that are being developed to support the Three Pillar outcome goals. Turbulence modeling research of necessity will have to focus on providing tools for addressing issues such as advanced high lift systems, Reynolds number scaling, flow control, noise reduction, wind tunnel data corrections, and reduced design cycle time and costs, and these

M. D. Salas et al. (eds.), Modeling Complex Turbulent Flows, 1–3.

issues will undoubtedly fall under the purview of different Pillar goals. This will require improved coordination and cooperation across the turbulence modeling community and across the Pillar goals.

Where do we stand regarding turbulence modeling research? There have been much resources expended to date on turbulence modeling, and although much progress has been achieved, there are still no turbulence models that are being used consistently throughout industry to provide the accuracy and confidence levels necessary for routine computations of flows about complex aerodynamic configurations. Despite the projections throughout the early 1980's, computational fluid dynamics (CFD) has not replaced the wind tunnel. In fact, CFD, wind tunnel testing, and flight testing currently form the system which provides the aerodynamic data that industry uses to make design and production decisions regarding future aerospace vehicles. Thus, industry would be expected to need turbulence modeling and CFD to provide engineering information from models and algorithms that are accurate and reliable, that have known limits of applicability, that are user friendly, and that are cost-effective to use. Since turbulence modelers, in general, tend to work more closely with turbulence researchers and CFD algorithm developers, there is the tendency to push for as much physics as possible in their models and this may not be the best approach if the ultimate users of turbulence models want good engineering tools. The question then is: how much physics is enough? One way to address what the turbulence modeling customers need is to conduct turbulence modeling users workshops; these workshops would showcase the user community rather than the turbulence modelers. The desired outcome for these workshops would be to identify how industry, government, and academia are applying and using existing turbulence models, what are their needs, where are the successes, and what lessons have been learned. This hopefully would help identify where research needs to focus and how best to maximize the return on investment.

Turbulence modeling research to date suffers from problems other than knowing who the customer is. Much of the turbulence modeling research has focused on modeling the effect of turbulence on mean flows rather than modeling the turbulence physics; therefore, much of the turbulence modeling effort has focused on tweaking or adding constants and terms in the models to predict the available experimental data, which too often is mean flow data and not turbulence data. Although much has been said over the years regarding the need for definitive turbulence modeling experiments, there remains a paucity of high quality dynamic turbulence data useful for modeling and validation for flows about complex geometries. Another problem is that the numerics of the models may be inconsistent and not compatible with the CFD algorithm numerics; since the numerical algorithms for turbulence models place a greater demand on numerics than does the Navier-Stokes equations, both CFD'ers and turbulence modelers need to work together more closely. The bottom line is that to meet the engineering requirements of industry, research in turbulence modeling should be refocused to develop a hierarchy of turbulence models with increasing physics and known applicability and variability, and an increased emphasis must be placed on modeling and verification experiments on geometries and configurations representative of those of practical interest to aerospace designers.

To accomplish what needs to be done to successfully model, predict, and control the flows of interest to industry will require more cooperation and coordination among the turbulence modeling community. It will also require the turbulence modeling community to work within outcome goals like those being proposed by the Aeronautics and Space Transportation Enterprise. This will ensure that the turbulence modeling effort is focused on engineering tools useful to industry.

ARMY TURBULENCE MODELING NEEDS

THOMAS L. DOLIGALSKI
U.S. Army Research Office
Research Triangle Park, North Carolina

1. Introduction

Many Army systems involve turbulent flow. In order to accurately predict the performance of these systems it is necessary to account for the turbulent flow physics. While turbulence modeling techniques such as the Baldwin-Lomax and $k-\varepsilon$ models are adequate for many of these computations, many flows of contemporary Army interest involve non-equilibrium turbulence processes which are not accounted for in these conventional approaches.

2. Conventional Models

One example of such a flow occurs during dynamic stall, which occurs on oscillating helicopter rotorblades. Previous attempts (Dindar & Kaynak, 1992; Dindar *et al.*, 1993; Srinivasan *et al.*, 1993; Ekaterinaris & Menter, 1994; Srinivasan *et al.*, 1995) to assess the efficacy of a variety of turbulence models for the prediction of this flow have demonstrated relatively poor correlation with experimental data for even integrated quantities such as lift, drag and pitching moment (which are less sensitive to turbulence modeling details than local quantities such as wall shear stress). In a recent review article Carr and McCroskey (1992) concluded that: "Turbulence modeling becomes of crucial importance when dynamic stall is considered. This is particularly true when the question of incipient separation and dynamic-stall-vortex development is to be represented by a single turbulence model; under these conditions, the use of a turbulence model based on equilibrium attached boundary layers in steady flow (e.g. eddy viscosity, Baldwin-Lomax) is open to serious question. The task of predicting separation by definition deals with boundary layers that have experienced very strong pressure gradients, often both positive and negative; the flow approaching unsteady separation contains high levels of vorticity induced by these pres-

M. D. Salas et al. (eds.), Modeling Complex Turbulent Flows, 5–7.

sure gradients, and is strongly unsteady. Recent study has shown that modification of the turbulence model can completely change the resultant flow results; at the same time, very little has been experimentally documented about the character of turbulence under these conditions." Similar difficulties occur when attempting to compute the flowfield in the base region of Army missiles and projectiles. Previous studies (Childs & Caruso, 1987; Tucker & Shyy, 1993; Sahu, 1994; Chuang & Chieng, 1994) have met with varying degrees of success in predicting these flows; for example, a recent review by Dutton *et al.* (1995) stated that "all of the turbulence models employed failed to correctly predict the shear layer spreading rate, which is a fundamental characteristic of the near-wake flow."

3. Direct and Large Eddy Simulations

Recent developments in the formulation and application of direct numerical simulation (DNS) and large eddy simulation (LES) have raised the possibility for the use of these techniques for Army applications: in fact, Tourbier and Fasel (1994) have applied these approaches to the computation of supersonic axisymmetric baseflow, albeit at relatively low Reynolds numbers. While DNS and LES are clearly useful tools to understand flow physics, it seems unlikely that these approaches will be routinely used to design and analyze Army systems. Direct numerical simulation requires extremely large computational resources, especially at the high Reynolds numbers typically encountered. Large eddy simulation reduces these requirements to some extent; however, much more sophisticated subgrid scale models will be necessary in order to compute many flows of Army interest (this is a similar closure problem to that encountered when solving the Reynolds-averaged Navier-Stokes equations). Advocates for DNS and LES approaches often argue that as computer power continues to increase the use of these techniques will become more and more routine. While the enhanced power of these computers will undoubtedly allow the application of these methods to a larger number of aerodynamic flows, it seems unlikely to this observer that such application will be routinely performed in the industrial design setting: the current industrial trend is towards more multidisciplinary calculations, where aerodynamics is coupled with combustion, structural dynamics and other disciplines (and perhaps these aeromechanics disciplines are in turn coupled with a design optimizer, which itself runs for many iterations). The net effect of this multidisciplinary trend is to actually reduce the computational resources available for the aerodynamic aspects of the design problem being considered. The problem is further exacerbated by the move by many industries from large supercomputers towards heterogeneous computing environments formed by network workstations.

4. Turbulence Modeling Needs

It seems clear that turbulence modeling has a continued important role to play for many Army aerodynamics applications. In view of the importance of this approach and the availability of new non-equilibrium models generated recently (as documented in this meeting), it is somewhat surprising to see the continued use of turbulence models developed over a decade ago for attached equilibrium boundary layers, even in situations where their use seems clearly inappropriate. This lack of recognition for the advantages offered by alternative approaches appears to stem largely from the lack of consensus within the turbulence modeling community about the efficacy of different models, and the lack of adequate and systematic validation for these models. Almost thirty years ago the first Stanford conference (Kline *et al.*, 1968) was held to achieve these goals with the models then available, and today seems a particularly appropriate time to hold a similar conference to examine the efficacy of today's generation of turbulence models.

References

Carr, L.W. and McCroskey, W.J., 1992. "A Review of Recent Advances in Computational and Experimental Analysis of Dynamic Stall," Proceedings of IUTAM Symposium on Fluid Dynamics of High Angle of Attack, Tokyo, Japan.

Childs, R.E. and Caruso, S.C., 1987. "On the Accuracy of Turbulent Base Flow Predictions," AIAA paper 87-1439.

Chuang, C.C. and Chieng, C.C., 1994. "Supersonic Base Flow Computations by Higher Order Turbulence Models," Proceedings of the 1994 International Symposium on Turbulence, Heat and Mass Transfer, Lisbon, Portugal.

Dindar, M. and Kaynak, U., 1992. "Effect of Turbulence Modeling on Dynamic Stall of a NACA0012 Airfoil," AIAA paper 92-0027.

Dindar, M., Kaynak, U., and Fuji, K., 1993. "Nonequilibrium Turbulence Modeling Study of Light Dynamic Stall of a NACA0012 Airfoil," AIAA Journal of Aircraft **30**(3), pp. 304-308.

Dutton, J.C., Herrin, J.L., Molezzi, M.J., Mathur, T., and Smith, K.M., 1995. "Recent Progress on High- Speed Separated Base Flows," AIAA paper 95-0472.

Ekaterinaris, J.A. and Menter, F.R., 1994. "Computation of Separated and Unsteady Flows with One- and Two-Equation Turbulence Models," AIAA paper 94-0190.

Kline, S.J., Morkovin, M.V., Sovran, G., and Cockrell, D.J., 1968. Proceedings of Turbulent Boundary Layers - 1968 AFOSR-IFP-Stanford Conference.

Sahu, J., 1994. "Numerical Computations of Supersonic Base Flow with Special Emphasis on Turbulence Modeling," Technical Report ARL-TR-438, Army Research Laboratory, Aberdeen Proving Ground, Maryland.

Srinivasan, G.R., Ekaterinaris, J.A., and McCroskey, W.J., 1995. "Evaluation of Turbulence Models for Unsteady Flows of an Oscillating Airfoil," *Computers in Fluids* **24**(7), pp. 833-861.

Srinivasan, G.R., Ekaterinaris, J.A., and McCroskey, W.J., 1993. "Dynamic Stall of an Oscillating Wing Part 1: Evaluation of Turbulence Models," AIAA paper 93-3403.

Tourbier, D. and Fasel, H.F., 1994. "Numerical Investigation of Transitional Axisymmetric Wakes at Supersonic Speeds," AIAA paper 94-2286.

Tucker, P.K. and Shyy, W., 1993. "A Numerical Analysis of Supersonic Flow Over an Axisymmetric Afterbody," AIAA paper 93-2347.

THE BEST TURBULENCE MODELS FOR ENGINEERS

PETER BRADSHAW
Stanford University
Mechanical Engineering Department
Stanford, California

Abstract. *"The best turbulence model"* is obviously the Navier-Stokes equations, which are almost universally accepted as an accurate description of turbulence in simple fluids. Solving them to obtain converged statistics (DNS) requires computer work proportional to the cube of the Reynolds number, approximately, so there will *always* be a limit to the attainable Reynolds number. At present this limit is of the order of the Reynolds number attained in small-scale laboratory experiments. Thus *"The best turbulence model for* aeronautical *engineers"* is – again obviously – not the Navier-Stokes (NS) equations. (DNS is and always will be a valuable research tool for developing and testing models based on simplifications of the NS equations: see Moin and Mahesh (1998).) The less-obvious point of this Abstract is that "best", in the title, is a time-dependent word and that it is short for "most cost-effective". This paper is a review of options, a brief history, and an assessment of current ideas and future possibilities. It seems likely that Reynolds-averaged models will continue to advance modestly, somewhat less likely that engineers will routinely use the highest-level models, and almost certain that large-eddy simulation will start to replace or assist Reynolds-averaged models in the next decade or so.

1. Introduction

Simplification of the NS equations usually involves Reynolds averaging. Applying it to the complete spectrum – the normal meaning of the phrase, used below – gives the RANS equations. In large-eddy simulation (LES), it is applied to the small "sub-grid-scale" (SGS) eddies only, by filtering in **x**-space or κ-space. Reynolds averaging is a brutal simplification: $O(10^{10})$ variables are needed to describe a DNS run to convergence at, say, given

M. D. Salas et al. (eds.), Modeling Complex Turbulent Flows, 9–28.

x in a 2D boundary layer, while a Reynolds-averaged calculation produces typically 50-point profiles of perhaps half a dozen variables. LES filtering is less severe but typically only a tiny fraction of the full (DNS) information remains. For recent reviews of LES, see Ferziger (1996) and Mason (1996).

Therefore, Reynolds-averaged turbulence modeling involves throwing away nearly all the information in the instantaneous NS equations and replacing it by empirical correlations. The consequence is that Reynolds-averaged models (and SGS models) often perform poorly in flows very different from those used to calibrate them. This does not rule out flow-to-flow changes, e.g. the zonal modeling suggested by Kline (Ferziger *et al.*, 1990), if controlled by pattern-recognition routines within the code. Nominally, the goal is a "universal" Reynolds-averaged model, giving acceptable engineering accuracy in all flows of engineering interest. It is not logically possible to prove that there can never be a universal model but certainly no current model is near the goal. This is a very important point even in the *development* of turbulence models: if empirical coefficients are chosen by using data from a range of different flow types, the implied assumption is that the final model will be valid in all such flow types.

Perhaps the most important defect of current engineering turbulence models – referred to repeatedly at this meeting – is that their *non-universality* (the boundaries of their range of acceptable engineering accuracy), cannot be estimated at all usefully a priori. Some models' general faults are known from experience, but, even so, very few codes output warning messages when the model is leaving its region of proven reliability.

LES, on the other hand, almost certainly does become substantially "universal" at large turbulence Reynolds number, such that the filter cut-off wave number can be chosen well above the energy-containing range and at least slightly below the viscous range. Then, the SGS model merely has to dissipate the energy cascaded from the larger eddies, and the details (e.g. contributions of the SGS motion to the Reynolds stresses, or accurate representation of backscatter) are unimportant. This is just another version of the principle of Reynolds-number independence. It is not quite the same thing as Kolmogorov's principle of universal small-scale structure at high local Reynolds numbers, but the latter also implies that SGS model coefficients should not be strongly flow-dependent at high local *Re*. Alas, any simulation of a wall-bounded flow must work at low turbulence *Re*(defined, say, as $k^2/(\epsilon\nu)$ which is about $4.5u_\tau y/\nu$ in the log law layer). Here, the Reynolds-stress-bearing range of the spectrum extends into the high-wave-number viscous-dependent range – so most of the stresses must be carried by the SGS model, or the LES just degenerates into a DNS, with the accompanying restriction to low bulk *Re*. This is the "pacing item" in LES.

As in RANS models, the alternative to integrating to the wall is to match the calculation to a "wall function" somewhere outside the viscous wall region so that the grid spacing and computer work do not depend on bulk Reynolds number. Unfortunately the simplest choice for LES, using the steady-state log law as an instantaneous boundary condition, is not reliable: Cabot (1996) obtained adequate results in simple flows but poor results in a backstep flow. Other possibilities will be discussed in Section 8, but it is clear that a real advance is needed before LES becomes a tool in high-*Re*fluids engineering. RANS is all we can afford at present.

This paper begins with a brief history of Reynolds-averaged turbulence modeling, followed by a section each on the two main types of current models – transport equations for eddy viscosity, and term-by-term modeling of the exact stress-transport equations – and their detailed difficulties, whose solution is the immediate goal of modeling. Most current models use a transport equation for dissipation: its insecure foundations, and a possible alternative, are discussed in Section 5. Section 6 deals with another weak point of nearly all models, the use of gradient diffusivity to model turbulent transport of stresses $-\overline{u_i u_j}$, energy $\overline{u_i^2}/2 \equiv k$ and dissipation ϵ. Section 7 discusses the restrictions/guides to modeling that arise from special cases, such as limiting behaviour at a solid surface: if these are not really important as restrictions, they should not be taken too seriously as guides. Section 8 is a guess at longer-term progress, in eddy-viscosity models (which may not have much mileage), stress-transport models (whose firmer foundations have not yet had much appeal in industry), and in large-eddy simulation (which seems to be the long-term hope, probably still in conjunction with cheaper Reynolds-averaged models for the less critical regions of a flow).

2. Reynolds-Averaged Models – History

We have already conceded that the best turbulence models for engineers, right now, are Reynolds-averaged models. This has been true since modeling started, but the complexity of the models actually used by engineers has increased greatly over the last 40 years. However a plateau seems to have persisted for some years past (in fact two plateaux – the persistence of the 1972-vintage k, ϵ model in industrial use, and the persistence of the framework of the 1975-vintage Launder-Reece-Rodi stress-transport model in model development). Weighting the NS equations before Reynolds averaging gives exact transport equations for the *rates of change* of the Reynolds stresses along a mean streamline, necessarily with further unknowns on the right-hand side. *All* turbulence models are approximations, not necessarily term by term, of these equations. The dimensions of the terms are [velocity]^2/[time] or [velocity]^3/[length], so modeling involves equations for time scale, length scale or an equivalent, to accompany velocity scales derived from the Reynolds stresses themselves. Stress-transport models and other PDE-based models were not seriously developed until the late 1960s: earlier models had to be simple enough to run on, at best, an electromechanical desk calculator.

The method of Head (1958) was by far the best of the so-called *integral methods*, the first being that of Gruschwitz (1931: see Schlichting (1979)). These solved ordinary differential equations (in 2D), namely the momentum integral equation for momentum thickness θ and an empirical equation for a shape parameter (usually $\delta^*/\theta \equiv H$) with an algebraic relation for skin-friction coefficient as a function of the shape parameter and the momentum-thickness Reynolds number. All the other integral methods effectively died at the 1968 Stanford "Olympics" meeting (Kline *et al.*, 1969), but Head's continued in engineering use for some years. The reason for its success is that it was based on a straightforward (though crude) physical assumption (that the dimensionless entrainment rate was a function of H only) rather than on pure data correlation like most of its predecessors.

In fact, Head's method used a special case of the assumption of direct connection between turbulence properties and mean-flow properties at a given point in space (or, in the case of 2D integral methods, at a given x position). Another special case is the *eddy-viscosity assumption.* This can be rigorously justified from the exact stress-transport equations only in one case: this is local (energy) equilibrium, where the mean and turbulent transport terms in the stress-transport equations are negligible and the remaining terms express a link between the turbulence quantities and the local mean velocity gradients. (Note that the word "equilibrium" is sometimes used in other senses, such as constant anisotropy of the Reynolds stresses.) The pragmatic justification for eddy viscosity is simply that the transport terms are usually fairly small, and the link between the Reynolds stresses and the mean strain rates is close enough that eddy viscosity often varies more slowly, or more simply, than the stresses themselves. One can always *define*, and indeed measure, an eddy viscosity as the ratio of a stress to a strain rate – the question is whether it is easier to find reliable model equations for eddy viscosity than for the stresses directly, bearing in mind that the eddy viscosity can be different for different components of the stress tensor.

Head's method was superseded, after electronic digital computers came into general engineering use in the late 1960s, by the "algebraic eddy viscosity" methods of Mellor and collaborators and of Cebeci and Smith (for these and other models not specifically referenced, see Wilcox (1998) (note also that throughout the present paper *recent*, rather than primary or exhaustive, references to advances in modeling are given). Here the eddy viscosity for the x, y-plane shear stress was related to mean velocity and length scales – in the outer layer of a boundary layer, for example, the eddy viscosity was assumed proportional to $U_e\delta^*$, while in the inner layer of any wall flow the eddy viscosity was chosen to yield mixing length $= \kappa y$. With careful choice of mean-flow scales, these models performed well in thin shear layers where the mean-flow scales were well-behaved. (The widely-used Baldwin-Lomax method is a fix-up of Cebeci-Smith to produce an answer in flows where the displacement thickness δ^* is not well defined, e.g. in boundary layers on curved surfaces where $\partial U/\partial y = -U/R$ in the external stream. It also runs without numerical breakdown in highly complex flows where *any* algebraic eddy-viscosity relation goes wildly wrong, and alas it is popular in some circles for this very reason.)

After algebraic eddy-viscosity models came a series of models in which the eddy viscosity was related to velocity and length scales of the *turbulence*, for which model transport equations were solved: the most popular eddy-viscosity-transport model, originally due to Jones and Launder in 1972, takes the eddy viscosity proportional to k^2/ϵ, where k and ϵ are the turbulent kinetic energy and its rate of dissipation, for which semi-empirical transport equations are solved.

The k, ϵ model has remained as the industry standard for longer than any other model: the more responsible users recognize, and work around, its defects (e.g. generally *poorer* performance than algebraic eddy-viscosity models in boundary layers in pressure gradients). However it seems to be an evolutionary dead end, necessarily trapped in the eddy-viscosity framework and therefore incapable of real improvement, although we shall see that k and ϵ are not the best choice for the variables of a "two-equation" method. Apparently nothing short of an asteroid impact will suffice to exterminate k, ϵ and the other eddy-viscosity models, so we need to examine their present state and what mileage is left in them. Stress-transport models, not yet widely used in industry, are discussed in Section 4.

3. Eddy Viscosity – Generalities

Using eddy viscosity is like keeping one foot on the bottom when learning to swim – a restriction, but helpful if one does not venture into deep water.

Eddy viscosity is the ratio of a turbulence quantity (Reynolds stress) to a mean-flow quantity (rate of strain) and it is no more plausible a priori to scale it on turbulence quantities only (as in the k, ϵ model) than on mean-flow quantities only. The state of local energy equilibrium, required to justify eddy viscosity rigorously, implies that the length and velocity scales of the mean flow and of the turbulence are proportional and *either* set can be used to scale the eddy viscosity. However, choosing turbulence scales, which can be legitimately defined by transport equations, goes some way towards allowing for departures from local equilibrium – what used to be called the "history effect" in the days of Head-type integral methods.

The k, ϵ model is the most popular of the two-equation group of eddy-viscosity-transport models, but various others have been developed. Nearly all are members of the family $k, k^m \epsilon^n$, but there is a real difference between them. One can always transform a given member of the family back to k, ϵ (say). However, all common models use gradient-transport assumptions in the terms expressing diffusion (turbulent transport) of the two variables; for dimensional correctness, the diffusivity of k and of the second variable in any model is taken proportional to the equivalent of k^2/ϵ. The diffusional flux is the product of this diffusivity and the gradient of the diffused quantity, and itself appears inside a derivative. Thus, when transformed back to k, ϵ variables, the equations contain squares and products of the derivatives of k and ϵ. Unlike the diffusion terms in the original equations, these transformed derivative-product terms do not integrate to zero over the flow volume. They therefore represent source terms which are not present in (say) the ordinary k, ϵ model. This difference between different members of the $k, k^m \epsilon^n$ family cannot be reduced to zero by adjustment of coefficients (short of ignoring diffusion altogether). Thus, as pointed out by Spalding (1991) there must be an optimum choice, according to some set of criteria, for the second variable. The k, ω model corresponds to $m = -1, n = 1$, i.e. $\omega \propto \epsilon/k$: it happens to give the right intercept for the law of the wall for velocity without any explicit viscous dependence but, more importantly, it is considerably better than k, ϵ in predicting boundary-layer flows and (Heyerichs & Pollard, 1996) heat transfer, and it reproduces the well-verified Van Driest transformation of the log law for compressible flow rather accurately (Huang *et al.*, 1994), which other two-equation models do not. There is no obvious physical reason for these successes of k, ω; a full optimization study might suggest non-integer values of m and n! However, k, ω seems to deserve more attention by engineers; any k, ϵ code could be easily converted.

Durbin and colleagues (e.g. Durbin (1995)) have developed a three-equation eddy-viscosity model. In the case of a simple shear flow, the third variable reduces to the Reynolds stress normal to the plane of the layer, but in more general cases it becomes a scale without explicit physical meaning.

3.1. FREE-STREAM BOUNDARY CONDITIONS

A general difficulty with free-stream edges was discussed by Cazalbou *et al.* (1994): they considered only two-equation eddy-viscosity methods, but similar effects are likely in stress-transport methods because the near-edge modeling of the dissipation and Reynolds stresses is nearly the same. Experiments and simulations show that near a free-stream edge the turbulent kinetic energy equation reduces to (advection = diffusion), *i.e.*, (mean transport = turbulent transport) or "pure transport" – production and dissipation being of smaller order – and model equations should reproduce this. (Typical model equations for the second variable, e.g. ϵ, are patterned so closely on the TKE equation that, rightly or wrongly, they behave in the same way.) The x-component momentum equation and the k and ϵ equations reduce to quasilinear diffusion equations, with diffusivities proportional to k^2/ϵ or equivalent: this means trouble, because the diffusivities go to zero at the edge, which is therefore at a finite value of y so that discontinuous derivatives can occur instead of smooth asymptotes.

Consider the general $k, k^m\epsilon^n$ model, with $k^m\epsilon^n \equiv \phi$ and with σ_k and σ_ϕ as the ratios of the diffusivity of momentum (eddy viscosity) to the diffusivities of k and ϕ respectively. Treating the (t, y) problem rather than the (x, y) problem, for simplicity, Cousteix *et al.* (1997) give the solution for an edge propagating in the y direction at velocity c (a "front"). Here k, ϕ, and the velocity defect vary as (different) powers of the distance inward from the edge, $ct - y$, for $ct - y > 0$, where $y = 0$ is the front position at $t = 0$. The eddy viscosity is linearly proportional to $ct - y$.

The condition for the model equations to reduce to pure transport near the edge is just $\sigma_k < 2$, and most models take $\sigma_k = 1$ – except the original k, ω model (Wilcox, 1998) which has $\sigma_k = \sigma_\omega = 2$ exactly and violates the condition. The condition for the exponent of $\partial U/\partial y$ to be positive (giving zero velocity gradient at the edge) is

$$L \equiv \sigma_k(2 + m/n) - \sigma_\phi/n < 1 \tag{1}$$

and the exponents of k and ϕ are positive (giving zero values at the edge) if $L > 0$ so the combined condition is simply $0 < L < 1$. The original k, ω model gives $L = \sigma_k - \sigma_\phi$ which is exactly zero and therefore k, ω again violates the condition. These shortcomings of k, ω merely lead to slightly worse near-edge velocity profiles than other models give, and could be cured by small adjustments. Note that this singular behaviour of k, ω has nothing to do with the choice $m = -1, n = 1$ as such. Also it has nothing to do with the sensitivity of the original k, ω model to free-stream values of ω: Wilcox (private communication) points out that the specific difficulty, shared with most other two-equation models, was that if coefficients were chosen to fit one of the canonical free shear layers (round jet, plane jet and plane mixing layer) poor results were obtained for the others. In order to get good results in all the free shear layers the free-stream values had (obviously) to be chosen in ranges where the answers were still sensitive to these choices, and this could not be done consistently.

3.2. CHOICES FOR THE SECOND VARIABLE

The only choices in widespread use appear to be the k, ϵ and k, ω models. Cousteix *et al.* (1997) used the second variable $\epsilon/\sqrt{k}$, i.e. $m = -1/2, n = 1$, so Eq. (1) is $0 < 3\sigma_k/2 - \sigma_\phi < 1$: they chose $\sigma_k = 1.5$ and $\sigma_\phi = 2$. Aupoix (private communication) states that this model gives better results than k, ω in the outer region, as one would expect from Section 3.1. Gibson and Harper (1997) also used $\epsilon/\sqrt{k}$, with $\sqrt{k}$ as the first variable: the above analysis does not apply to this case because $\sigma_{/sqrtk}$ cannot be simply related to σ_q.

Of the other choices of $k^m \epsilon^n$, those with $n < 0$, notably the time scale k/ϵ and the length scale $k^{3/2}/\epsilon$, have the operational disadvantage that the usual propagation of information away from a solid surface implies negative gradient diffusivity (according to simple law-of-the-wall analysis, k is independent of y in the log law layer and $\epsilon \propto 1/y$). Some at least go to infinity in a free stream (e.g. k/ϵ). There is no objection to formulating a model with one choice of m and n and then transforming it to another choice, without further approximation, for numerical convenience: Kalitzin *et al.* (1996) transformed ω into $\sqrt{1/\omega}$. The source term that arises in such transformations is fully nonlinear and may lead to numerical difficulties.

3.3. ONE-EQUATION MODELS

Several "one-equation" models have been developed, though they are as yet less widely used than two-equation models. In 1965, Glushko followed Prandtl in solving a model TKE equation to generate an eddy viscosity $k^{1/2}/l$, where l/δ was specified as a function of y/δ. In 1967 Bradshaw, Ferriss and Atwell used Townsend's (1961) anisotropy parameter $a_1 \equiv -\overline{uv}/(2k)$ to convert the TKE equation into a shear-stress transport equation, avoiding the use of eddy viscosity but again with $l/\delta = f(y/\delta)$: a_1 was treated as an empirical function but only a constant value was ever used. The more recent model of Johnson and King can be regarded as a simplified version of this model, using an eddy viscosity to give the shear-stress profile shape but an ODE for $-\overline{uv}_{max}$ to specify the shear-stress *level.* These models all use an algebraic correlation for length scale and are therefore restricted to shear layers with a well-behaved thickness – with no guarantee that the same correlation will work for different types of shear layer. However the Johnson-King model has been used successfully in the initial region of separated flows, especially shock/boundary-layer interactions.

Nee and Kovasznay, Secundov (see Shur *et al.* (1995)), Baldwin and Barth and most recently Spalart and Allmaras (S-A), have proposed transport equations for eddy viscosity as such, length-scale (or time-scale) information being implied by the presence of the mean velocity gradients in the source terms of the transport equation. Baldwin-Barth was formally derived as a simplification of the k, ϵ model, and seems not to be very satisfactory: the others are empirical. S-A uses the mean vorticity, rather than the mean strain rate, in the eddy-viscosity formula, and this minimizes unphysically rapid response of the Reynolds stresses in regions of rapid deformation where the strain rate changes rapidly but vorticity does not, at least in incompressible 2D flow. This use of vorticity is of course itself unphysical but at no deeper level of wrongdoing than the use of eddy viscosity itself – the near-equilibrium flows in which eddy viscosity is a justifiable simplification are mostly simple shear flows, where mean strain rate and mean vorticity are numerically nearly equal. The S-A transport equation was assembled term by term, by considering flows of increasing complexity. This straightforward approach seems to have been very successful – as was Head's, 40 years ago – and in at least some cases S-A is superior to k, ϵ.

4. Stress-Transport Models – Generalities

The most complicated models currently considered by engineers are "stress-transport" or "second-moment" models in which the above-mentioned exact transport equations for some or all of the Reynolds stresses are modeled term by term. The turbulent kinetic energy or some other combination of Reynolds stresses again provides a velocity scale, and a length scale (or equivalently a time scale) is usually deduced from a model dissipation transport equation minimally different from that used in two-equation models: the ω equation has also been used, but according to Launder (private communication, 1996) it does not seem to be a consistent improvement.

It is so obvious that stress-transport models are more realistic in principle than eddy viscosity models that the modest improvements in performance they give are very disappointing – and most engineers have decided that the increased numerical difficulties (complexity of programming, expense of calculation, occasional instability) do not warrant changing up from eddy-viscosity models at present. Even stress-transport models often give very poor predictions of complex flows – notoriously, the effects of streamline curvature are not naturally reproduced, and empirical fixes for this have not been very reliable.

The weak physical points of stress-transport models are the pressure-strain terms in the stress-transport equations themselves, and all aspects of the dissipation equation. The weak numerical points are the large number of equations as such, and the greater stiffness. The latter arises because eddy-viscosity models lead to well-behaved quasi-laminar mean-flow equations with the stresses expressed in terms of mean-velocity gradients, while the stresses derived explicitly from a Reynolds-stress model count as source terms, not diffusive terms, in the mean-flow equations.

4.1. THE PRESSURE-STRAIN "REDISTRIBUTION" TERM

The conventional division of the pressure-strain term into a "fast" part containing the mean strain rates and a "slow" part containing only turbulence quantities is just a consequence of Reynolds averaging and does not imply a two-part physical process: one tends to forget that individual eddies do not see the Reynolds-averaged flow. In turn, modellers usually forget that the NS (Poisson) solution for the pressure is an integral over the whole flow volume, so that parameterizing the pressure-strain term model with local quantities is a crude approximation. Bradshaw *et al.* (1987) showed that approximating the integral for the "fast" term by putting the mean strain rate outside the integral, instead of inside, gave results within roughly 20% of those obtained by integrating DNS data in a channel flow – except in the viscous wall region where very large and erratic discrepancies appeared. This strongly suggests that difficulties in modeling the viscous wall region are not attributable solely to viscosity and the no-slip boundary condition as such, but are related to the no-permeability condition (acting via the pressure fluctuations) and the strong inhomogeneity of the flow.

Durbin (1993) developed a semi-empirical elliptic PDE for the pressure-strain term, specifically to allow for inhomogeneity and avoid near-wall "damping functions". In thin shear layers where mean x- and z-derivatives are small, it reduces to an ODE in y, and solution costs are small. Bradshaw and Strigberger (1995) used an informal parametrization of the Poisson equation to obtain an integral equation (in y) for mean-square surface pressure fluctuation. These two approaches could possibly be merged.

Another addition to pressure-strain modeling is the use of terms which are nonlinear in the Reynolds stresses. In the "fast" term, the constraints of invariance, and of proper limiting behaviour at a solid surface and in the case of rapid distortion, limit the number of nonlinear terms and provide links between their coefficients. The conventional linear form of the "slow" term, due to Rotta in 1951, implies that the tendency to isotropy is governed by a simple time constant, which is not likely to be an accurate model of turbulence. To date, the only alternatives suggested are empirical powers of anisotropy. Clearly, nonlinear models are more likely to exhibit near-singular behaviour in unusual cases but by the same token may be better able to represent the curious behaviour of turbulence.

4.2. ALGEBRAIC STRESS MODELS (ASM)

These are an attempt to preserve the best features of stress-transport models in an eddy-viscosity format (see the paper by Speziale in this volume). Typically, the mean and turbulent transport terms in the transport equation for a given Reynolds stress are assumed proportional to that Reynolds stress, so that all these terms can be deduced from the modeled TKE equation. This can easily be seen to fail for the mean transport terms, because $D\overline{u_i u_j}/Dt = (\overline{u_i u_l}\partial U_j/\partial x_l + \overline{u_i u_l}\partial U_j/\partial x_l) + \ldots$ so the change in mean transport due to a change in mean strain rate is not proportional to $\overline{u_i u_j}$ in general. (The ASM assumption implies constant anisotropy of the stress tensor.) Fu *et al.* (1988) found that the ASM assumption for turbulent transport was also unsatisfactory in free shear layers (which grow more rapidly than wall flows and thus have larger rates of turbulent transport near free-stream edges). Craft *et al.* (1993) point out an additional, numerical, disadvantage in rapidly-changing flows needing the full RANS equations: the ASM assumptions lead to a very stiff system. ASM produces an anisotropic effective eddy viscosity and can thus simulate stress-induced secondary flows, e.g. in noncircular ducts and jets, which isotropic eddy-viscosity models cannot. However stress-induced secondary flows are less common than *skew*-induced secondary flows such as vortices in curved ducts, and it is not clear that ASM is significantly better in predicting the decay of the latter. It seems unlikely that Algebraic Stress Models based on the above assumptions will replace two-equation models in industry, but perhaps more realistic assumptions about transport terms could be fitted into the ASM framework.

5. Dissipation Transport Equations

Most two-equation and stress-transport models use either the dissipation, or a closely-related quantity like $\omega \propto \epsilon/k$, to provide a length or time scale. Here we discuss the dissipation equation alone, for clarity and because the right route to a model transport equation for another quantity is probably through consideration of the dissipation.

Typical modeled dissipation transport equations are not recognizable term-by-term models of the exact equation. The variable in the exact equation is the viscous dissipation rate, $\epsilon = 2\nu\overline{s_{ij}s_{ij}}$, and the terms on the right-hand side represent processes in the viscous, dissipating eddies. Model equations usually resemble the turbulent energy equation, with the variable changed from k to ϵ and an empirical coefficient in front of each dimensionally-correct term.

The dissipation rate is actually set by the larger eddies, which carry the Reynolds stresses and especially the turbulent energy, and for which we want a length or time scale. Therefore we should consider, not the dissipation, but the rate of energy transfer from the larger eddies up the wave-number spectrum to the dissipating eddies. This rate is nominally equal to ϵ but is governed by a very different equation, essentially independent of viscosity. Note that the rate of energy transfer is finite but the cascade process of transfer from the top of the energy-containing range of wave number κ to the much higher dissipating range is in principle instantaneous – the energy spectral density in the intervening inertial subrange is small by definition so that energy can only pass through and not reside there. Therefore there is no extra time scale associated with the cascade process. The same argument shows that the energy-transfer rate is equal to the dissipation rate appearing in the turbulent energy equation at the same point in Reynolds-averaged space. This energy transfer rate from the top of the energy-containing range, at given time, does indeed depend on processes over a preceding interval of the order of one energy-containing eddy turnover time, say k/ϵ – but the same applies to almost any other property of the energy-containing range that could be used to give a length or time scale. The transport equation for energy-transfer rate is an integral, over the energy-containing range of κ, of the Fourier transform of the transport equation for the spatial derivatives of certain two-point triple correlations. It is of course algebraically intractable but groups of terms are recognizable as sources, sinks and transport, and could be evaluated numerically from DNS results.

6. Turbulent Diffusion Modelling

It is of course inconsistent to abandon eddy viscosity (turbulent diffusivity of momentum) in favor of transport equations for the Reynolds stresses, but still use eddy diffusivity to model turbulent transport ("diffusion") of Reynolds stress and of dissipation rate. The hope is that the turbulent transport terms are small enough for gradient diffusion to be a good enough approximation. This seems to be the case in most shear layers: diffusion is only a small part of the Reynolds-stress budgets except near free-stream edges (where the unphysical consequences of gradient diffusion have been discussed in Section 3.1). In most current models, diffusion of dissipation (or other length-scale quantity) is the main mechanism by which the presence of a solid wall constrains eddy size, but fortunately the diffusivity coefficient in the model dissipation equation is determined, in the inner layer, by the need to reproduce the right Karman constant, and the same value of the coefficient is commonly used throughout the flow for want of anything better. In compressible flows, modeled diffusion terms typically look like

$$\frac{Dk}{Dt} = \ldots + \frac{1}{\rho}\frac{\partial}{\partial y}\left(\frac{c_\mu \rho k^2}{\sigma_\epsilon \epsilon}\frac{\partial k}{\partial y}\right) + \ldots \tag{2}$$

Now in the inner layer, where $\tau \approx \tau_w$ implying $\rho k \approx$ constant, not only does $\partial\rho/\partial y$ appear explicitly but it implies $\partial k/\partial y \neq 0$, so the modeled diffusion is not negligible as it is in constant-density flow – but it is by no means obvious that real diffusion processes are significantly altered by density gradients! As mentioned above only the k, ω model satisfactorily, though coincidentally, reproduces the Van Driest transformed log law (which corresponds to local equilibrium – *i.e.*, neglect of k diffusion, with $\epsilon \propto 1/y$).

7. Realizability and Limiting Cases

Schumann (1977) first discussed the need to avoid unphysical values of variables (negative normal stresses or dissipation, correlation coefficients numerically greater than unity, etc.). Most current models give "realizable" results in this sense: for a review see Shih (1997). The problem is not too severe in practice, since unrealizable results usually occur at free-stream edges where variables are supposed to tend to zero from above – so numerical clips in the code would remove spurious negative values without significant effect on predictions. However realizability is widely regarded as necessary on principle, like rotational and translational invariance. This is a fundamental question in modeling, whether of turbulence or of anything else: a model is approximate (by definition), so to what extent should we force it to have the *mathematical* properties of the exact equations?

7.1. THE SURFACE BOUNDARY CONDITION

7.1.1. *The Limit $y \to 0$*

Models intended for integration to the wall should in principle be able to reproduce the limiting forms of the different variables at $y = 0$, notably $\overline{u^2} \propto y^2$, $\overline{w^2} \propto y^2$, $\overline{v^2} \propto y^4$, $\overline{uv} \propto y^3$, and $\epsilon \approx$ constant. However, acceptable prediction of the mean velocity field in attached wall flows merely means generating the right additive constant in the logarithmic law, details of the Reynolds shear stress profile in the viscous wall region being unimportant. (Correct prediction of heat transfer over a range of Prandtl number is more demanding.) Forcing the model to reproduce the "two-component" limit ($v \ll u, w$) and calibrating it for the right log law constant is probably more realistic than making model coefficients functions of local Reynolds number. However, it is by no means certain that building in the proper limiting behaviour as $y \to 0$ improves the accuracy of predictions at larger y, or even the near-wall predictions in separating or reattaching flows.

It is now clear that the coefficients of proportionality in the above-mentioned limiting forms are strongly flow-dependent because of pressure fluctuations induced by the outer flow (equivalently, "blocking" or "splat effect"). Therefore accuracy in detail is not possible with any local model, although better might be expected from a non-local model for pressure terms (e.g. Durbin (1993)).

7.1.2. *The Law of the Wall*

Even if the law of the wall is not used as a "boundary" condition somewhere outside the viscous wall region, most models are calibrated to reproduce the logarithmic law for a constant-stress layer, or the mixing-length formula with $l = \kappa y$ for the more general case where shear stress varies with y. This is the strength and weakness of most models, not only the algebraic eddy viscosity models which enforce $l = \kappa y$ explicitly. In slowly-changing boundary layer or duct flows the law of the wall is valid for $y/\delta \leq 0.2$ approximately so the velocity at the outer edge of the wall-law region is 0.6-0.7 of free stream – and quite large errors in predicting the velocity profile in the outer part of the flow will have little effect on bulk quantities like skin friction coefficient. In rapidly-changing flows, particularly boundary layers in strong adverse pressure gradient, the region of applicability of the law of the wall will shrink and the outer-layer calculation becomes more critical.

Both the mixing-length formula and the log law analysis neglect streamwise changes, and it seems to be good luck that the log law works better than the mixing-length formula in boundary layers in pressure gradient. The derivations of the law of the wall for velocity and for temperature are closely analogous, but the law of the wall for temperature behaves very badly in flows in pressure gradient, which is not only the corresponding *bad* luck but also a warning against excessive trust in the law of the wall for velocity! (See Bradshaw and Huang (1995).)

Alternatives to the logarithmic law have been proposed by George and Castillo (1993) and Barenblatt *et al.* (1997). These seem to deal with higher level changes in flow structure than are normally considered in turbulence modeling, but it is certainly true that the standard law-of-the-wall assumptions – e.g. complete independence of shear-layer thickness and streamwise gradients – are very crude. Zagarola *et al.* (1997) discuss the position and present recent high-Reynolds-number data, and Oberlack (1997) reviews mathematically-possible forms of similarity analysis, based on Lie group theory.

7.2. RAPID DISTORTION

Another limiting case is rapid distortion (where generation terms resulting from large strain rates are large compared to other terms on the right-hand sides of the stress-transport equations). Eddy-viscosity methods, including ASM, cannot handle this limit. In the case of stress-transport models, analytical results for rapid distortion impose a constraint on at least one of the coefficients in the pressure-strain model, and many modellers have used this for calibration. However it is not clear that requiring the correct behaviour as the ratio of TKE production to dissipation tends to infinity will improve results in real flows, where the ratio rarely exceeds two.

7.3. THERE IS NO UNIVERSAL MODEL

Constraints of realizability, invariance, limiting behaviour etc., are very powerful in reducing the number of free empirical coefficients in a turbulence model *if it is assumed that the coefficients are the same in all parts of all flows* – i.e. if the model is assumed to be universal. However it is quite clear that no model is universal, giving good results for all flows of interest – so why should a model intended for shear layers exactly satisfy the predictions of rapid-distortion theory, and why should the limiting behaviour at a wall be used to fix the coefficients in a model whose predictions are important only in the body of the flow? Over-use of limiting cases encourages the development of over-complex models with insufficient empirical input from real flows. Some time ago, Aristotle remarked "It is the mark of an educated man to look for precision in each class of things just so far as the nature of the subject admits" – an imprecise statement itself, but a useful warning. Using the known physics of turbulence in model development is all that saves us from complete empiricism (a.k.a. witch-doctoring) but we must remember that Reynolds-averaged models will always be an extremely crude and incomplete representation of turbulence, and an insecure foundation for mathematical analysis.

8. New Approaches

The NS equations express Newton's second law of motion and a simpler exact alternative (what could really be called a "theory" of turbulence) is unlikely to be found. Therefore the present paradigm of turbulence modeling, *Average the NS equations and use data or simulations empirically to recover a closed, simpler, system of equations*, is unlikely to change. A form of averaging intermediate (in computing cost!) between complete Reynolds averaging and conventional LES would be a great step forward. A spectrum partition like LES but at much lower wave number ("VLES") does not look hopeful; some form of conditional averaging might be a better prospect.

A variety of nonlinear eddy-viscosity relations has appeared in recent years, some (Gatski & Speziale, 1993) based on ASM-like truncations of stress-transport models, some, e.g. Myong and Kasagi (1990), Craft *et al.* (1993) and Nagano *et al.* (1997), more empirical. In most cases the standard k and ϵ model equations are used. To deal with the case of system rotation, which undoubtedly alters turbulence behaviour, models include the vorticity as well as the strain rate – but only in the higher-order terms, not in the main linear term as in the one-equation model of Spalart and Allmaras. A detailed review by Apsley *et al.* (1997) comments that "The mechanisms by which [improvements in results are obtained] are, however, quite different from the actual physical interactions which are represented principally through groups of stress-strain products in the evolution equations for the Reynolds stresses...In effect, non-linear models are made to mimic the physical behaviour by means of mathematical artifacts and careful calibration." The resulting danger is that these models may fail spectacularly in extreme cases, the nonlinear terms being largest when rates of strain are large. A numerical danger is that the more elaborate eddy-viscosity models may lead to increasingly stiff systems and may not have much advantage in computational speed and stability over stress-equation models.

The reason for the survival of eddy diffusivity in modeling turbulent transport of stress or dissipation, even in stress-transport models, is the lack of a plausible alternative. Townsend pointed out in 1956 that turbulent transport is mainly transport by the large eddies – so that a gradient-transport model is unrealistic – and defined a "transport velocity" as the rate of transport divided by the mean value of the quantity being transported, e.g. the transport velocity of turbulent energy $\overline{u_i^2}/2$ in the x_j direction is $(\overline{p'u_j}/\rho + \overline{u_i^2 u_j}/2)/(\overline{u_i^2}/2)$. Unfortunately transport velocities seem to be less simply behaved than diffusivities. What is wanted is some kind of nonlocal model, generally producing transport down the gradient but evaluating the gradient as an average over a large-eddy volume. Nakayama and Vengadesan (1993) have taken a step in this direction. Various authors, the latest being Craft *et al.* (1997), have considered transport equations for the triple products: for reasons of computational economy only fairly simple models are acceptable, and of course equilibrium assumptions are likely to lead back to gradient diffusion. Nagano and Tagawa (1990) found empirical correlations between the different triple products, in simple shear layers, which would in principle allow all to be predicted from a transport equation for one only. However such correlations may not be reliable in complex flows. Eddy diffusivity is a recurring embarrassment (see Section 6).

Model transport equations for dissipation and other variables of the form $k^m \epsilon^n$ are also an embarrassment, because they are not term-by-term models of exact equations in the way that the stress-transport model equations are. The transport equation for energy transfer rate (Section 5) may provide better insight. Perneix and Durbin (1996) suggested that the dissipation equation was often blamed for the faults of the stress-transport equations, but certainly far more effort has been devoted to the stress-transport equations, particularly the pressure-strain term, than to the insecurely-based dissipation equation.

8.1. ADVANCES OVER STRESS-TRANSPORT MODELING

Reynolds and collaborators (see paper in this volume) have developed a stress-transport model which employs an "eddy axis tensor" to improve on the crudity of the conventional scalar eddy length scale. The model is constructed to limit to rapid-distortion theory for very large strain rates and to the k, ϵ model in the opposite extreme of near-equilibrium flow.

Models based on the probability distribution of the velocity have been used for combustion calculations, with pdf equations for species, and Pope (1994), see also this volume) shows that there is a one-to-one relation between stochastic Lagrangian methods and stress-transport models. The advantages of pdf models over stress-transport models in single-species flows are not decisive, and they pose modeling difficulties of their own.

Higher-level models that have been suggested are mainly multi-point (two-point) closures: these become very cumbersome in inhomogeneous flows and seem unlikely to come into widespread engineering use.

8.2. LARGE-EDDY SIMULATION – SUB-GRID-SCALE MODELS

At present, LES development seems to have slowed down after the impetus of the "dynamic" SGS model (for recent published work see Najjar and Tafti (1996) and Wang and Squires (1996)), and the Reynolds-number limitation implied by uncertainties about near-wall performance remains. It seems inescapable that an LES sub-grid-scale model should limit to a complete Reynolds-averaged model as the filter wave number becomes small. Then the simulation can operate at indefinitely high bulk Re: the near-wall region is covered by a coarse grid and an almost-Reynolds-averaged SGS model, using some form of wall function to bridge the viscous wall region. Much work is implied by the "almost" and the "some form"! The steady-state law of the wall is not acceptable as an instantaneous boundary condition for the full energy-containing range, but it presumably is acceptable for the lowest part of the wave number spectrum: if the LES filter cutoff is chosen within this acceptable range, and if the SGS model can represent the upper part of the Reynolds-stress spectrum, we have an LES scheme with no Reynolds number limit. For Reynolds-number independence the distance from the surface at which the wall function is applied to the resolved motion must be chosen as a fraction of the flow width, not a multiple of the viscous length. Critical regions, e.g. the neighborhood of separation and reattachment points, may have to be treated by full DNS.

Present SGS models are mostly crude and not easy to extend to the whole spectrum. The "dynamic model", which calibrates the SGS model itself from the properties of the high end of the resolved spectrum, is justifiable with any rigor only in the case of very high local *Re*, with the filter wave number somewhere in the middle of the inertial subrange. It is obviously incapable of extension to the full spectrum, for then there would be no calibration range left. The dynamic model is even suspect when the energy-containing and dissipating ranges overlap, so that all structural parameters are changing rapidly with wave number near the filter cutoff.

9. Conclusions

This writer's view is that Reynolds-averaged models will continue to improve, and their use in engineering will certainly become more widespread, irrespective of any significant improvements. Eddy-viscosity models are likely to remain popular despite their basic shortcomings. It is difficult to see what improvements in stress-transport models would make them more attractive to engineers: a basic advantage which has not been exploited is that their greater realism could at least lead to worthwhile accuracy/applicability estimates. The wrong answer may be preferable to no answer at all, but it is nice to have some idea of how far wrong it may be.

At present, LES is not the best turbulence model for engineers, and a breakthrough in SGS models for the near-wall region is needed, but there is much more hope for LES than for DNS, and engineers and funding agencies need to keep it in mind as the longer-term solution. In 25 years, LES will probably be in widespread use in high-tech industry, almost certainly for difficult sub-regions imbedded in Reynolds-averaged domains.

Probably the most unrealistic model assumption ever made (by members of a respected research group, many years ago) is that "pressure transport" $\overline{p'u_i}$ is related, by an eddy diffusivity, to the *mean pressure gradient.* Modellers uncertain of the plausibility of their ideas can take comfort from this long-standing record; they are unlikely to break it.

10. Acknowledgements

I am grateful to Prof. F.T.N. Nieuwstadt, Dr. P.R. Spalart and Dr. D.C. Wilcox for helpful comments on a draft of this paper.

References

Apsley, D.D., Chen, W.-L., Leschziner, M.A. and Lien, F.-S., 1997. "Non-Linear Eddy-Viscosity Modelling of Separated Flows," *J. Hydraulic Res.* **35**, p. 723.

Barenblatt, G.I., Chorin, A.J., and Prostokishin, V.M., 1997. "Scaling Laws for Fully Developed Turbulent Flow in Pipes: Discussion of Experimental Data," *Proc. Nat Acad. Sci. USA* **94**, p. 773.

Bradshaw, P. and Huang, P.G., 1995. "The Law of the Wall in Turbulent Flow," *Proc. Roy. Soc. London* **A451**, p. 165.

Bradshaw, P. and Strigberger, J., 1995. "Modeling of Surface Pressure Fluctuations," Final Report on Boeing P.O. HY7422.

Bradshaw, P., Mansour, N.N., and Piomelli, U., 1987. "On Local Approximations of the Pressure-Strain Term in Turbulence Models," *Studying Turbulence Using Numerical Simulation Databases*, Ames/Stanford Center for Turbulence Research CTR-S87, p. 159.

Cabot, W., 1996. "Near-Wall Models in Large Eddy Simulations of Flow Behind a Backward-Facing Step," NASA Ames/Stanford Center for Turbulence Research, Annual Research Briefs, p. 199.

Cazalbou, J.B., Spalart, P.R., and Bradshaw, P., 1994. "On the Behavior of Two-Equation Models at the Edge of a Turbulent Region," *Phys. Fluids* **6**, p. 1797.

Cousteix, J., Saint-Martin, V., Messing, R., Bezard, H., and Aupoix, B., 1997. "Development of the k, ϕ Turbulence Model," presented at *11th International Symposium on Turbulent Shear Flows.*

Craft, T.J., Kidger, J.E., and Launder, B.E., 1997. "Importance of Third-Moment Modeling in Horizontal, Stably-Stratified Flows," presented at *11th International Symposium on Turbulent Shear Flows.*

Craft, T.J., Launder, B.E., and Suga, K., 1993. "Extending the Applicability of Eddy Viscosity Models Through the Use of Deformation Invariants and Non-Linear Elements," *Proc. 5th IAHR Symposium on Refined-Flow Modelling and Turbulence Measurements.*

Durbin, P.A., 1993. "A Reynolds Stress Model for Near-Wall Turbulence," *J. Fluid Mech.* **249**, p. 465.

Durbin, P.A., 1995. "Separated Flow Computations with the $k - \epsilon - v^2$ Model," *AIAA J.* **33**, p. 659.

Ferziger, J.H., Kline, S.J., Avva, R.V., Bordalo, S.N., and Tzuoo, K.-L., 1990. "Zonal Modelling of Turbulent Flows - Philosophy and Accomplishments," *Near-wall Turbulence - 1988 Zaric Memorial Conference*, S.J. Kline and N.H. Afgan, eds., Hemisphere, p. 800.

Ferziger, J.H., 1996. "Recent Advances in Large-Eddy Simulation," *Engg. Turb. Modelling and Expt. 3*, W. Rodi and G. Bergeles, eds., Elsevier, p. 163.

Fu, S., Huang, P.G., Launder, B.E., Leschziner, M.A., 1988. "A Comparison of Algebraic and Differential Second-Moment Closures for Axisymmetric Turbulent Shear Flows With and Without Swirl," *J. Fluids Engg.* **110**, p. 216.

Gatski, T.B. and Speziale, C.G., 1993. "On Explicit Algebraic Stress Models for Complex Turbulent Flows," *J. Fluid Mech.* **254**, p. 59.

George, W.K. and Castillo, L., 1993. "Boundary Layers with Pressure Gradient - Another Look at the Equilibrium Boundary Layer," *Near-Wall Turbulent Flows*, R.M.C. So, C.G. Speziale, and B.E. Launder, eds., Elsevier, p. 901.

Gibson, M.M. and Harper, R.D., 1997. "Calculation of Impinging Jet Heat Transfer with the Low-Reynolds Number q-ζ Turbulence Model," *Int. J. Heat and Fluid Flow* **18**, p. 80.

Head, M.R., 1958. "Entrainment in the Turbulent Boundary Layer," *British Aero. Res. Counc.* R. & M. 3152.

Heyerichs, K. and Pollard, A., 1996. "Heat Transfer in Separated and Impinging Turbulent Flows," *Int. J. Heat Mass Transfer* **39**, p. 2385.

Huang, P.G., Bradshaw, P., and Coakley, T.J., 1994. "Turbulence Models for Compressible Boundary Layers," *AIAA J.* **32**, p. 735.

Kalitzin, G., Benton, J.J., and Gould, A.R.B., 1996. "Application of Two-Equation Turbulence Models in Aircraft Design," *AIAA* paper 96-0327.

Kline, S.J., Morkovin, M.V., Sovran, G., and Cockrell, D.J., eds., 1969. *Proceedings, Computation of Turbulent Boundary Layers – 1968 AFOSR-IFP-Stanford Conference*, Thermosciences Division, Stanford University.

Mason, P.J., 1996. "Large-Eddy Simulation: A Critical Review of the Technique," *Quart. J. Roy. Met. Soc.* **120**, p. 1.

Moin, P. and Mahesh, K., 1998. "Direct Numerical Simulation: A Tool in Turbulence Research," *Ann. Rev. Fluid Mech.* **30**, p. 539.

Myong, H.K. and Kasagi, N., 1990. "Prediction of Anisotropy of the Near-Wall Turbulence With an Anisotropic Low-Reynolds-Number k, ϵ Turbulence Model," *J. Fluids Engg.* **112**, p. 521.

Nagano, Y. and Tagawa, M., 1990. "A Structural Turbulence Model for Triple Products of Velocity and Scalar," *J. Fluid Mech.* **215**, p. 639.

Nagano, Y., Hattori, H., and Abe, K., 1997. "Modeling the Turbulent Heat and Momentum Transfer in Flows Under Different Thermal Conditions," *Fluid Dynamics Research* **20**, p. 127.

Najjar, F.M. and Tafti, D.K., 1996. "Study of Discrete Test Filters and Finite Difference Approximations for the Dynamic Subgrid-Scale Stress Model," *Phys. Fluids* **8**, p. 1076.

Nakayama, A. and Vengadesan, S., 1993. "A Non-Local Turbulent Transport Model," presented at *9th Symposium on Turbulent Shear Flows*, paper 26-4.

Oberlack, M., 1997. "Unified Theory for Symmetries in Plane Parallel Turbulent Shear Flows," *Stanford/Ames Center for Turbulence Research* Manuscript 163, submitted to J. Fluid Mech.

Parneix, S. and Durbin, P.A., 1996. "A New Methodology for Turbulence Modelers Using DNS Database Analysis *Stanford/Ames Center for Turbulence Research, Annual Research Briefs*, p. 17.

Pope, S.B., 1994. "On the Relationship Between Stochastic Lagrangian Models of Turbulence and Second-Moment Closures," *Phys. Fluids* **6**, p. 973.

Schlichting, H., 1979. *Boundary Layer Theory*, McGraw-Hill, New York.

Schumann, U., 1977. "Realizability of Reynolds-Stress Turbulence Models," *Phys. Fluids* **20**, p. 721.

Shih, T.-H., 1997. "Developments in Computational Modeling of Turbulent Flows," *Fluid Dyn. Res.* **20**, p. 67.

Shur, M., Strelets, M., Zaikov, L., Gulyaev, A., Kozlov, V., and Secundov, A., 1995. "Comparative Numerical Testing of One- and Two-Equation Turbulence Models for Flows with Separation and Reattachment," *AIAA* paper 95-0863.

Spalding, D.B., 1991. "Kolmogorov's Two-Equation Model of Turbulence," *Proc. Roy. Soc. London* **A434**, p. 211.

Townsend, A.A., 1961. "Equilibrium Layers and Wall Turbulence," *J. Fluid Mech.* **11**, p. 97.

Wang, Q. and Squires, K.D., 1996. "Large Eddy Simulation of Particle-Laden Turbulent Channel Flow," *Phys. Fluids* **8**, p. 1207.

Wilcox, D.C., 1998. *Turbulence Modeling for CFD* (2nd Edition) DCW Industries, La Cañada, CA.

Zagarola, M.V., Perry, A.J., and Smits, A.J., 1997. "Log Laws or Power Laws: The Scaling in the Overlap Region," *Phys. Fluids* **9**, p. 2094.

THE MODELING OF TURBULENT FLOWS WITH SIGNIFICANT CURVATURE OR ROTATION

B.E. LAUNDER
UMIST
Manchester, England

1. Introduction

The present contribution aims to provide an impression of the successes and weaknesses of different closure levels when applied to the calculation of flows with appreciable curvature or rotation and some directions for improvement.

Twenty-five years have elapsed since Bradshaw's (1973) definitive AGARDograph alerted the world to the fact that in near-wall flow, eddy-viscosity models underestimated the effects of secondary strain associated with curvature by at least an order of magnitude. Similar weaknesses arise with rotation. Yet still, eddy-viscosity models are widely — indeed, almost exclusively — used for industrial CFD calculations involving flow over or around curved or rotating surfaces. What are the reasons for this state of affairs? A major reason, discussed more fully in Section 2, is that in most circumstances arising in industry, curvature or rotation leads to appreciable secondary mean flow. As an example, Figure 1 shows a diagram prepared by Rolls Royce plc to highlight the complexities of flow through a blading stage. As is evident, numerous secondary strains are generated. While in principle, these secondary strains might have been expected to expose further weaknesses in the isotropic viscosity concept, in practice, in three-dimensional flows, eddy viscosity models often do better than would be supposed from considering their performance in two-dimensional boundary layers.

A further reason for the modest take-up rate of second-moment closures for industrial CFD problems involving curvature and rotation is that the most widely used approaches employ 'wall-reflection' terms designed for flow over plane surfaces whose extent is very large compared with the length scale of the turbulence flowing past it. Understandably, it makes

M. D. Salas et al. (eds.), Modeling Complex Turbulent Flows, 29–51.

no sense to switch from an eddy viscosity scheme if, in the situations of interest, one has to redesign special-purpose wall-reflection models to suit the particular geometry considered. A final reason is that curvature often provokes separation and this is an area where no turbulence model has a distinguished track record due principally to currently employed versions of the ε equation returning too high levels of shear stress in adverse pressure gradients.

Modelers must certainly be sensitive to all these currents if they are to advance the quality of industrial CFD modeling for curved and rotating shear flows.

The remainder of the paper is organized in three further sections followed by summarizing conclusions: Section 2 examines computations of a number of three-dimensional curved and rotating flows which have been published over the past decade. They serve to provide direct evidence of the ambivalent position noted above. Section 3 considers non-linear eddy viscosity models - a closure level that has become popular in the Nineties. The aspiration is that such schemes should retain most of the desirable numerical features of linear eddy-viscosity models while capturing much of the observed sensitivity to streamline curvature and rotation which the former miss. Finally, Section 4 reports newer strategies in second-moment closure and further modeling refinements. In all these areas, the writer especially draws on computations from his group at UMIST thereby to enable more assured comparisons between one modeling level and another. In any event, the volume includes contributions from most other leading turbulence-closure specialists who are well able to advocate the rationale for their own strategies.

2. Relative Performance of EVM's and SMC's in Curved and Rotating Flows

To begin our consideration, we note that the transport equation for the streamwise vorticity, Ω_x, for an incompressible, three-dimensional flow may be written:

$$\begin{aligned}\frac{D\Omega_x}{Dt} &= \underbrace{\Omega_x\frac{\partial U}{\partial x}}_{(A)}+\underbrace{\Omega_y\frac{\partial U}{\partial y}}_{(B)}+\underbrace{\Omega_z\frac{\partial U}{\partial z}}_{(C)}\\ &\quad+\underbrace{\frac{\partial^2}{\partial y\partial z}\left(\overline{w^2}-\overline{v^2}\right)+\frac{\partial^2\overline{vw}}{\partial y^2}-\frac{\partial^2\overline{vw}}{\partial z^2}}_{(D)}+\underbrace{\nu\left(\frac{\partial^2\Omega_x}{\partial y^2}+\frac{\partial^2\Omega_x}{\partial z^2}\right)}_{(E)} \quad (1)\end{aligned}$$

The viscous terms, (E), are purely diffusive in character as are, likewise, the turbulent stress terms (D) if they are approximated by a linear eddy-

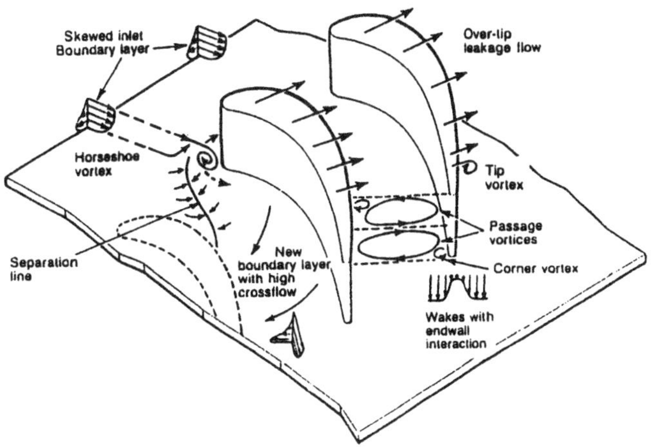

Figure 1. Secondary flow arising in turbomachinery blading (Courtesy Rolls Royce plc.).

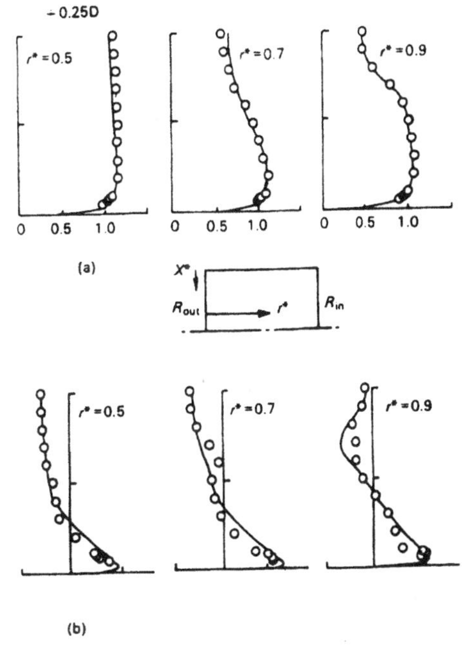

Figure 2. Flow through a 90° bend of square cross section: velocity profile 0.25 diameter downstream of flow exit. Symbols: Experiments, Taylor *et al.* (1982); Lines: EVM computations, Iacovides *et al.* (1982); a) steamwise profiles; b) radial profiles.

viscosity model (EVM):

$$\overline{u_i u_j} = +\frac{2}{3}\delta_{ij}k - \nu_t\left(\frac{\partial U_i}{\partial x_j} + \frac{\partial U_j}{\partial x_i}\right) \tag{2}$$

where ν_t is the turbulent viscosity.

This modeling strategy can be fatal to any attempt at predicting fully-developed flows in straight ducts of non-axisymmetric cross-section; for then, terms (D) provide the only mechanism for creating streamwise vorticity while their representation by Eq. (2) transforms the source into a diffusion term with the result that no secondary flows are generated. In curved ducts, however, the mean-strain terms in Eq. (1) are also present and may be so large as to render the limitations of Eq. (2) unimportant. However, it is important to note that mean velocity gradients are largest very close to the wall in the viscous sublayers. This means that it is vital to employ a model of turbulence that is used right up to the wall rather than adopt a high-Reynolds number model matched to wall functions. It was the failure to respect this requirement that led to a number of transport models of turbulence achieving spectacularly bad predictive performance for the 90°-bend test case at the 1980-1 Stanford 'Olympics' (Kline *et al.*, 1982). Yet, as Figure 2, taken from Iacovides *et al.* (1987), illustrates, Eq. (2) leads to virtually complete agreement with the mean velocity field of Taylor *et al.* (1982). In this example, while the $k-\varepsilon$ EVM had been adopted for the main flow, across the vital near-wall sublayer just Van Driest's (1956) form of the mixing-length hypothesis (MLH) had been used (in order - in 1986 - to keep mesh requirements within the core capability of the then available computers).

In fact, the above computations were originally planned to be just the initial set, to be followed shortly thereafter by second-moment closure (SMC) computations. However, there was evidently no scope for improving on the agreement already achieved with the EVM. Attention was therefore shifted to the square-sectioned U-bend flow of Chang *et al.* (1983). Besides the EVM calculations, an algebraic truncation of the SMC of Gibson & Launder (1978) (GL) was adopted. In applying this model to the flow in a square-sectioned bend, a decision was needed on how to handle wall-reflection effects in the pressure-strain process, ϕ_{ij}. For the case of an infinite flat wall for which the model had been calibrated, ϕ_{ij} takes the form:

$$\phi_{ij} = -1.8\frac{\varepsilon}{k}\left(\overline{u_i u_j} - \frac{2}{3}\delta_{ij}k\right) - 0.6\left(P_{ij} - \frac{1}{3}\delta_{ij}P_{kk}\right) + \phi_{ijw} \tag{3}$$

where the wall reflection part took the form:

$$\phi_{ijw} = \left[\begin{array}{l} 0.5\frac{\varepsilon}{k}\left(\overline{u_k u_m} n_k n_m \delta_{ij} - \frac{3}{2}\overline{u_k u_i} n_k n_j - \frac{3}{2}\overline{u_k u_j} n_k n_i\right) \\ +0.2\left(\tilde{\phi}_{km} n_k n_m \delta_{ij} - \frac{3}{2}\tilde{\phi}_{ik} n_k n_j - \frac{3}{2}\tilde{\phi}_{jk} n_k n_i\right) \end{array} \right] \frac{0.4k^{3/2}}{\varepsilon x_n} \tag{4}$$

Here P_{ij} is the generation rate of stress by mean shear, n_k denotes the unit vector normal to the wall, x_n is the distance from the wall, and $\tilde{\phi}_{ij}$ denotes $-0.6\left(P_{ij} - \frac{1}{3}\delta_{ij}P_{kk}\right)$. Now, in flow through a square duct one may construct four wall-normal vectors each with their associated wall distance. While the importance of these terms diminishes with distance from the wall, in a fully developed duct flow the effects of the walls are felt right to the duct center. Thus, a major question is how 'wall-reflection' should be interpreted in such multi-wall situations. In their computations Choi *et al.* (1989) supposed ϕ_{ijw} could be summed in turn for all four walls assuming that the GL proposal, which was devised for an infinite plane wall, could be applied directly to each of the four sides, two of which were curved. Figure 3 compares streamwise velocity profiles 90° around the U-bend at different distances from the bend's symmetry axis for three different model treatments: two EVM's and the SMC discussed above. Figures 3a and 3b both employ the linear $k - \varepsilon$ EVM over most of the flow but the former adopts wall functions while the latter employs the MLH. As discussed earlier, the use of wall functions leads to an underestimate in streamwise vorticity, which in turn means that the "hole" in the axial velocity is entirely missed. There is significant improvement when the MLH is employed across the sublayer but there evidently remains much scope for improvement, some of which is provided by ASM computations in Figure 3c. As a sequel, Figure 4 shows the more recent computations of Li (1995) (see also (Iacovides *et al.*, 1996a)) of this flow comparing algebraic and differential versions of the same closure. The scales are not the same between figures but careful comparison indicates that the broken line in Figure 4 is not identical with the solid line in Figure 3c even though the models are the same. This is because the elapse of seven years between the two computations enabled the number of grid nodes to be increased by a factor of 3. Switching to the full differential stress transport model, *i.e.*, the Differential Stress Model (DSM) version improves further the level of agreement.

Of course, in terms of modeling, it may not make sense to merge a full second-moment closure with the mixing-length hypothesis. Other forms of sublayer modeling will be presented later. First, however, let us include the effects of rotation. When one rotates one's coordinate axes to make a spinning object - perhaps a turbine disc - appear stationary to an observer riding on the reference frame, it is necessary to add Coriolis terms both to the mean momentum equation and to the stress transport equation. In the

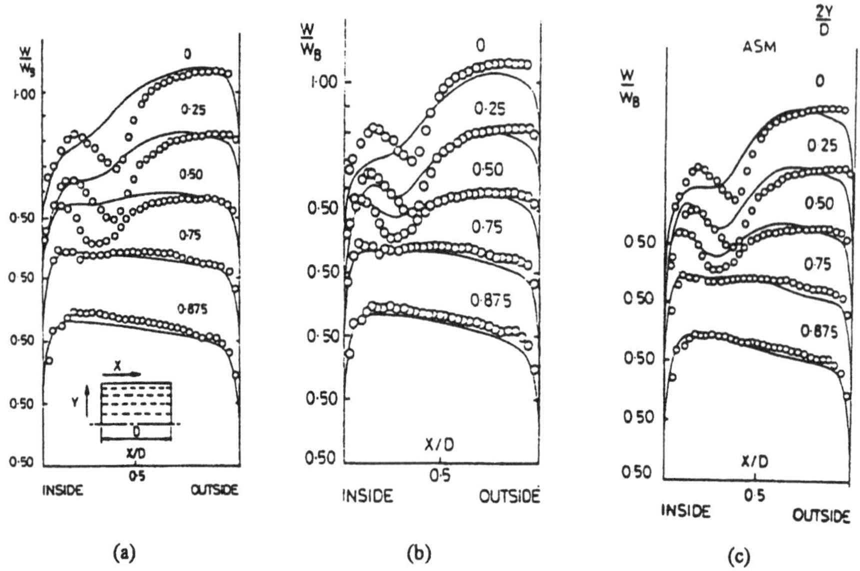

Figure 3. Flow in an unseparated U-bend of square cross section 90° after bend entry. Symbols: Experiments, Chang *et al.* (1983); Lines: Computations, Choi *et al.* (1989). a) $k - \varepsilon$ EVM with wall functions; b) $k - \varepsilon$ EVM with MLH sublayer treatment; and c) ASM with MLH sublayer treatment.

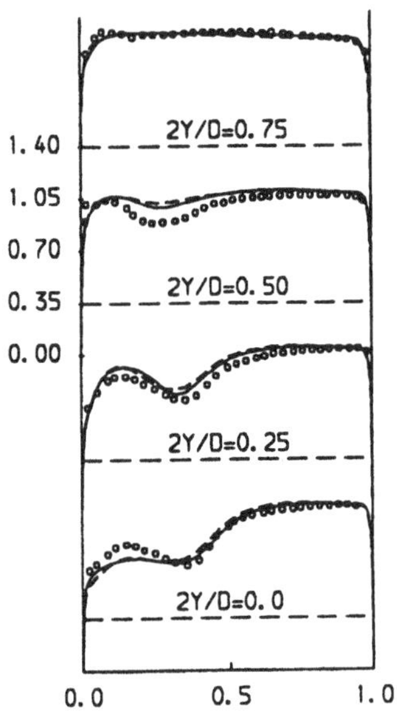

Figure 4. Further consideration of square U-bend using 'Basic' DSM plus MLH sublayer treatment Symbols: Experiments, Chang *et al.* (1983). —— DSM; - - - - - ASM, Li (1995).

latter, the direct extra source F_{ij} is:

$$F_{ij} = -2\Omega_k \left(\overline{u_j u_m}\varepsilon_{ikm} + \overline{u_i u_m}\varepsilon_{jkm}\right) \tag{5}$$

Let us note first that if Eq. (5) is contracted the result is:

$$F_{kk} = 0 \tag{6}$$

Consequently, if one adopts a linear $k - \varepsilon$ EVM, there is no effect of rotation on the turbulence energy level (unless some *ad hoc* term is added to the ε equation). A popular test for examining success in accommodating the effects of rotation on turbulence is flow through a plane channel in orthogonal-mode rotation, Figure 5a. The SMC calculation of Launder *et al.* (1987) extended the GL model by replacing, in the pressure-strain term, P_{ij} by $\left(P_{ij} + \frac{1}{2}F_{ij}\right)$. (The factor $\frac{1}{2}$ multiplying F_{ij} is suggested by an examination of the exact correlation for ϕ_{ij}, Cousteix & Aupoix (1981), Bertoglio (1982)). The rotation leads to an extra generation of shear stress of $\left(\overline{u^2} - \overline{v^2}\right)\Omega$ (x being the stream direction, y normal to walls) which, assuming $\overline{u^2}$ to be larger than $\overline{v^2}$, leads to an augmentation of shear stress near the pressure surface and a diminution near the suction surface. The consequent effects on the mean velocity profile are well-mimicked by the model as are the effects on wall shear stress (Ro being the rotation number $\Omega D/\overline{U}$).

As indicated above, a linear $k - \varepsilon$ EVM would return identical profiles to that found in a non-rotating channel. This might seem to indicate that modeling at this level could not be adequate in rotating flows. However, if one shifts to the 'real-world' of three-dimensional flows, the rotational terms in the mean momentum equations will then not vanish. Accounting for the consequent induced secondary flow may then be resolved adequately with an EVM closure. Indeed, the situation is essentially analogous to that of a curved duct. By way of illustration, Figure 6 shows the average Nusselt number on the pressure and suction surfaces of a rotating square duct from Bo *et al.* (1995). At a rotation number of 0.12 (not shown) both EVM and ASM schemes return nearly the same variation in satisfactory accord with experiment; when the rotation rate is doubled, however, the experiments of Wagner *et al.* (1991) indicate an irregular variation of Nu on the suction surface that is better mimicked by the ASM computations. The ASM velocity-field predictions indicate the probable cause of this variation: at a distance of 6D into the duct there is a marked second vortex near the suction surface, which thereafter collapses to give a single vortex mode. It is foreseeable that such a change in structure would give rise to an abrupt reduction in Nusselt number.

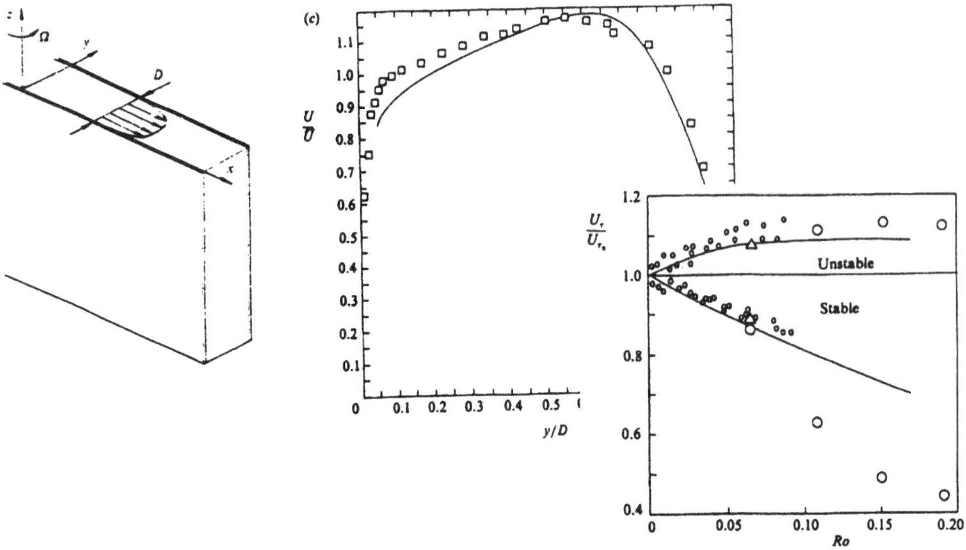

Figure 5. Flow through a plane channel in orthogonal-mode rotation, Launder *et al.* (1987). a) Flow configuration, b) Effect of curvature on mean velocity profile, Ro=0.21 and c) Dependence of wall shear stress on rotation.

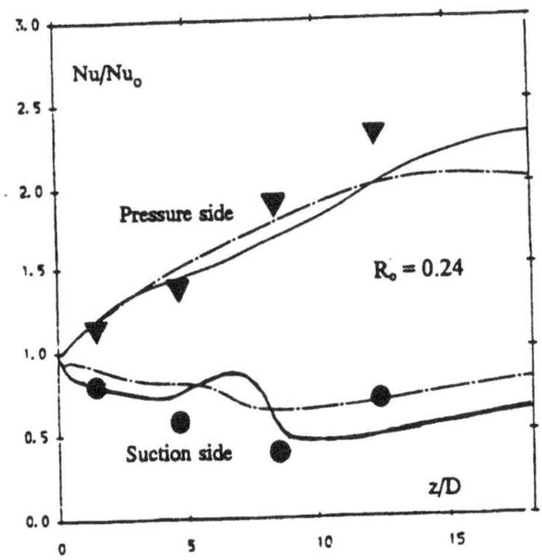

Figure 6. Nusselt-number development in entry region of a square duct rotating in orthogonal mode. Symbols: Experiments, Wagner *et al.* (1991); Computations: Bo *et al.* (1995). - — - $k-\varepsilon/1-eq$ EVM; — Low-*Re* ASM.

Finally, in reviewing comparative successes of EVM's and SMC's in curved and rotating ducts, attention is turned to the flow in a hairpin bend such as is employed inside turbine blades for cooling, using air that has by-passed the combustion chamber. Figure 7, drawn from Iacovides *et al.* (1996b), compares for the non-rotating limit predictions for three models: a two-equation EVM interfaced to a one-equation EVM in the sublayer and two algebraic SMC's: one that is matched to the one-equation EVM and another spliced to a one-equation ASM across the sublayer. This last model, the most elaborate of the treatments, achieves the best agreement though separation is occurring too late, a behavior in accord with remarks in Section 1.

Finally, in drawing comparisons between the behavior of a linear EVM and the "Basic" SMC, reference is made to two other major studies of three-dimensional flow by Malecki (1994) and Sotiropoulos & Patel (1995), both of whom used the extension of the "Basic" model to apply up the wall, proposed by Launder & Shima (1989). In the former, computations were made of the flow over the suction surface of a wind-tunnel model of a swept wing at an angle of attack. A substantial streamwise vortex is generated by the 35° sweep which in turn leads to appreciable secondary shear stress ($\overline{vw}$) by the most downstream station shown in Figure 8. The DSM computations are evidently in markedly better agreement with the data for both shear stresses ($\overline{vw}$ and $\overline{uv}$) than the three EVM's tested (the mixing length hypothesis, the $k - \varepsilon$ model and the $k - \omega$ scheme of Wilcox (1988)) as are, correspondingly, the mean velocity profiles (not shown). Sotiropoulos & Patel (1995) computed the downstream half of a ship's hull and noted a much stronger streamwise vortex shed from near the trailing edge with the SMC. The most striking feature of their computations, however, was an analysis the flow's harmonic character in the plane of the propeller shown in Figure 9 where, unlike the EVM, the SMC gives results virtually coincident with the data.

This section has provided an overview of the performance of turbulence models based on twenty-year-old methodology at predicting 3D flows with curvature and/or rotation. Linear eddy viscosity modeling is not as bad as one might anticipate from considering two-dimensional shear flows; nevertheless, the basic SMC or its low-Reynolds number refinement does better. The main weakness of these SMC's is the need to prescribe wall-reflection corrections; and no model does assuredly well when curved walls provoke separation. The next two sections consider how well much newer approaches to closure fare.

3. Non-Linear Eddy Viscosity Models (NLEVM's)

If one seeks to capture the sensitivity of the turbulent stress field to streamline curvature while avoiding the perceived difficulties of handling the solution of a further system of intricately coupled transport equations, the idea of providing an explicit non-linear connection between the stress and deformation fields is certainly appealing. In the context of modeling turbulence, the first general proposal of this type was by Pope (1975). While his form comprised up to fifth-order products of the strain and vorticity tensors, most of the schemes actually tested in the past decade have contained just quadratic products (for example, Speziale, 1987; Nisizima and Yoshizawa, 1987; Rubinstein and Barton, 1990; Myong and Kasagi, 1990; Choi *et al.*, 1989).

If, however, one converts an implicit algebraic truncation of an SMC into an NLEVM by successive substitution, one arrives at cubic or higher order products of stress and strains. This conclusion very much accorded with the findings of Suga (1995) (See also Craft *et al.* (1993; 1995; 1996; 1997) and Horiuti (1996)) that, to capture sensitivity to streamline curvature, any NLEVM needed to include cubic products of the deformation field.

The postulated stress-deformation relation in Suga's work was:

$$\begin{aligned}
a_{ij} &\equiv \frac{\overline{u_i u_j}}{k} - \frac{2}{3}\delta_{ij} = -\frac{\nu_t}{k}S_{ij} \\
&+ c_1\frac{\nu_t}{\tilde{\varepsilon}}\left(S_{ik}S_{kj} - \frac{1}{3}S_{kl}S_{kl}\delta_{ij}\right) \\
&+ c_2\frac{\nu_t}{\tilde{\varepsilon}}\left(\Omega_{ik}S_{kj} + \Omega_{jk}S_{ki}\right) \\
&+ c_3\frac{\nu_t}{\tilde{\varepsilon}}\left(\Omega_{ik}\Omega_{jk} - \frac{1}{3}\Omega_{ik}\Omega_{lk}\delta_{ij}\right) \\
&+ c_4\frac{\nu_t k}{\tilde{\varepsilon}^2}\left(S_{ki}\Omega_{ij} + S_{kj}\Omega_{ii}\right)S_{kl} \\
&+ c_5\frac{\nu_t k}{\tilde{\varepsilon}^2}\left(\Omega_{il}\Omega_{lm}S_{mj} + S_{il}\Omega_{lm}\Omega_{mj} - \frac{2}{3}S_{lm}\Omega_{mn}\Omega_{nl}\delta_{ij}\right) \\
&+ c_6\frac{\nu_t k}{\tilde{\varepsilon}^2}S_{ij}S_{kl}S_{kl} + c_7\frac{\nu_t k}{\tilde{\varepsilon}^2}S_{ij}\Omega_{kl}\Omega_{kl} \qquad (7)
\end{aligned}$$

where $S_{ij} \equiv (\partial U_i/\partial x_j + \partial U_j/\partial x_i)$, $\Omega_{ij} \equiv (\partial U_i/\partial x_j - \partial U_j/\partial x_i)$, and $\tilde{\varepsilon}$ is the so called "homogeneous" dissipation rate, $\varepsilon - 2\nu\left(\partial k^{1/2}/\partial x_j\right)^2$.

To mimic the stress field in simple shear at high Reynolds number for arbitrary levels of the strain parameter, $S\left(\equiv \frac{k}{\varepsilon}\sqrt{\frac{1}{2}\left(\frac{\partial U_i}{\partial x_j} + \frac{\partial U_j}{\partial x_i}\right)^2}\right)$ the following relations were imposed on the coefficients:

$$c_2 = 0.1;\ (c_1 + c_3) = 0.16;\ (-c_5 + c_6 + c_7) = 0$$

while c_μ was made a function of S.

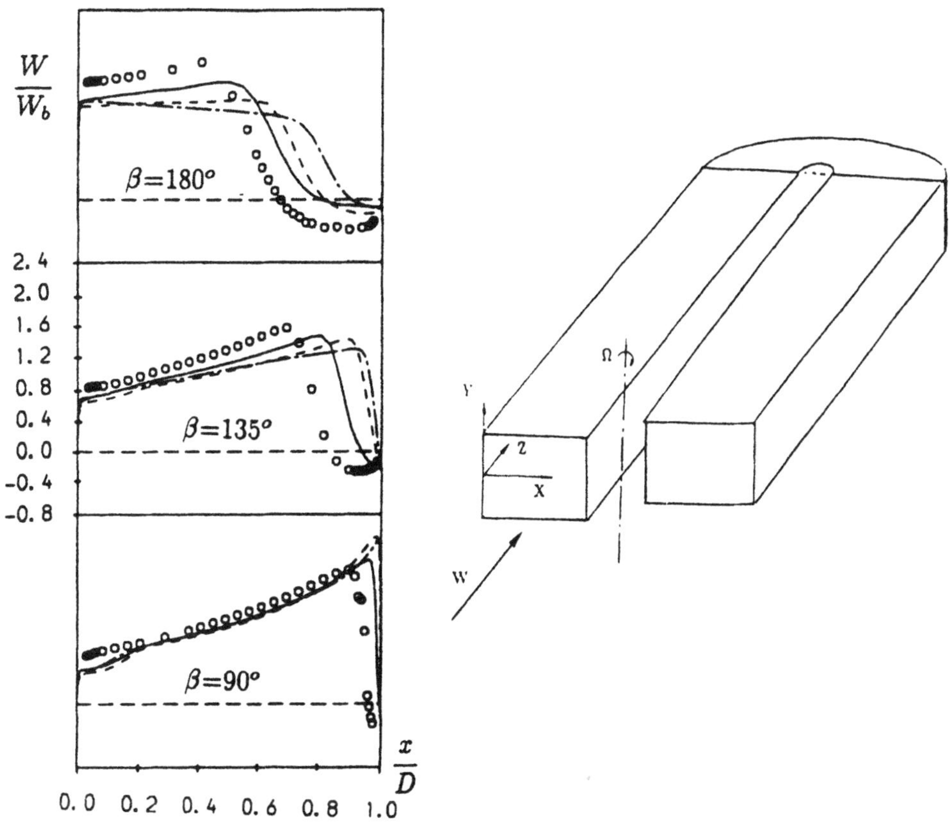

Figure 7. Streamwise velocity on center plane of square sectioned U-bend. Symbols: Experimental data; Computations: Iacovides *et al.* (1996b). — . — $k-\varepsilon/1-eq$ EVM; - - ASM/$1-eq$ EVM; —— ASM/$1-eq$ ASM.

Turning to turbulent flows with curved streamlines, for flow through a pipe rotating about its own axis, it is known that, in the fully-developed state, the swirl velocity displays a roughly parabolic variation between the axis and the pipe wall. This is in marked contrast with the solid-body rotation observed in laminar flow (and predicted in turbulent flow too whenever a linear EVM is used in modeling). When the above NLEVM constitutive relation is adopted one finds that in solid-body rotation the turbulent transport of angular momentum is non-zero only if c_4 or c_5 is non zero; thus, any

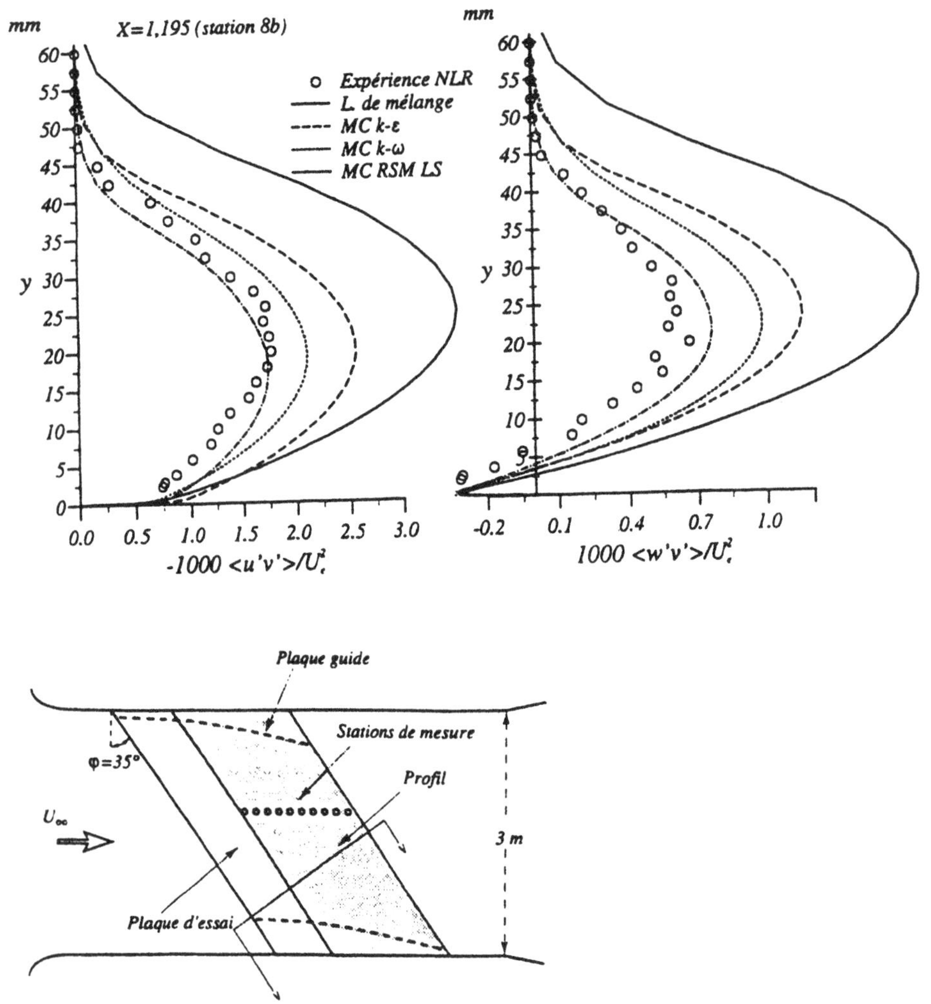

Figure 8. Streamwise vortex on suction surface of inclined swept wing: from Malecki (1994). Prediction of primary and secondary shear stress. Symbols: NLR Experiments; Lines: Computations (note: 'RSM LS' = Launder-Shima Reynolds stress model; 'L de mélange' = mixing length hypothesis.).

quadratic scheme would return solid-body rotation as the fully-developed state.

As a second case, if one considers a two-dimensional boundary layer

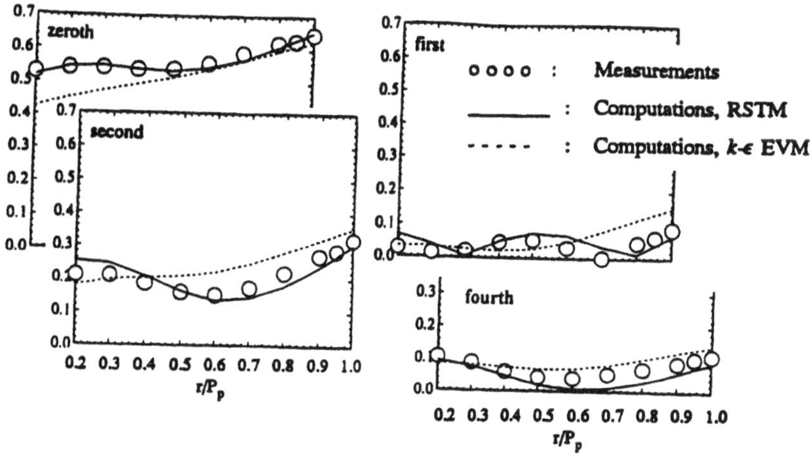

Figure 9. Boundary layer harmonics at propeller plane of shear flow detaching from the hull of a "mystery tanker"; from Sotiropoulos & Patel (1996). Symbols: Experiments; Lines: Computations.

developing on a curved surface, the only elements, besides the conventional linear term, affecting the shear-stress level are the cubic terms with coefficient c_5, c_6 and c_7. In the UMIST work, therefore, the coefficient c_5 is set to zero and, to match the simple-shear requirement above, c_6 is set equal to the negative of c_7. The coefficients c_4 and c_7 are then chosen to optimize agreement with the two types of curved flow. As would be expected, accord with these two curved flows is very satisfactory (Craft *et al.*, 1996; Craft *et al.*, 1997) and is not reproduced here. Figure 9, however, shows the case of an axisymmetric impinging jet. This involves a quite different type of strain field in which, on the axis, only normal straining occurs. Evidently, the scheme is much more successful than the linear EVM in reproducing both the dynamic field (Reynolds stresses and mean flow) and the consequent distribution of heat-transfer coefficient along the wall.

As a second example, Figure 11 shows the distribution of heat flux around a heated blade in a turbine cascade (experiments of Nicholson *et al.*, (1982)) in which the turbulence intensity level in the approaching stream was 4%. The linear EVM predicts much too high levels of heat flux at the stagnation point as well as over the pressure surface of the blade. Strikingly better results are obtained, however, with the NLEVM though it is noted that transition to turbulent flow on the suction surface still takes place rather more abruptly than in the experiments.

Finally, in this section, a few words of caution are in order. Following the considerable success of the NLEVM over a range of unseparated flows, it

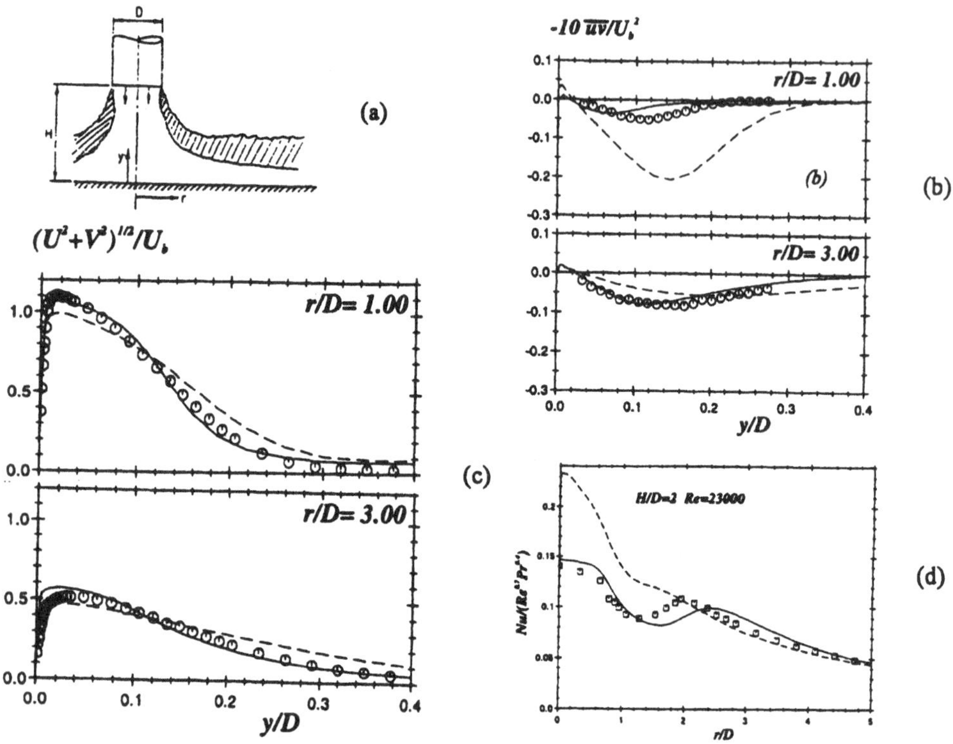

Figure 10. Prediction of turbulent impinging axisymmetric jet (from Craft *et al.*, 1997), H/D = 2.0. Symbols: Experiments, Cooper *et al.* (1993); —— NLEVM; - - - - $k-\varepsilon$ EVM, Craft *et al.* (1997). a) Configuration; b) shear-stress; c) mean velocity; and d) Nusselt number.

was decided to apply it to flow over rib-roughened surfaces where separation zones formed both downstream of the rib and just upstream. Iacovides (1997) has found, however, that the non-linear model predicts considerably too high stress levels and this leads to too high Nusselt numbers behind the rib. It is likely that careful recalibration can remove this anomalous behavior without significantly worsening agreement for the other test cases already considered in Suga's study. It does underline, however, that models of this type should be used only for interpolation.

4. Further Developments in Second-Moment Closure

This section discusses improvements - both actual and potential - in modeling the unknown processes on the right-hand side of the stress-transport equation: ϕ_{ij} - the pressure-strain term; ε_{ij}, the stress dissipation rate; and d_{ij} the turbulent diffusion process. Where possible, relevant applications in

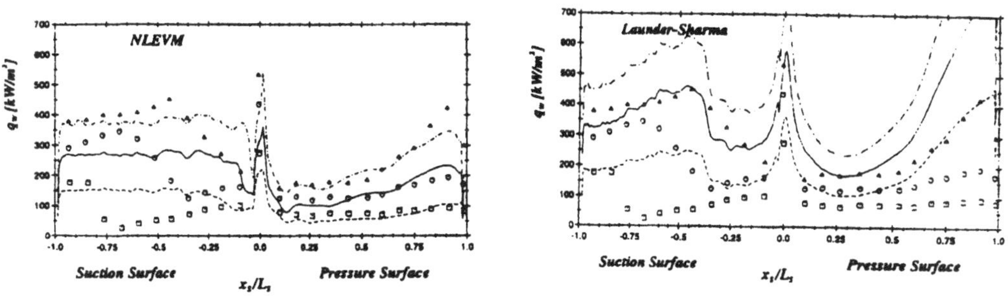

(a) (b)

Figure 11. Local heat flux on turbine-blade at three Reynolds numbers. Symbols: Experiments (Nicholson *et al.*, 1982); Lines: Computations (Suga *et al.*, 1997). a) NLEVM; b) $k-\varepsilon$ EVM.

curved or related flows will be given.

The weakest element of the closures adopted in Section 2 is the employment in ϕ_{ij} of 'wall-reflection' corrections based on the use of wall-normal vectors and distance from the wall. These corrections were designed and calibrated for flat surfaces that were very large in extent. Alas, these are not the types of configuration to which industry usually seeks predictions; the flow through the turbine-blade cooling passage shown in Figure 1 is more representative. Thus, at a practical level, the requirement must be to eliminate wall-reflection processes from the model or to devise a more general method of accounting for them. The writer's group has focused on the former approach though progress is also being made on the latter by Durbin (1993) and co-workers, Wizman *et al.* (1996).

With regard to the former, the strategy, first advocated by Lumley (1978), is to make the closure comply with the constraints of the two-component limit (TCL) to which real turbulence must collapse at a rigid surface (or, indeed, at a free surface). In the case of the pressure-strain process, one requires that, if one normal stress should vanish - $\overline{u_n^2}$, say - then ϕ_{nn} should also equal zero. The first specific 'two-component' forms of ϕ_{ij} were given by Shih & Lumley (1985); the UMIST model adopts the same broad philosophy but employs, at several points, different closure constraints, Fu *et al.* (1987), Craft *et al.* (1996). While the forms that result from applying the two-component limit are algebraically very complicated (see, for example, Table 1 for the model of mean-strain effects on ϕ_{ij}), the number of free coefficients is small - indeed, in some cases, zero. This is not necessarily a virtue, for it is comforting to have a few disposable coefficients with which to ensure good agreement for a number of key flows. Neverthe-

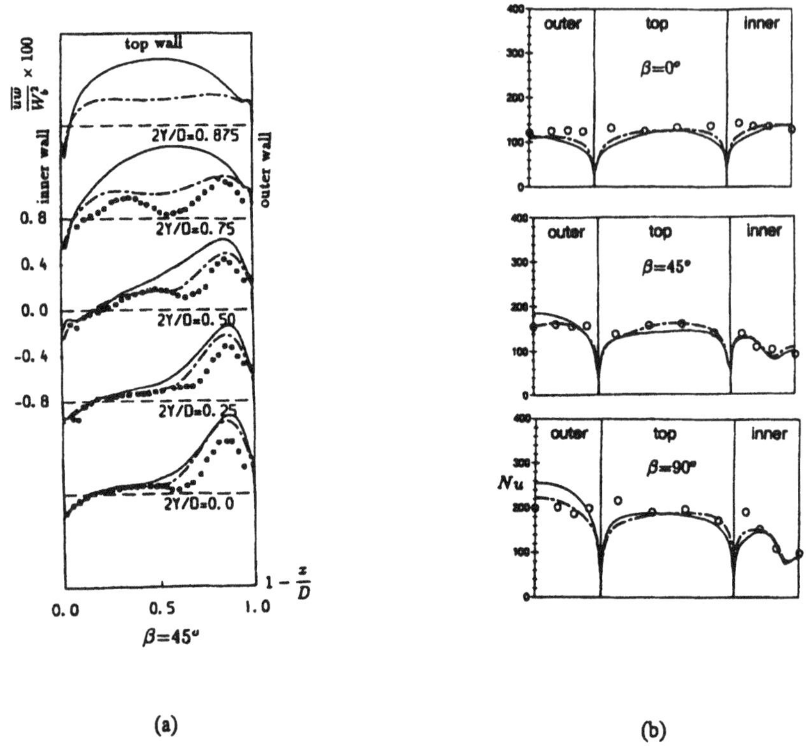

Figure 12. Application of TCL model to flow around U-bend. Symbols: Experiments: - — - TCL Model; — Basic Model (from Iacovides *et al.*, 1996). a) shear-stress distribution at 45°; b) Nusselt-number development.

less, the approach has proved successful for a number of quite different classes of flows. A wide range of computations in free shear flows and buoyant flows are respectively summarized in the reviews by Launder (1989) and Craft *et al.* (1996). In each of these topic areas flows may approach the two-component limit, in the former case by very rapid shearing and in the latter by the damping of vertical fluctuations in a stably stratified flow. For the case of flow parallel to a wall, Launder & Li (1994) showed that, with the use of the TCL model for ϕ_{ij}, it was possible to predict the variation of turbulent stresses throughout the flow without employing wall-reflection agencies. This point was underlined by their showing that flow through straight rectangular ducts could be predicted much more successfully with the TCL version than with the 'Basic Model' even though wall-reflection terms were employed in the latter case. The same model has more recently been applied by Iacovides *et al.* (1996a) to the flow around the previously considered U-bend of square cross section from which two features are noted here. Figure 12a presents, for the 45° station, the $\overline{uw}$ shear stress profiles along lines parallel to the symmetry plane at different distances (2Y/D) from this plane. On the symmetry plane itself note first that, as would be expected, the experimental data of Chang *et al.* (1983) exhibit much reduced shear stresses near the convex wall and amplified levels near the concave surface. Concerning the second-moment predictions, both the TCL model and the Basic Model with wall reflection capture this asymmetry rather well at this symmetry-plane position. As one moves towards the upper wall, however, the TCL model is evidently much more successful at following the measured variation. Likewise, the Nusselt number, whose variation around the perimeter is shown in Figure 12b is certainly more faithfully reproduced by the TCL scheme than the Basic Model.

TABLE 1. TCL Model of Mean-Strain Effects on ϕ_{ij}
Fu *et al.* (1987), Craft *et al.* (1996)

$$\begin{aligned}
\phi_{ij2} = \; & -0.6\left(P_{ij}-\frac{1}{3}\delta_{ij}P_{kk}\right)+0.3\varepsilon a_{ij}\left(P_{kk}/\varepsilon\right) \\
& -0.2\left[\frac{\overline{u_k u_j}\,\overline{u_l u_i}}{k}\left(\frac{\partial U_k}{\partial x_l}+\frac{\partial U_l}{\partial x_k}\right)-\frac{\overline{u_l u_k}}{k}\left(\overline{u_i u_k}\frac{\partial U_j}{\partial x_l}+\overline{u_j u_k}\frac{\partial U_i}{\partial x_l}\right)\right] \\
& -c_2\left[A_2\left(P_{ij}-D_{ij}\right)+3a_{mi}a_{nj}\left(P_{mn}-D_{mn}\right)\right] \\
& +c_2'\left[\left(\frac{7}{15}-\frac{A_2}{4}\right)\left(P_{ij}-\frac{1}{3}\delta_{ij}P_{kk}\right)\right. \\
& +0.1\varepsilon\left[a_{ij}-\frac{1}{2}\left(a_{ik}a_{kj}-\frac{1}{3}\delta_{ij}A_2\right)\right]\left(P_{kk}/\varepsilon\right)-0.05a_{ij}a_{lk}P_{kl} \\
& +0.1\left[\left(\frac{\overline{u_i u_m}}{k^2}P_{mj}+\frac{\overline{u_j u_m}}{k}P_{mi}\right)-\frac{2}{3}\delta_{ij}\frac{\overline{u_l u_m}}{k}P_{ml}\right] \\
& +0.1\left[\frac{\overline{u_l u_i u_k u_j}}{k^2}-\frac{1}{3}\delta_{ij}\frac{\overline{u_l u_m u_k u_m}}{k^2}\right]\left[6D_{lk}+13k\left(\frac{\partial U_l}{\partial x_k}+\frac{\partial U_k}{\partial x_l}\right)\right] \\
& \left.+0.2\,\frac{\overline{u_l u_i u_k u_j}}{k^2}\left(D_{lk}-P_{lk}\right)\right]
\end{aligned}$$

where $D_{ij} \equiv -\left(\overline{u_i u_k}\partial U_k/\partial x_j+\overline{u_j u_k}\partial U_k/\partial x_i\right)$

The model is not quite the panacea that the above comparisons may suggest. In the foregoing example, despite the appreciable secondary flow, the velocity vectors did not make an impingement angle greater than about

25° with any of the walls. The model is not adequate, however, for computing normally impinging jets. Work has begun on reformulating the closure for this and other 'difficult' flows, Craft & Launder (1996), Craft (1997a).

With the above developments in modeling ϕ_{ij}, it is our view that it is time to place a greater effort on modeling ε_{ij}. The stress dissipation rate is conventionally obtained by solving a transport equation for the kinetic energy dissipation rate:

$$\frac{D\varepsilon}{Dt} = \frac{c_{\varepsilon 1}}{2}\frac{P_{kk}\varepsilon}{k} - c_{\varepsilon 2}\frac{\varepsilon^2}{k} + \mathit{diff}\,(\varepsilon) + \text{low Reynolds number terms} \qquad (8)$$

then invoking some algebraic relation to inter-relate ε_{ij} and ε:

$$\frac{\varepsilon_{ij}}{\varepsilon} = f\left(\delta_{ij}, a_{ij}, A, A_2, R_t\right) \qquad (9)$$

A frequently noted weakness of the above form of ε equation is that it is insensitive to effects of rotation since, in a rotating flow, the energy generation rate associated with rotation, $\frac{1}{2}F_{kk}$, is zero. Yet, there have been a number of DNS and experimental studies of decaying grid turbulence which indicate a marked sensitivity to rotation. To the writer it isn't clear that the omission of a rotation-dependent source in the ε equation is important in most practical flows (after all, the computations of the rotating channel in Figure 5 employed the same ε equation as in a non-rotating flow). Nevertheless, the modeling of the source term in the ε equation has been the subject of criticism and speculation for over twenty years (Lumley and Khajeh-Nouri, 1974) and it is certainly high time to bring the treatment of this variable to a more consistent state. The problems with modeling ε mainly stem from the fact that it is expected to serve both as the true energy dissipation rate as well as the energy transfer rate across the inertial subrange. Aside from the fact that, increasingly, CFD is called on to handle flows far from local equilibrium, it also needs to be acknowledged that, while the true dissipation rate may be isotropic, the energy entering the inertial subrange rarely is. The most sensible route thus seems to be to distinguish between the energy transfer rate, e, (say) and the dissipation rate, ε, solving transport equations for each. While a practical formulation still awaits development, one may note that:

- Oberlack (1995) has gone some way to developing an equation for e_{ij} which, on contraction, provides a transport equation for e. The principal source in such an equation is proportional to $e_{ij}\partial U_i/\partial x_j$ while in rotating flows there is an exact Coriolis term equal to $-4\Omega_k e_{il}\varepsilon_{kli}$.
- The transport equation for ε must be largely empirical but should not contain mean strain or rotation terms. Its source would, of course, be e (Hanjalic *et al.*, 1980).

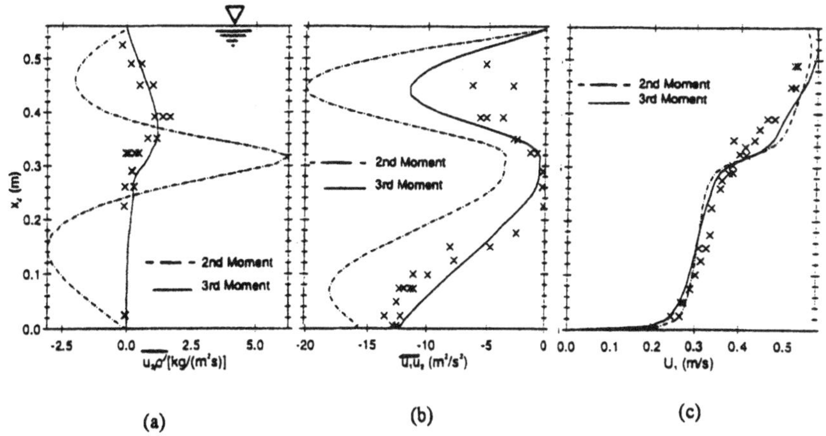

Figure 13. Effect of 3rd moments in a stably stratified mixing layer. Symbols: Experiments, Uittenbogaard (1988); Lines: Computations, Kidger (1997), a) density flux; b) shear stress; and c) mean velocity.

- The quantity (e/ε) provides a dimensionless measure of the non-equilibrium of the turbulence spectrum and, as such, may be appropriately used as an argument in the various empirical coefficients.
- The quantities e_{ij}/e and $\varepsilon_{ij}/\varepsilon$ would (contrary to Oberlack's proposal) be prescribed algebraically, the former depending on a_{ij} and invariants of the stress field, while the latter may be expected to be a function of the turbulent Reynolds number, the spectral imbalance (e/ε), and e_{ij}/e.
- In the stress transport equation, e would be used to provide a turbulent time scale, k/e, in the pressure-strain and diffusion terms.

If the above strategy seems too complicated, one should reflect that, even for an isothermal flow, a CFD analysis at 2nd-moment level with a single scale equation requires 11 transport equations to be solved for a three-dimensional flow. Adding a second turbulent scale equation will thus have a barely detectable effect on computing time - yet can be expected to offer the prospect of greatly improving the realism of the computations.

In all computations of flow over curved and rotating surfaces known to the writer, the stress diffusion has been handled by models designed from the standpoint of cheapness and stability rather than fidelity in modeling the processes. The rationale is that the diffusion terms are not of decisive importance in the stress budgets, especially in flow near walls; so, the overall effect of the diffusion modeling on the mean flow is rather small. One needs to be wary of accepting this viewpoint, however, in flow over highly curved convex surfaces. The strong curvature will in all probability create a separated shear layer in which diffusion rates may grow rapidly while dissipation rates will fall due to the rapid growth in length scale following detachment. Moreover, due to the stabilizing streamline curvature, stress generation will fall abruptly. These are circumstances where the accurate modeling of diffusive transport may be essential to capture the flow's evolution.

In fact, so far as flow over curved or rotating surfaces is concerned, the above is speculation: plausible, but no incontrovertible cases to cite. However, the analogy between streamline curvature and stratified flows is frequently invoked and with that precedent, a recent example of a stably stratified mixing layer makes the point rather effectively. Figure 13 shows predictions by Kidger (1997) of Uittenbogaard's (1992) open-channel flow in which mixing is occurring between fast-moving fresh water and a (lower) slower-moving saline stream. Computed profiles are shown arising from two treatments of the third-moment correlations: approximation via the generalized gradient diffusion hypothesis (GGDH) of Daly & Harlow (1970); and a partial 3rd-moment closure where transport equations are solved for all third moments containing salinity fluctuations (but, to keep computing time at a reasonable level, where the GGDH was still used for triple velocity correlations). For all other terms in the second-moment and dissipation equations, identical approximations are applied. Figure 13 shows what a dominant effect the treatment of the triple moments has on the resultant density fluxes and, indeed, on the consequent shear-stress profile too. It is true that the resultant mean velocity profile is not greatly affected but this simply indicates that the low level of shear stress in the highly stratified region takes a considerable time to modify the mean velocity. An inference that seems reasonable to draw from the above is that, in separated or free flows subject to appreciable streamline curvature or effects of rotation, simple gradient-diffusion approximations for second-moment transport are inadequate. While one may decide to stop short of third-moment closure, ways must be found to incorporate the effects of the source terms in the triple-moment equations associated with curvature and rotation.

5. Concluding Remarks

The main points may be summarized as follows:

- In really complex flows involving curvature or rotation, use of an eddy viscosity model is often less of a disaster than would be inferred from corresponding simple two-dimensional shears.
- It is, however, vitally important in near-wall flows with strong streamwise vorticity (which may be induced by curvature or rotation) to

employ a grid that extends all the way to the wall; matching to a logarithmic law beyond the sublayer means that the vorticity source is greatly underestimated.

- Non-linear EVM's can capture a much wider range of curved flows than linear models provided that cubic level deformation terms are retained in the constitutive equations. (As was demonstrated, quadratic terms play no role in sensitizing the stress field to streamline curvature).
- Notwithstanding, second-moment closure offers the most reliable closure approach for reproducing with fidelity the effects of streamline curvature and rotation on turbulence.
- However, the use, in most models, of 'wall-reflection' terms and wall-normal vectors make them unsuited for application to the complex geometries that arise in practical problems. As has been demonstrated, TCL closures can avoid these weaknesses.
- The principal remaining weakness in second-moment closure seems to be linked with the ε or other scale-determining equation and, in particular, with the reliance on a single turbulent time or length scale. It has been argued that distinguishing the energy transfer rate from the dissipation rate provides a better physical basis for closure, while adding negligibly to the overall cost of the CFD computation.
- In separated or free flows, gradient models of second-moment transport are inadequate when generation is weak (as it often will be near a convex surface).

6. Acknowledgements

The evolution of TCL modeling at UMIST has been especially helped by the research of Drs. T.J. Craft, S. Fu, N.Z. Ince, S.-P. Li and D.P. Tselipidakis and by Mr. J. Kidger. Mrs. C. King prepared the camera-ready version of this paper for publication.

References

Bertoglio, J.P. (1982). *AIAA J.* **20**, pp. 1175.

Bo, T., Iacovides, H., and Launder, B.E. (1995). *Trans. ASME J. Turbomachinery* **117**, pp. 474-484.

Bradshaw, P. (1973). Effects of streamline curvature on turbulent flow. AGARDograph No. 169.

Chang, S.M., Humphrey, J.A.C., and Modavi, A. (1983). *Physico-Chemical Hydrodynamics* **4**, pp. 243.

Choi, Y.-D., Iacovides, H., and Launder, B.E. (1989). *ASME J. Fluids Engineering* **111**, pp. 59-69.

Cooper, D., Jackson, D.C., Launder, B.E., and Liao, G.X. (1993). *Int. J. Heat Mass Transfer* **36**, pp. 2675-2684.

Cousteix, J. and Aupoix, B. (1981), *La Recherche Aérospatiale*, No. 1981-4, p. 275.

Craft, T.J. (1997a). Personal communication.
Craft, T.J. (1997b). Computations of separating and reattaching flows using a low-Reynolds number second-moment closure. *Proc. 11th Symp. Turbulent Shear Flows*, Paper 30.4, Grenoble.
Craft, T.J., Ince, N.Z., and Launder, B.E. (1996). *Dynamics of Atmospheres and Oceans* **25**, pp. 99-114.
Craft, T.J., Kidger, J., and Launder, B.E., (1997). Importance of third-moment modeling in horizontal, stably stratified flows. *Proc. 11th Symp. Turbulent Shear Flows*, Paper 20.3, Grenoble.
Craft, T.J. and Launder, B.E. (1996). *Int. J. Heat & Fluid Flow* **17**, pp. 245-254.
Craft, T.J., Launder, B.E., and Suga, K. (1993). Extending the applicability of eddy viscosity models through the use of deformation invariants and non-linear elements. *Proc. 5th Int. Symp. Refined Flow Modeling and Turbulence Measurements*, Presses Ponts et Chaussées, Paris.
Craft, T.J., Launder, B.E., and Suga, K. (1995). A non-linear eddy viscosity model including sensitivity to stress anisotropy. *Proc. 10th Symp. Turbulent Shear Flows*, Session 23, Pennsylvania State University, pp. 19-25.
Craft, T.J., Launder, B.E., and Suga, K. (1996). *Int. J. Heat Fluid Flow* **17**, pp. 108-115.
Craft, T.J., Launder, B.E., and Suga, K. (1997). *Int. J. Heat Fluid Flow* **18**, pp. 15-28.
Daly, B.J. and Harlow, F.H. (1970). *Phys. Fluids* **13**, pp. 2634-2649.
Durbin, P. (1993). *J. Fluid Mech.* **249**, pp. 465-498.
Fu, S., Launder, B.E., and Tselipidakis, D.P. (1987). Accommodating the effects of high-strain rates in modeling the pressure-strain correlation. *UMIST Mech. Eng. Dept.*, Rep. TFD/87/5.
Gibson, M.M. and Launder, B.E. (1978). *J. Fluid Mech.* **86**, p. 491.
Hanjalic, K., Launder, B.E., and Schiestel (1980). Turbulent Shear Flows 2, pp. 36-49, Springer Verlag, Heidelberg.
Horiuti, K. (1996). Personal communication.
Iacovides, H. (1997). Personal communication.
Iacovides, H., Launder, B.E., and Li, H.-Y. (1996a). *Experimental Thermal and Fluid Science* **13**, pp. 419-429.
Iacovides, H., Launder, B.E., and Li, H.-Y. (1996b). *Int J. Heat & Fluid Flow* **17**, pp. 22-33.
Iacovides, H., Launder, B.E., and Loizou, P.A. (1987). *Int. J. Heat & Fluid Flow* **8**, pp. 320-325.
Kidger, J. (1997). Personal communication.
Kline, S.J., Cantwell, B.J., and Lilley, G.M., editors (1982). Complex turbulent flows: 3, Comparison of computation with experiment, Thermo-Sciences Division, Stanford University, Stanford, CA.
Launder, B.E. (1989). *Int. J. Heat & Fluid Flow* **10**, pp. 282-340.
Launder, B.E. and Li, S.-P. (1994). *Phys. Fluids* **6**, p. 999.
Launder, B.E. and Shima, N. (1989). *AIAA J.* **27**, pp. 1319-1325.
Launder, B.E., Tselipidakis, D.P., and Younis, B. (1987). *J. Fluid Mech.* **183**, pp. 63-75.
Li, H.-Y. (1995). Ph.D. Thesis, Faculty of Technology, University of Manchester.
Lumley, J.L. and Khajeh-Nouri, B.J. (1974). *Adv. Geophysics* **18A**, pp. 169.
Lumley, J.L. (1978). *Adv. Applied Mechanics* **18**, Academic, New York.
Malecki, P. (1994). Etude de modéles de turbulence pour les couches limites tridimensionnelles, Thése de Docteur de l'ENSAE, Toulouse, France.
Myong, H.K. and Kasagi, N. (1990). *Trans. ASME, J Fluids Eng.* **112**, pp. 521-524.
Nicholson, J.H., Forest, A.E., Oldfield, M.L.G., and Schultz, D.L. (1982). Heat-transfer optimized turbine rotor blades - an experimental study using transient techniques, ASME Paper 82-GT-304.
Nisizima, S. and Yoshizawa, A. (1987). *AIAA J.* **25**, pp. 414-420.
Oberlack, M. (1995). Turbulent Shear Flows - 9, pp. 33-52, Springer, Berlin.
Pope, S.B. (1975). *J. Fluid Mech.* **72**, pp. 331-340.

Rubinstein, R. and Barton, J.M. (1990). *Phys. Fluids A2*, pp. 1472-1476.

Shih, T.-H. and Lumley, J.L. (1985). Modeling of pressure correlations terms in Reynolds stress and scalar-flux equations Rep FDA-85-3, Sibley School of Mech. & Aerosp. Engg., Cornell Univ.

Shih, T.-H., Zhu, J., and Lumley, J.L. (1993). NASA TM-105993.

Sotiropoulos, F. and Patel, V.C. (1995). Application of Reynolds-stress transport models to stern and wake flows. *J. Ship Research* **39**, No. 4, pp. 263-283.

Speziale, C. (1987). *J. Fluid Mech.* **178**, pp. 459-475.

Suga, K. (1995). Development and application of a non-linear eddy viscosity model sensitized to stress and strain invariants. Ph.D. Thesis, Department of Mechanical Engineering, UMIST.

Taylor, A.M.K.P., Whitelaw, J.H., and Yianneskis, M. (1982). *J. Fluids Eng.* **104**, p. 350.

Uittenbogaard, R. (1992). Measurement of turbulent fluxes in a steady, stratified mixing layer. Proc. 3rd Int. Symp. on Refined Flow Modeling and Turbulence Measurement, Tokyo.

Van Driest, E.R. (1956). *J. Aero. Sci.* **23**, p. 1007.

Wagner, J.H., Johnson, B.V., and Hajek, T.J. (1991). *Trans. ASME J. Turbomachinery* **113**, pp. 42-51.

Wilcox, D.C. (1988). *AIAA J.* **26**, pp. 1299-1310.

Wizman, V., Laurence, D., Kanniche, M., Durbin, P., and Demuren, A. (1996). *Int. J. Heat & Fluid Flow* **17**, pp. 255-266.

A PERSPECTIVE ON TURBULENCE MODELING

STEPHEN B. POPE
Sibley School of Mechanical & Aerospace Engineering
Cornell University
Ithaca, New York

1. Introduction

In this paper, we speculate on the usage of turbulence models over the next 10 or 20 years. To this end, some insights are gained by reflecting on the developments over the last 25 years, since the introduction of the k-ε turbulence model (Jones *et al.*, 1972).

It is emphasized (in Section 2) that there currently exist a broad range of models and a broader range of applications. In 10 or 20 years, computer power will have advanced substantially. Will this lead to an abandonment of the simpler models (based on the Reynolds-averaged Navier-Stokes (RANS) equations), in favor of large-eddy simulations (LES)? The consequences of increased computer power are considered in Section 3. It is argued that, in most applications, there are more pressing demands for the increased computer power than the use of computationally-intensive approaches such as LES. Consequently, while the use of LES will undoubtedly increase, it is argued that the bulk of turbulent flow calculations will still be based on RANS (or unsteady RANS) approaches. The whole range of turbulence models will remain useful.

The general opinion (e.g., Bradshaw *et al.* (1996)) is that the progress towards accurate and general RANS models (Reynolds-stress models (RSM) in particular) has been disappointingly slow. As discussed in Section 4, it is important to distinguish between the *level of closure* and *particular models*. The level of closure defines a class of models; and a particular model is a member of one such class. If, in some application, a particular turbulence model is found to be inaccurate, is this because the turbulence physics cannot be adequately represented at the level of closure? Or is it because the coefficients in the particular model have been poorly chosen? While every level of closure has its intrinsic limitations, it is suggested that the some-

M. D. Salas et al. (eds.), Modeling Complex Turbulent Flows, 53–67.

what disappointing performance of RANS models is due more to the choice of model coefficients. Specifically, current models are some distance from the *optimal model* at that level of closure.

Optimal models are defined and discussed in Section 5. Since models at different levels of closure will be used for many years to come, it would be extremely valuable to develop a general methodology to determine optimal models. (It is remarkable how little effort over the last 25 years has been applied to this objective.)

2. Current Range of Models and Applications

Nearly all modeling approaches in use and envisioned fall into one of the following three classes:

1. RANS models.
2. Partial resolution of unsteady turbulent motions.
3. Full resolution of unsteady turbulent motions.

RANS models can be applied to statistically stationary flows, for which they predict the stationary fields of the mean velocity and some turbulence properties. These models include:

- algebraic models (e.g., Baldwin *et al.* (1978))
- one-equation models (e.g., Spalart *et al.* (1992))
- two-equation models (e.g., Jones *et al.* (1972))
- Reynolds-stress models (e.g., Launder *et al.* (1975))
- elliptic relaxation models (e.g., Durbin (1993))
- PDF models (e.g., Haworth *et al.* (1986a)).

Even though they lack generality, it is important to appreciate that the simplest models can perform quite satisfactorily for narrow—but technologically important—classes for flows (see, e.g., Wilcox (1993)).

RANS models represent the turbulence in terms of one-point one-time statistics, and they do not explicitly represent time-dependent turbulent motions. In some circumstances—when there are large-scale unsteady turbulent motions that have a dominant effect upon the flow—it may be preferable to account explicitly for some of these unsteady motions. One way is through large-eddy simulation: but it should be appreciated that there is again a range of possibilities for the *partial resolution of unsteady turbulent motions*. Such possibilities include:

- unsteady RANS modeling—in which only the largest (usually 2D) unsteady motions are resolved
- very large eddy simulations (VLES)—in which some fraction of the turbulent energy-containing motions are resolved

- LES—in which essentially all of the energy-containing motions are resolved.

Durbin (1995) provides an excellent example of the success of unsteady RANS for a flow with vortex shedding, for which the corresponding steady RANS model calculations are grossly inaccurate. The computational cost and difficulty of such unsteady RANS calculations is orders of magnitude less than LES calculations.

In LES (and other partial resolution approaches) modeling is required. Only by fully resolving all scales of turbulent motion through direct numerical simulation (DNS) can all modeling be avoided.

DNS is a powerful research tool that the author has used extensively (e.g., Yeung *et al.* (1989); Juneja *et al.* (1996).) But DNS should not be considered as a possible approach to engineering turbulent flow calculation (at moderate or high Reynolds number, within the twenty-year time frame being considered here). Not only is the computational cost outrageous and prohibitive, but the computational effort is misplaced. From an engineering perspective, in nearly all circumstances it is the energy-containing turbulent motions that are important, whereas in DNS over 99.9% of the effort is devoted to the smallest dissipative motions.

To substantiate this observation, consider a DNS of homogeneous isotropic turbulence using a pseudo-spectral method. Spatial resolution requires the highest wavenumber $k_{\max}$ to satisfy

$$k_{\max}\eta \geq 1.5,$$

where η is the Kolmogorov lengthscale. The dissipative range can be taken to be wavenumbers greater than k_D, where

$$k_D\eta = 0.1.$$

In the three-dimensional wavenumber space, the fraction of the wavenumbers that are in the dissipative range is

$$\frac{k_{\max}^3 - k_D^3}{k_{\max}^3} \geq \frac{15^3 - 1}{15^3} = 0.9997.$$

Thus, 99.97% of the computational effort is devoted to the dissipative motions.

Not only is there a broad range of turbulence models, but there is also a broad range of uses to which they are put. The following are some of the distinguishing characteristics of the different uses:

- geometric complexity. From simple 2D airfoils, to the flow in the ports and cylinders of internal combustion engines.

- complexity of physical processes. Applications consisting solely of single-phase fluid mechanics are the exception rather than the rule. More often, the applications involve heat or mass transfer, chemical reactions, multi-phase flow etc.
- accuracy required. In many aerodynamic applications, calculations are required to be accurate to less than 1%. In more complex, less well understood flows, 20% accuracy may be very useful.
- number of calculations. In some applications, considerable effort can be justified to make a calculation of a single flow. In others, such as design optimization, possibly thousands of flow calculations are required.
- user's background and support. In some industrial settings, turbulence-model calculations are performed by highly-qualified (e.g., Ph.D. degree) and experienced personnel; and there may be a parallel experimental program to validate the models for the range of flows considered. Increasingly though, commercial CFD codes incorporating turbulence models are being used by less qualified (e.g., M.S. degree) and less experienced personnel, with less experimental validation.

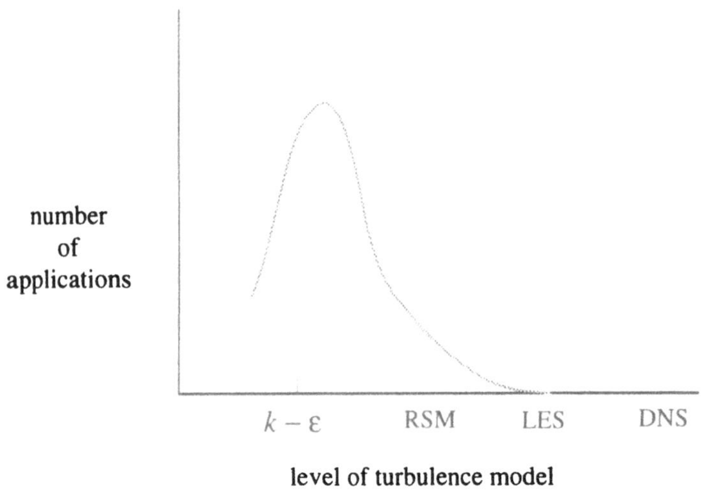

Figure 1. Distribution of current turbulence model usage.

Figure 1 provides a rough picture of current turbulence model usage in engineering applications. The center of the distribution is squarely at the two-equation level (e.g., k-ε): this is the simplest level at which a *complete closure* is possible (without requiring the specification of a mixing length, for example); and the k-ε model is implemented in nearly all commercial CFD codes. But the range extends to yet simpler models (one-equation and algebraic) and also to Reynolds stress models, with a handful of LES applications.

3. Impact of Increasing Computer Power

Figure 2 shows the speed (measured in flops—floating point operations—per second) of the fastest computers over the last 30 years. It may be seen that this speed has increased quite consistently at a rate of a factor of 30 per decade. While there is no sound basis for extrapolation beyond a few years, it is nevertheless generally supposed that this trend will continue.

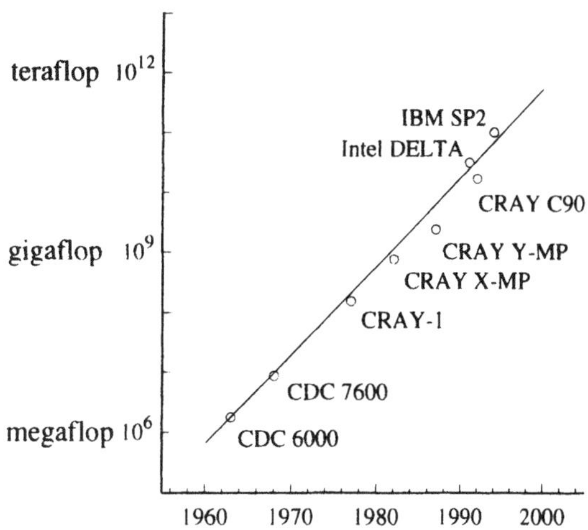

Figure 2. Speed (flops per second) of the fastest supercomputers against year of their introduction. The line shows a growth rate of a factor of 30 per decade. (Adapted from Foster (1995).)

Clearly, an increase in computer power of 30 in 10 years or 900 in 20 years will have a major impact on how turbulent flow calculations are performed. But the nature of this impact requires careful consideration.

3.1. RANGE OF COMPUTATIONS

There is, of course, a large range in the scale of turbulent flow computations performed. It is useful to characterize this range by three types of calculations.

1. large-scale research computations requiring of order 200 CPU hours on the most powerful supercomputer. The channel-flow DNS of Kim

(1987) required 250 hours. Such calculations are generally much too expensive for engineering applications.
2. large engineering calculations requiring 15 minutes CPU time on a supercomputer, or equivalently 25 hours CPU time on a workstation.
3. repetitive engineering calculations—as may be required in a design optimization study—requiring 1 minute of CPU time on a workstation.

The relative magnitudes of these computations is summarized in Table 1. For use in a "large engineering calculation" a methodology that is currently a "large-scale research computation" requires an increase in computer speed of about 1,000. With a speed increase rate of 30 per decade, this corresponds to 20 years. To use the same methodology for "repetitive engineering computations" requires a further factor of 1,000 in speed, and hence a further 20 years of hardware development. (These estimates of 20 and 40 years may be pessimistic by a factor of 2, say, because they do not take into account algorithmic developments, which typically keep pace with hardware developments.)

TABLE 1. CPU time requirements of different types of computations.

	large-scale research computation	large engineering calculation	repetitive engineering computation
CPU time on supercomputer	200 hrs	15 min	—
CPU time on workstation	—	25 hrs	1 min
relative CPU time	1.2×10^6	1.5×10^3	1
	now	20 years	40 years
	—	now	20 years

These considerations suggest that relatively simple models, that currently require significantly less CPU time than "large-scale research computations," will continue to be dominant in engineering usage over the next 20 years.

3.2. USE OF INCREASED COMPUTER POWER

In considerations of turbulent flow calculations, it is often assumed that all of the future increases in computer power can be used for more demanding turbulence approaches—LES in place of RANS, for example. In fact, there are many demands on the computer resources, and turbulence modeling may be far from the most pressing. Examples of the uses of increased computer power are:

1. improved spatial resolution in the computations to achieve numerical accuracy
2. more complete and accurate representation of the boundary geometry
3. use of larger solution domains so that boundary conditions can be specified more easily
4. modeling of other processes such as: heat and mass transfer; chemical reactions; multi-phase flow; acoustics
5. use of more versatile, general and user-friendly codes (which are less efficient in terms of CPU time)
6. reduce run times—to enable repetitive engineering calculations, for example
7. use of more demanding turbulence models.

3.3. DETERMINANTS OF COMPUTER REQUIREMENTS

In order of importance, the primary factors determining the computer requirements of a turbulence modeling approach are:

1. length and time scales that have to be resolved
2. number (and nature) of independent variables
3. number (and nature) of equations to be solved.

DNS requires the resolution of the smallest Kolmogorov scales, which is why it is prohibitively demanding for moderate and high Reynolds number flows. RANS approaches, on the other hand, require resolution only of mean fields. In approaches with partial resolution of the unsteady turbulent motions, the computer requirements rise rapidly as the scales of the resolved motions decreases. For statistically-stationary flows, statistical approaches (e.g., RANS) have no temporal resolution requirements, and converged solutions can be obtained in of order 100 iterations. In contrast, for LES and DNS time-marching methods are required, and typically tens of thousands of time steps are needed.

Not all independent variables are the same. The computer requirements increase rapidly going from OD to 1D to 2D and to 3D flows. But in PDF methods, the computer work rises only linearly (to a first approximation) with the number of independent sample-space variables. For example, Saxena *et al.* (1998) performed PDF calculations with 18 independent variables (2 positions, 3 velocities, 1 turbulence frequency, and 14 thermochemical compositions) for a piloted methane jet flame. This required just 25 CPU hours on a workstation.

In comparison with items 1 and 2 in the list above, the impact of item 3—the number of equations solved—is relatively minor.

3.4. COMPUTATIONAL COSTS

We have discussed CPU times extensively above. But this is just one component of the cost of performing a turbulent flow calculation. The primary components of the cost are

1. code development costs (paid directly by the developer, or indirectly by the user through license fees)
2. manpower costs in performing the calculations (salary, education and training)
3. hardware costs (CPU time, etc.).

At commercial rates, large DNS calculation can cost well over $1,000,000, and so for this case item 3 dominates. And in such applications, computational efficiency (in terms of CPU time) is of prime importance.

With simpler models, the other components of cost may be more significant. Indeed, as illustrated in Fig. 3, as CPU speeds increase, eventually the CPU costs become small compared to the other costs. In these circumstances, rather than CPU efficiency, the important issues become ease of use, reliability and accuracy.

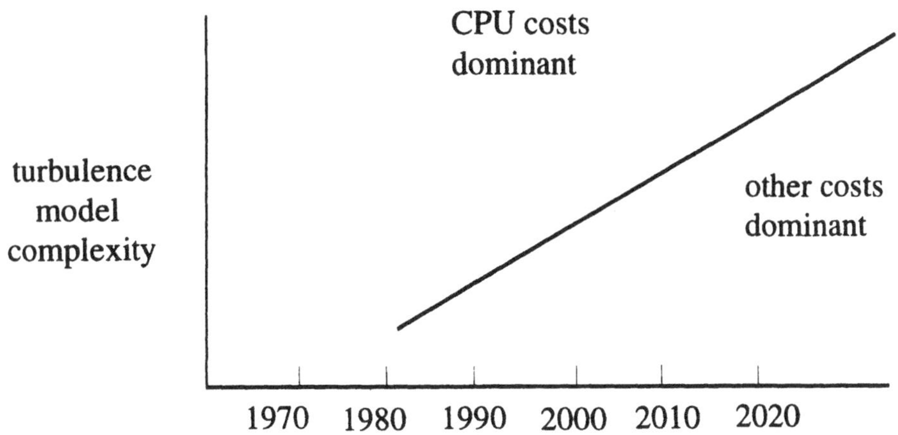

Figure 3. Relative importance of computation costs for turbulence models of different complexities. With increasing time, as CPU speeds increase, CPU costs become less dominant.

4. Future Usage of Turbulence Models

Figure 4 shows the conjectured distribution of turbulence model use 10 or 20 years from now. The main points are

1. The total usage will increase
   (a) because of the increased availability and ease-of-use of commercial CFD codes
   (b) because the increased computer power allows the treatment of more applications (most with complex geometry and complex physical processes)
   (c) because of model improvements, the accuracy (and hence usefulness) of the calculations will increase
2. Most of the increases in computer power will be used for items 1–6 enumerated in Section 3.2, not for more complex turbulence modeling. Hence the distribution shown in Fig. 4 moves only slightly in the direction of more demanding models.
3. There will be more use of models which partially resolve the unsteady turbulent motions (e.g., unsteady RANS and LES), but RANS methods remain dominant.

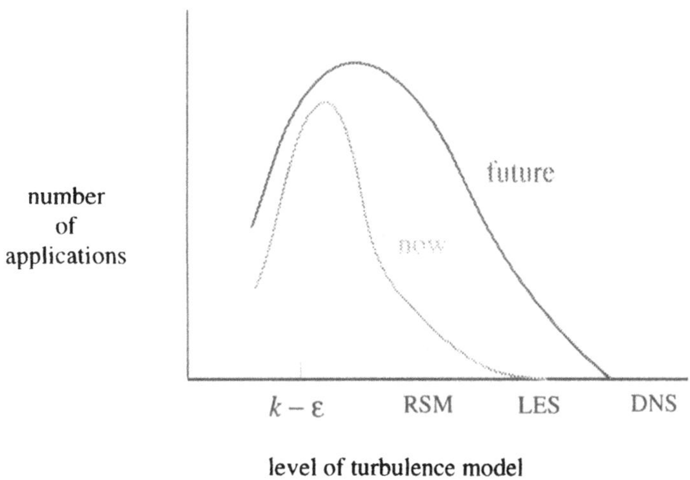

Figure 4. Distribution of current and future (10–20 years) turbulence model usage.

Further to (3) above, we can ask: how will the increased usage in engineering of "partial resolution" models take place? And, in this context, what developments in modeling methodology will be useful? On the first question, one route that may be followed in some applications is the abandonment of current RANS-based methods, with a jump to LES. More prevalent, however, in the industrial setting will be a more gradual transition from

RANS, to unsteady RANS, to VLES, etc. At present, there are distinct model equations solved at the different levels of resolution (RANS–LES–DNS). A useful aim of modeling research in this area is to develop a unified model, which can be applied (and is accurate) at any specified level of resolution, from RANS to DNS.

As discussed in Section 2, there is a broad range of turbulent flow applications, most involving more complex physical and chemical processes than the constant-property Newtonian flows that form the focus of current LES research. The development of "partial resolution models" for such complex processes is necessary if this methodology is to be broadly applied. As an example, Colucci *et al.* (1998) extend the PDF methodology to treat chemically reactive flows by LES.

5. Levels of Closure and Particular Models

Based on a collaborative testing of turbulence models, Bradshaw *et al.* (1996) summarize the status of RANS modeling thus: "Our conclusion is, alas, much the same as that of the 1980-81 meeting: no current Reynolds-averaged turbulence model can predict the whole range of complex turbulent flows to worthwhile engineering accuracy." "Stress-transport models ... did appear somewhat better than eddy-viscosity methods, but not enough to warrant the abandonment of eddy-viscosity models." This disappointing state of affairs prompts a number of questions, primarily:

- why has more progress not been made in the last 25 years?
- have RANS models reached their full potential, or are there RANS models (yet to be discovered) with markedly superior performance?

As a first step in a discussion of these questions we emphasize the distinction between the *level of closure* (or *class of models*) and *particular models*.

Figure 5 shows an incomplete classification of RANS models. The models are classified (and sub-classified) in terms of the *forms* of their constituent equations. A particular model—a point in Fig. 5—is characterized by the complete specification of all of the coefficients appearing in the model equations.

Consider, for example, the class (not the most general) of k-ε models defined by the equation:

$$\langle u_i u_j \rangle = k \mathcal{F}_{ij}(\widehat{\mathbf{S}}, \widehat{\boldsymbol{\Omega}}), \tag{1}$$

$$\frac{\bar{D}k}{\bar{D}t} = \nabla \cdot \left(\frac{C_\mu k^2}{\sigma_{k\varepsilon}} \nabla k \right) - \langle u_i u_j \rangle \frac{\partial \langle U_i \rangle}{\partial x_j} - \varepsilon, \tag{2}$$

$$\frac{\bar{D}\varepsilon}{\bar{D}t} = \nabla \cdot \left(\frac{C_\mu}{\sigma_\varepsilon} \frac{k^2}{\varepsilon} \nabla \varepsilon \right) + \frac{\varepsilon^2}{k} S_\varepsilon(\widehat{\mathbf{S}}, \widehat{\boldsymbol{\Omega}}), \tag{3}$$

where $\langle \mathbf{U} \rangle$ is the mean velocity, $\bar{D}/\bar{D}t$ is the mean substantial derivative $(\partial/\partial t + \langle \mathbf{U} \rangle \cdot \nabla)$ and $\widehat{S}_{ij}$ and $\widehat{\Omega}_{ij}$ are the normalized mean rates of strain and rotation:

$$\widehat{S}_{ij} \equiv \frac{1}{2}\frac{k}{\varepsilon}\left(\frac{\partial\langle U_i\rangle}{\partial x_j} + \frac{\partial\langle U_j\rangle}{\partial x_i}\right), \tag{4}$$

$$\widehat{\Omega}_{ij} \equiv \frac{1}{2}\frac{k}{\varepsilon}\left(\frac{\partial\langle U_i\rangle}{\partial x_j} - \frac{\partial\langle U_j\rangle}{\partial x_i}\right). \tag{5}$$

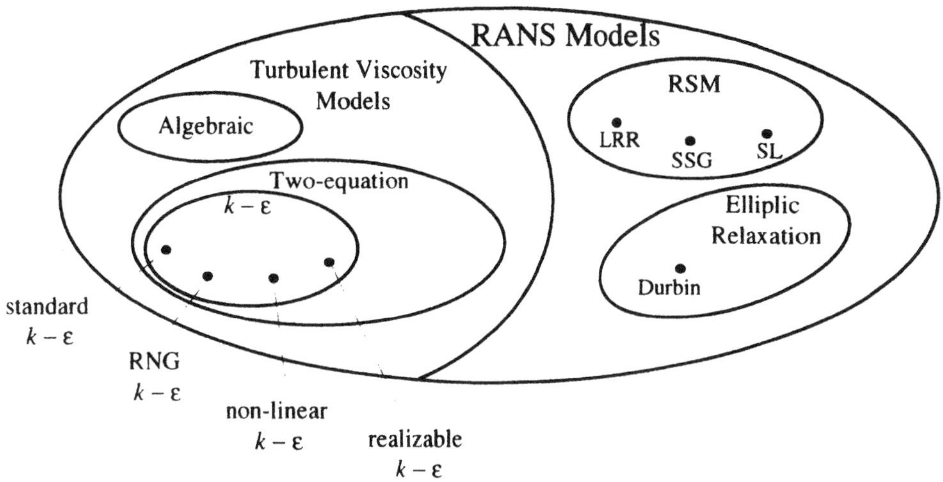

Figure 5. Sketches of particular models as members of subclasses and classes of models.

A particular model within the class corresponds to a particular specification of the constants C_μ, σ_k and σ_ε, of the non-dimensional scalar function $S_\varepsilon(\widehat{\mathbf{S}}, \widehat{\mathbf{\Omega}})$, and of the symmetric non-dimensional tensor function $\mathcal{F}_{ij}(\widehat{\mathbf{S}}, \widehat{\mathbf{\Omega}})$. For example, the specifications for the standard k-ε model are:

$$C_\mu = 0.09, \quad \sigma_k = 1.0, \quad \sigma_\varepsilon = 1.3, \tag{6}$$

$$\mathcal{F}_{ij} = \frac{2}{3}\delta_{ij} - 2C_\mu \widehat{S}_{ij}, \tag{7}$$

$$S_\varepsilon = 2C_\mu C_{\varepsilon 1}\widehat{S}_{ij}\widehat{S}_{ij} - C_{\varepsilon 2}, \tag{8}$$

with

$$C_{\varepsilon 1} = 1.44 \text{ and } C_{\varepsilon 2} = 1.92. \tag{9}$$

It is important to appreciate that the assessment of *particular models* and of *classes of models* must be based on qualitatively different arguments.

For example, a comparison between experimental data and model calculations may reveal that (for the flow considered) the standard k-ε model calculation of the mean velocity is in error by up to 40%. Nothing can be deduced from this observation about the performance of *all* k-ε models.

In the absence of mean velocity gradients, the function $\mathcal{F}_{ij}$ is inevitably isotropic (e.g., $\mathcal{F}_{ij} = \frac{2}{3}\delta_{ij}$). Consequently *no* k-ε model (indeed, no turbulent viscosity model) can represent the Reynolds-stress anisotropy in decaying, anisotropic grid turbulence.

A quest for better models can proceed in two distinct directions. One is to seek higher levels of closure, so that the class of model considered can represent more turbulence phenomena. For example, a Reynolds-stress level may be chosen to overcome the turbulent viscosity models' inability to represent anisotropic decaying turbulence. Or a structure-based level (Reynolds *et al.*, 1995; Van Slooten *et al.*, 1997) may be chosen to overcome Reynolds-stress models' inability to calculate rapid distortions with mean rotation. Valuable though quests in this direction may be, it should be appreciated that:

a) any level of statistical closure provides only a partial representation of the turbulence, and therefore it cannot account for all turbulence behaviors
b) even though a level of closure is known to have limitations, a model at that level may nevertheless be able to calculate a range of flows to useful accuracy.

The second direction that can be taken in the quest for better models is to seek particular models within each class which are more accurate than current models. In the context of engineering turbulence model calculations (for most inert flows), essentially all of the progress that has been made in the last 25 years has been in this direction. The class of two-equation, k-ε, and Reynolds stress models had certainly been identified by 1972 (e.g., Jones *et al.* (1972); Hanjalic *et al.* (1972)). The newer k-ε models (e.g., RNG, non-linear, realizable) and Reynolds-stress models (e.g., LRR, Shih-Lumley, SSG) represent different particular models within the same class.

6. Towards Optimal Turbulence Models

It is argued above that the full range of models will be in use for the next 20 years—from two-equation to LES, with RANS models being dominant. It would be extremely valuable, therefore, to determine the *optimal model* within each class, and to have a general methodology for doing so. We offer the following four-step definition of an optimal model.

1. The *class of models* considered is defined by a set of equations (e.g., Eqs. 1–3) involving *free constants* (e.g., $C_\mu, \sigma_k, \sigma_\varepsilon$) and *free coefficients*

(e.g., $\mathcal{F}_{ij}$ and S_ε).

2. A *range of flows* is considered for which there are accurate experimental data.
3. For every model within the class, the *error* ϵ is defined as a specified weighted difference between the model calculation and the experimental data.
4. The *optimal model* corresponds to the specification of the free constants and coefficients[1] that minimizes the error ϵ.

It is all but certain that all current models are sub-optimal. Taking the standard k-ε model as an example, it has been known for 25 years (Rodi, 1972) that C_μ is more appropriately a non-trivial function of $\widehat{\mathbf{S}}$ and $\widehat{\mathbf{\Omega}}$ than a constant. While functional forms of C_μ have been proposed (e.g., Pope (1975)) there is no consensus on a near-optimal form. A methodology for determining optimal models would be able to determine the appropriate specifications, and would determine how far from optimal current models are.

Model coefficients are often specified by assuming them to be constant, and by determining the value of the constant by reference to a single observation. For example $C_\mu = 0.09$ stems from the observation (in turbulent shear flow) $C_\mu^{\frac{1}{2}} = |\langle u_1 u_2 \rangle|/k \approx 0.3$. Alternatively, a simple functional form is assumed, which is constructed or constrained to behave correctly in extreme limiting circumstances, such as rapid distortion or two-component or two-dimensional turbulence. Again, the numerical parameters in the assumed functional form are typically determined by reference to a few simple observations, often in homogeneous turbulence. Arguably, the relatively slow progress in Reynolds-stress modeling is due to overly restrictive functional forms for the coefficients, and to the lack of a methodology for determining near-optimal values of the free parameters.

There are strong theoretical arguments showing that some classes of models are superior to others: non-linear viscosity models are superior to linear viscosity models; Reynolds stress models are superior to turbulent viscosity models; models with more than one scale equation are superior to those with a single scale equation (e.g., the ε equation). This superiority is manifest in that the error ϵ incurred by the optimal model in the "superior" class, is less than that incurred by the optimal model in the "inferior" class. But of course, through an abysmal choice of the free coefficients, a model in the "superior" class can perform much worse than a reasonable model in the "inferior" class. Indeed, it has often been the case that in the early years of their development, particular models in a "superior" class are found

[1] With coefficients—as opposed to constants—the minimization problem is underdetermined, and additional smoothness conditions on the coefficients are appropriate.

not to be convincing improvements over the best models in the "inferior" class. In the assessment cited above—"stress-transport models ... appear somewhat better than eddy-viscosity models"—we believe that the qualifier is "somewhat" rather than "decisively" because current Reynolds-stress models are far from optimal.

The development of a methodology to determine optimal turbulence models (as defined above) is a challenge for future research. The constants in the generalized Langevin model (GLM) of Haworth *et al.* (1986a) were determined by the four-step optimization procedure described above; but no work on a general methodology for determining optimal turbulence-model coefficients is known to the author.

7. CONCLUSIONS

We have considered the future of turbulent flow computations in the context of engineering problems. The principal points of this paper—many of them conjectures—are as follows:

1. There is a very broad range of turbulent flow problems, and it is valuable (now and into the future) to have a broad range of turbulence modeling approaches.
2. Over the next 20 years there will be increased usage of turbulence models, dominantly RANS models.
3. There will be more usage of models that partially represent unsteady turbulent motions; more by a transition from RANS to unsteady RANS and beyond rather then by a jump to LES.
4. The projected increase in computer power (a factor of 30 in 10 years, 1000 in 20 years) will be used more on the first 6 items enumerated in Section 3.2 than on much more computationally-demanding approaches such as LES.
5. There would be great value in a methodology for determining optimal turbulence models at different levels of closure—as defined in Section 6.
6. The observed marginal superiority of Reynolds stress models over turbulent viscosity models—in contrast to their marked theoretical superiority—is attributed to existing Reynolds stress models being far from optimal.
7. Challenges for future research include
   (a) developing a general methodology for determining optimal turbulence models
   (b) developing unified models that can be applied (and are accurate) at any specified level of resolution of the unsteady turbulent motions (from RANS to LES to DNS)

(c) developing models for complex physical and chemical processes (e.g., combustion, multi-phase flow) for use in "partial resolution models."

References

Baldwin, B. S. and Lomax, H., 1978. "Thin-layer approximation and algebraic model for separated turbulent flows," AIAA Paper 78-257, Huntsville, AL.

Bradshaw, P., Launder, B. E., and Lumley, J. L., 1996. "Collaborative testing of turbulence models," *Trans. ASME I: J. Fluids Engng.* **118**, pp. 243–247.

Colucci, P. J., Jaberi, F. A., Givi, P., and Pope, S. B., 1998. "Filtered density function for large eddy simulation of turbulent reacting flows," *Phys. Fluids* **10**, pp. 499–515.

Durbin, P. A. Durbin, 1993. "A Reynolds stress model for near-wall turbulence," *J. Fluid Mech.* **249**, pp. 465–498.

Durbin, P. A., 1995. "Separated flow computations with the k-ϵ-v^2 model," *AIAA J.* **33**, pp. 659–664.

Foster, I., 1995. *Designing and building parallel programs.* Addison-Wesley.

Hanjalic, K. and Launder, B. E., 1972. "A Reynolds stress model of turbulence and its application to thin shear flows," *J. Fluid Mech.* **52**, pp. 609–638.

Haworth, D. C. and Pope, S. B., 1986. "A generalized Langevin model for turbulent flows," *Phys. Fluids* **29**, pp. 387–405.

Jones, W. P. and Launder, B. E., 1972. "The prediction of laminarization with a two-equation model of turbulence," *Int. J. Heat Mass Transfer* **15**, pp. 301–314.

Juneja, A. and Pope, S. B., 1996. "A DNS study of turbulent mixing of two passive scalars," *Phys. Fluids* **8**, pp. 2161–2184.

Kim, J., Moin, P., and Moser, R., 1987. "Turbulence statistics in fully developed channel flow at low Reynolds number" *J. Fluid Mech.* **177**, pp. 133–166.

Launder, B. E., Reece, G. J., and Rodi, W., 1975. "Progress in the development of a Reynolds-stress turbulence closure," *J. Fluid Mech.* **68**, pp. 537–566.

Pope, S. B., 1975. "A more general effective-viscosity hypothesis," *J. Fluid Mech.* **72**, pp. 331–340.

Reynolds, W. C. and Kassinos, S. C., 1995. "A one-point model for the evolution of the Reynolds stress and structure tensors in rapidly deformed homogeneous turbulence," *Proc. R. Soc. Lond. A* **451**, pp. 87–104.

Rodi, W., 1972. *The prediction of free turbulent boundary layers using a two-equation model of turbulence*, Ph.D. thesis, Imperial College, London.

Saxena, V. and Pope, S. B., 1998. "PDF calculations of major and minor species in a turbulent piloted jet flame," Twenty-seventh Symp. (Int'l) on Combust., The Combustion Institute, in press.

Spalart, P. R. and Allmaras, S. R., 1992. "A one-equation turbulence model for aerodynamic flows," AIAA Paper 92-439, Reno, NV.

Van Slooten, P. R. and Pope, S. B., 1997. "PDF modeling of inhomogeneous turbulence with exact representation of rapid distortions," *Phys. Fluids* **9**, pp. 1085–1105.

Wilcox, D. C., 1993. *Turbulence modeling for CFD*, DCW Industries, La Cañada, CA.

Yeung, P. K. and Pope, S. B. Pope, 1989. "Lagrangian statistics from direct numerical simulations of isotropic turbulence,: *J. Fluid Mech.* **207**, pp. 531–586.

DEVELOPMENTS IN STRUCTURE-BASED TURBULENCE MODELING

Particle and one-point modeling of deformations of homogeneous turbulence

S. C. KASSINOS AND W. C. REYNOLDS
Department of Mechanical Engineering
Stanford University, Stanford, California

1. Introduction

The role of turbulence models in engineering codes is to provide the turbulent (Reynolds) stresses for the mean momentum equations. The Reynolds stresses themselves are seldom of any other use and this motivates the desire to minimize computational resources devoted to their computation. Economy in predictive efforts has another side of course, and that is the ability to compute reliably as many diverse flow regimes as possible without the need to modify the model in the engineering code. The choice of the appropriate level of sophistication in turbulence modeling is then one of a balance between minimizing computational effort per specific application on one hand, and maximizing reliability of results while minimizing code modifications on the other.

In simple flows, where the mean deformation rates are mild and the turbulence has time to come to equilibrium with the mean flow, the Reynolds stresses are determined by the applied strain *rate*. Hence in these flows, it is often adequate to use an eddy-viscosity representation. Modern k-ϵ models, which use transport equations for the turbulent kinetic energy k and dissipation rate ϵ, have been very useful in predicting near equilibrium turbulent flows, where the rms deformation rate S is small compared to the reciprocal time scale of the turbulence (ϵ/k).

Turbulent flows of importance in aircraft and propulsion system design involve complex three-dimensional time-dependent mean flow, where the turbulence is often subjected to very rapid deformation and strong mean rotation or curvature effects. In these non-equilibrium flows the structure of the turbulence plays an important role in determining the transport of the Reynolds stresses, and because the turbulence structure takes sometime to

M. D. Salas et al. (eds.), Modeling Complex Turbulent Flows, 69–87.

respond to the imposed deformation, an eddy-viscosity representation is not appropriate. Eddy-viscosity theory cannot possibly account in any general fashion for these structure-induced effects, because the basic assumptions on which the theory is based ignore the role of the turbulence structure. As a result, the ability of aerospace engineers to predict these complex flows is limited by the current state-of-the-art in one-point turbulence modeling, which invariably relies on an eddy-viscosity approach.

The response of turbulence to the kind of deformation that is often encountered in engineering applications (rapid rates and/or strong rotation) is described well by Rapid Distortion Theory (RDT). RDT is a closed two-point theory, but engineering models require one-point formulation. Therefore, what is needed is a one-point model that matches eddy viscosity models for weak deformation rates and RDT for rapid deformations.

Reynolds stress transport (RST) equations have been added to the PDE system used in turbulence models in an attempt to deal with the weakness of eddy-viscosity. While RST models have enjoyed some success, they are not yet widely used in industry because they have not proven reliably better than simpler models in dealing with the more challenging types of complex flows. The problem with RST models is that, even though they incorporate more of the correct physics, they still ignore the role of the turbulence structure. When structure-induced effects dominate, RST models are bound to fail as does eddy viscosity theory. The need to include structure information in turbulence models is most clearly demonstrated by considering the effects of strong mean rotation (or curvature effects): standard RST models as well as eddy-viscosity models fail to distinguish between turbulence fields with widely different responses to the same mean rotation.

Our goal has been the development of an engineering one-point model of turbulence that incorporates structure information in order to achieve the correct viscoelastic character. This led us to the introduction of new one-point turbulent tensors that carry the information missing from the Reynolds stresses. Kassinos and Reynolds (1994) and Reynolds and Kassinos (1995)[1] have already used these new tensors to construct a successful Particle Representation Model (PRM) and a one-point structure-based model (SBM) of RDT.

Here we outline extensions to these models that allow them to deal with nonlinear effects, important whenever the time scale of the mean deformation is large compared to that of the turbulence. An in-depth account of these new modeling ideas will be presented in a manuscript currently in the review process. In Section 2 we first introduce the new one-point structure tensors and explain notation, and then we lay down the basic ideas

[1]Hereafter denoted by KR94 and RK95 respectively.

of a Particle Representation Model (PRM) and show how it can emulate RDT. In Section 2.5 we show how a structure-based model of the nonlinear interactions can be included in the PRM. In Section 3 we outline a one-point model that follows from the extended PRM. An evaluation of the two models for several deformations of homogeneous turbulence is given in Section 4. Finally, Section 5 is a concluding summary.

2. Particle Representation Models

In Sections 2.1 and 2.2 we introduce the basic RDT equations and the key ideas behind the PRM for the exact emulation of RDT. Then in Section 2.5, we present the formulation of the extended model, including the effects of the nonlinear particle-particle interactions (IPRM).

2.1. THE BASIC RDT EQUATIONS

The discussion is restricted to inviscid RDT because in the large eddies, which contribute most of the Reynolds stresses, viscous effects are usually negligible, but this restriction can be removed. For the inviscid RDT of homogeneous turbulence, the fluctuating continuity and momentum equations are given by,

$$\frac{\partial u_i'}{\partial x_i} = 0 \tag{1}$$

and

$$\frac{\partial u_i'}{\partial t} + U_j \frac{\partial u_i'}{\partial x_j} = -G_{ij} u_j' - \frac{1}{\rho} p'_{,i}. \tag{2}$$

Here standard tensor notation is employed (subscripts after commas denote differentiation), U_i is the mean velocity, $G_{ij} = U_{i,j}$ is the mean velocity gradient tensor, and p' is the rapid part of the pressure fluctuations

$$\frac{1}{\rho} p'_{,kk} = -2 G_{mn} u'_{n,m} \,. \tag{3}$$

We introduce the turbulent stream function vector defined by

$$u_i' = \epsilon_{its} \Psi'_{s,t} \qquad \Psi'_{i,i} = 0 \qquad \Psi'_{i,nn} = -\omega_i' \,. \tag{4}$$

We require Ψ_i' to be divergence-free so that the last equality in (4) is valid. This choice is important for the physical meaning of the resulting structure tensors introduced by Kassinos and Reynolds (see KR94 and RK95). Here ω_i' denotes the components of the turbulent vorticity vector. Note that Ψ_i' satisfies a Poisson equation and hence like the pressure fluctuation carries non-local information.

2.1.1. *One-point structure tensors*

The Reynolds stress tensor and the associated non-dimensional and anisotropy tensors are defined by

$$R_{ij} = \overline{u_i' u_j'} = \epsilon_{ipq}\epsilon_{jts}\overline{\Psi'_{q,p}\Psi'_{s,t}}\,, \qquad r_{ij} = R_{ij}/q^2, \qquad \tilde{r}_{ij} = r_{ij} - \frac{1}{3}\delta_{ij}\,. \quad (5)$$

Here $q^2 = 2k = R_{ii}$. Introducing the isotropic tensor identity (Jeffreys 1931, Mahoney 1985)

$$\epsilon_{ipq}\epsilon_{jts} = \delta_{ij}\delta_{pt}\delta_{qs} + \delta_{it}\delta_{ps}\delta_{qj} + \delta_{is}\delta_{pj}\delta_{qt} - \delta_{ij}\delta_{ps}\delta_{qt} - \delta_{it}\delta_{pj}\delta_{qs} - \delta_{is}\delta_{pt}\delta_{qj} \quad (6)$$

and assuming homogeneity, one finds

$$R_{ij} + \underbrace{\overline{\Psi'_{k,i}\Psi'_{k,j}}}_{D_{ij}} + \underbrace{\overline{\Psi'_{i,k}\Psi'_{j,k}}}_{F_{ij}} = \delta_{ij}q^2\,. \quad (7)$$

The constitutive equation (7) shows that for a proper characterization of non-equilibrium turbulence the *componentality* information found in r_{ij} must be supplemented by *structure* information found in the one-point turbulent *structure* tensors D_{ij} and F_{ij} introduced by KR94. In addition to the basic definitions of these tensors that appear in (7), one can use equivalent representations for homogeneous turbulence in terms of the velocity spectrum tensor $E_{ij}(\mathbf{k})$ and vorticity spectrum tensor $W_{ij}(\mathbf{k})$. These are as follows:

- Structure *dimensionality* tensor

$$D_{ij} = \int \frac{k_i k_j}{k^2} E_{nn}(\mathbf{k})\, d^3\mathbf{k} \quad d_{ij} = D_{ij}/q^2 \quad \tilde{d}_{ij} = d_{ij} - \frac{1}{3}\delta_{ij} \quad (8)$$

- Structure *circulicity* tensor

$$F_{ij} = \int \mathcal{F}_{ij}(\mathbf{k})\, d^3\mathbf{k} \quad f_{ij} = F_{ij}/q^2 \quad \tilde{f}_{ij} = f_{ij} - \frac{1}{3}\delta_{ij}\,. \quad (9)$$

Here $\mathcal{F}_{ij}(\mathbf{k}) = k^2\overline{\hat{\Psi}_i\hat{\Psi}_j^*}$ is the circulicity spectrum tensor, which is related to the vorticity spectrum tensor $W_{ij}(\mathbf{k}) = \overline{\hat{W}_i\hat{W}_j^*}$ through the relation

$$\mathcal{F}_{ij}(\mathbf{k}) = \frac{W_{ij}(\mathbf{k})}{k^2}.$$

The familiar rapid pressure–strain-rate term is given by

$$T_{ij} = 2G_{ts}(M_{istj} + M_{jsti}) \quad (10)$$

where the fourth-rank tensor $\mathbf{M}$ is given by

$$M_{ijpq} = \int \frac{k_p k_q}{k^2} E_{ij}(\mathbf{k})\, d^3\mathbf{k}\,. \tag{11}$$

For homogeneous turbulence $D_{ii} = F_{ii} = q^2$ and it is possible to normalize (7) so that

$$r_{ij} + d_{ij} + f_{ij} = \delta_{ij} \tag{12}$$

where

$$r_{ij} = R_{ij}/q^2 \quad d_{ij} = D_{ij}/q^2 \quad f_{ij} = F_{ij}/q^2\,. \tag{13}$$

The tensor anisotropies $\tilde{r}_{ij} = r_{ij} - \frac{1}{3}\delta_{ij}$, $\tilde{d}_{ij} = d_{ij} - \frac{1}{3}\delta_{ij}$ and $\tilde{f}_{ij} = f_{ij} - \frac{1}{3}\delta_{ij}$ satisfy

$$\tilde{r}_{ij} + \tilde{d}_{ij} + \tilde{f}_{ij} = 0\,. \tag{14}$$

2.2. A PARTICLE REPRESENTATION OF THE RDT OF HOMOGENEOUS TURBULENCE

In a particle representation method, a number of key properties and their evolution equations are assigned to hypothetical particles. The idea is to follow an ensemble of "particles", determine the statistics of the ensemble and use those as the representation for the one-point statistics of the corresponding field. It is important to appreciate that these particles do not have to be physical elements of fluid. The idea of representing the turbulent flow by a large number of particles, each having its own set of properties, has been used over the past ten years by the combustion community in the form of Lagrangian PDF methods (for example see Pope, 1994). In these traditional approaches, a stochastic model is used, which can be chosen so that upon taking moments of the governing stochastic evolution equations one recovers one of the standard Reynolds stress transport (RST) models. This approach, however, uses modeling where it is not required, *i.e.* in emulating RDT. More recently, Vanslooten and Pope (1997) have incorporated the PRM of KR94 in their PDF approach. A more detailed discussion of the rapid PRM can be found in KR94.

2.2.1. *Particle properties*

We start with a discussion of the properties assigned to each of the hypothetical particles. The assigned properties are:

- $\mathbf{V}$ velocity vector
- $\mathbf{W}$ vorticity vector
- $\mathbf{S}$ stream function vector
- $\mathbf{N}$ gradient vector
- P pressure.

Here we consider a representation method using non-physical "particles" that correspond most closely to a vortex sheet (or 1D-1C flow). The single axis of dependence lies normal to the vortex sheet and parallel to the N_i vector, which provides a measure of gradients normal to the plane. The remaining vectors lie in the plane of independence.

The evolution of the vector properties assigned to each particle are governed by ordinary differential equations based on the Navier-Stokes equations. For example, a kinematic analysis leads to the RDT evolution equation for $\mathbf{N}$

$$\dot{N}_i = -G_{ki} N_k \,, \tag{15}$$

where

$$\dot{()} \equiv \frac{d()}{dt}$$

and $G_{ij} = U_{i,j}$ is the mean velocity gradient tensor. Equation (15) shows that $\mathbf{N}$ plays a role similar to that of the wavenumber vector $\mathbf{k}$. The unit vector $n_i = N_i/\sqrt{N_k N_k}$ satisfies

$$\dot{n}_i = -G_{ki} n_k + G_{km} n_k n_m n_i \,. \tag{16}$$

The PRM evolution equation for $\mathbf{V}$ is

$$\dot{V}_i = -G_{ik} V_k + 2G_{km} \frac{V_m N_k N_i}{N^2} . \tag{17}$$

The familiar Poisson equation (3) for the rapid pressure is the basis for the analogous definition

$$P = -2G_{km} \frac{V_m N_k}{N^2} \,. \tag{18}$$

Using (17) and (18), one obtains

$$\dot{V}_i = -G_{ik} V_k - P N_i \tag{19}$$

by analogy to the fluctuation momentum equation (2) under RDT.

2.3. REPRESENTATIONS FOR THE ONE-POINT STATISTICS

The Reynolds stress $R_{ij} = \overline{u_i' u_j'}$ is represented as

$$R_{ij} = \langle V_i V_j \rangle = \langle V^2 v_i v_j \rangle \,, \tag{20}$$

where the angle brackets denote averaging over an ensemble of particles. In the IPRM formulation the structure tensors, defined in (8) and (9), are represented as

$$D_{ij} = \langle S_n S_n N_i N_j \rangle = \langle V^2 n_i n_j \rangle \quad \text{and} \quad F_{ij} = \langle N_n N_n S_i S_j \rangle = \langle V^2 s_i s_j \rangle \tag{21}$$

where $s_i = S_i/\sqrt{S_k S_k}$ and $v_i = V_i/\sqrt{V_k V_k}$. A consequence of the orthogonality of the three vectors n_i, v_i, and s_i is that the IPRM representations in (20) and (21) satisfy the constitutive equations (7), (12) and (14), valid for homogeneous turbulence.

2.4. CLUSTER-AVERAGED EQUATIONS

The cluster-averaged implementation of the PRM offers a better computational efficiency. The idea is to do the averaging in two steps, the first step being done analytically. First, an averaging is done over particles that have the same $\mathbf{n}(t)$, followed by an averaging over all particles with different $\mathbf{n}(t)$. The one-point statistics resulting from the first (cluster) averaging are conditional moments, which will be denoted by

$$R_{ij}^{|\mathbf{n}} \equiv \langle V_i V_j | \mathbf{n} \rangle \qquad D_{ij}^{|\mathbf{n}} \equiv \langle V^2 n_i n_j | \mathbf{n} \rangle = \langle V^2 | \mathbf{n} \rangle n_i n_j \tag{22}$$

and

$$F_{ij}^{|\mathbf{n}} \equiv \langle V^2 s_i s_j | \mathbf{n} \rangle . \tag{23}$$

The conditionally-averaged stress evolution equation

$$\dot{R}_{ij}^{|\mathbf{n}} = -G_{ik} R_{kj}^{|\mathbf{n}} - G_{jk} R_{ki}^{|\mathbf{n}} + 2G_{km}(R_{im}^{|\mathbf{n}} n_k n_j + R_{jm}^{|\mathbf{n}} n_k n_i) \tag{24}$$

is obtained by using the definition (22) along with (17). Note that (16) and (24) are *closed* for the conditional stress tensor $R_{ij}^{|\mathbf{n}}$ and n_i. That is, they can be solved without reference to the other conditioned moments.

2.5. PARTICLE REPRESENTATION MODEL FOR SLOW DEFORMATIONS

Whenever the time scale of the mean deformation is large compared to that of the turbulence the nonlinear turbulence-turbulence interactions become important in the governing field equations. In the context of the Interacting Particle Representation Model (IPRM), these nonlinear processes are represented by a model for the particle-particle interactions.

Direct numerical simulations (Lee and Reynolds, 1985) show that under weak strain the structure dimensionality D_{ij} remains considerably more isotropic than does the Reynolds stress R_{ij}. This leads to counter-intuitive R_{ij} behavior in axisymmetric expansion flows (see Section 4.1.1), supported by experiments (Choi, 1983). Hence we modify the basic evolution equations (16) and (24) to account for these effects. The resulting cluster-averaged evolution equations are

$$\dot{n}_i = -G_{ki}^n n_k + G_{kr}^n n_k n_r n_i \tag{25}$$

$$\begin{aligned}\dot{R}^{|n}_{ij} &= -G^v_{ik}R^{|n}_{kj} - G^v_{jk}R^{|n}_{ki} - C_r[2R^{|n}_{ij} - R^{|n}_{kk}(\delta_{ij} - n_i n_j)] \\ &\quad +[G^n_{km} + G^v_{km}](R^{|n}_{im}n_k n_j + R^{|n}_{jm}n_k n_i)\,. \end{aligned} \tag{26}$$

Note that the mean velocity gradient tensor G_{ij} that appeared in (16) and (24) has been replaced by the *effective* gradient tensors G^v_{ij} and G^n_{ij}. These are defined by

$$G^n_{ij} = G_{ij} + \frac{C_n}{\tau} r_{ik} d_{kj} \qquad G^v_{ij} = G_{ij} + \frac{C_v}{\tau} r_{ik} d_{kj}\,. \tag{27}$$

Here $r_{ij} = R_{ij}/q^2$ and $d_{ij} = D_{ij}/q^2$ where $q^2 = 2k = R_{ii}$. The two constants are taken to be $C_n = 2.2, c_V = 2.2$. The different values for these two constants account for the different rates of return to isotropy of **D** and **R**. The time scale of the turbulence τ is evaluated so that the dissipation rate in the IPRM

$$\epsilon^{\text{PRM}} = q^2 \frac{C^v}{\tau} r_{ik} d_{km} r_{mi} \tag{28}$$

matches that obtained from a modified model equation for the dissipation rate,

$$\dot{\epsilon} = -C_0(\epsilon^2/q^2) - C_s S_{pq} r_{pq}\epsilon - C_\Omega \sqrt{\Omega_n \Omega_m d_{nm}}\,\epsilon\,. \tag{29}$$

The last term in (29) accounts for the suppression of ϵ by mean rotation. Here Ω_i is the mean vorticity vector, and the constants are taken to be

$$C_0 = 3.67 \qquad C_s = 3.0 \qquad \text{and} \qquad C_\Omega = 0.01\,. \tag{30}$$

Mean rotation acting on the particles tends to produce rotational randomization of the **V** vectors around the **n** vectors (Mansour *et al.*, 1991, KR94). The third (bracketed) term on the RHS of (26), is the *slow rotational randomization model*, which assumes that the effective rotation due to nonlinear particle-particle interactions, $\Omega^*_i = \epsilon_{ipq} r_{qk} d_{kp}$, should induce a similar randomization effect, while leaving the conditional energy unmodified. Based on dimensional considerations, and requirements for material indifference to rotation (Speziale, 1981, 1985), we take

$$C_r = \frac{8.5}{\tau}\,\Omega^* f_{pq} n_p n_q, \quad \Omega^* = \sqrt{\Omega^*_k \Omega^*_k}, \quad \Omega^*_i = \epsilon_{ipq} r_{qk} d_{kp}\,. \tag{31}$$

The rotational randomization coefficient C_r is sensitized to the orientation of the **n** vector, so that the slow rotational randomization vanishes whenever the large-scale circulation is confined in the plane normal to **n**.

The pressure P is determined by the requirement that $R^{|n}_{ik} n_k = 0$ is maintained by (25) and (26). This determines the effects of the slow pressure strain–rate-term without the need for further modeling assumptions

$$P = \underbrace{-2G_{mk}\frac{V_k N_m}{N^2}}_{\text{rapid}} - \underbrace{\frac{(C^v + C^n)}{\tau} r_{mt} d_{tk} \frac{V_k N_m}{N^2}}_{\text{slow}} \,. \tag{32}$$

3. A One-Point Structure-Based Model Based on the IPRM

A one-point structure-based model for the deformation of homogeneous turbulence can be derived directly from the IPRM formulation. At the one-point level, additional modeling assumptions must be introduced in order to deal with the non-locality of the pressure fluctuations. Here we limit the formulation of the one-point model to the case of irrotational deformation of homogeneous turbulence and discuss briefly the more general case. The complete one-point formulation will be presented separately .

As a result of the constitutive equation (7), a *structure-based* one-point model must carry the transport equations for only two of the three second-rank tensors. Here we propose a model based on the R_{ij} and D_{ij} equations, which are the one-point analogs of (25) and (26) . Using the definitions (20) and (21) and the evolution equations (25) and (26), and averaging over all clusters, one obtains

$$\dot{D}_{ij} = -D_{ik}G^n_{kj} - D_{jk}G^n_{ki} + 2q^2 G^n_{km} Z^d_{kmij} - 2G^v_{km} M_{mkij} \tag{33}$$

and

$$\begin{aligned} \dot{R}_{ij} &= -G^v_{ik} R_{kj} - G^v_{jk} R_{ki} \\ &-\hat{C}_r f_{pq}[2M_{ijpq} - (\delta_{ij} D_{pq} - q^2 Z^d_{ijpq})] \\ &+[G^n_{km} + G^v_{km}](M_{imkj} + M_{jmki}) \,. \end{aligned} \tag{34}$$

Here G^n_{ij} and G^v_{ij} are as defined for the IPRM in (27), and $\hat{C}_r = 8.5\Omega^*/\tau$ where Ω^* is given in (31). The fourth-rank tensors

$$Z^d_{ijkm} = \langle V^2 n_i n_j n_k n_m \rangle / q^2 \quad \text{and} \quad M_{ijpq} = \langle V^2 v_i v_j n_p n_q \rangle \tag{35}$$

must be modeled. Note that **Z** is the fully symmetric, energy-weighted fourth moment of a single vector, for which we have been able to construct a good model (see KR94 and RK95). What is more, one can use an exact decomposition[2] based on group theory (see KR94) to express M_{ijpq} in terms

[2]In the presence of mean rotation this exact decomposition involves additional terms (not given here) that require modeling of the *stropholysis* effects (see KR94 and RK95).

of fourth moments of a single vector and the second-rank tensors R_{ij} and D_{ij}:

$$M_{ipqj} = \tfrac{1}{2}q^2(Z^f_{ipqj} - Z^r_{ipqj} - Z^d_{ipqj})$$
$$+\tfrac{1}{6}[-3\delta_{ip}\delta_{qj}q^2 + 4(\delta_{qj}R_{ip} + \delta_{ip}D_{qj}) + 2(\delta_{qj}D_{ip} + \delta_{ip}R_{qj})]\,. \tag{36}$$

Here

$$Z^r_{ijpq} = \langle V^2 v_i v_j v_p v_q\rangle/q^2 \quad\text{and}\quad Z^f_{ijpq} = \langle V^2 s_i s_j s_p s_q\rangle/q^2. \tag{37}$$

Substituting (36) in (33) and (34) and using the definitions (13), one obtains

$$\begin{aligned}\dot{d}_{ij} &= -d_{jk}G^n_{ki} - d_{ik}G^n_{kj} + 2G^v_{km}r_{km}(d_{ij} - \tfrac{2}{3}\delta_{ij})\\ &\quad -\tfrac{2}{3}G^v_{km}d_{mk}\delta_{ij} + G^v_{kk}(\delta_{ij} - \tfrac{4}{3}d_{ij} - \tfrac{2}{3}r_{ij})\\ &\quad +(2G^n_{km} + G^v_{km})Z^d_{kmij} + G^v_{km}Z^r_{mkij} - G^v_{km}Z^f_{mkij}\end{aligned} \tag{38}$$

and

$$\begin{aligned}\dot{r}_{ij} = \;& \tfrac{1}{3}(G^v_{mj} + G^n_{mj})(2d_{mi} + r_{mi}) + \tfrac{1}{3}(G^v_{mi} + G^n_{mi})(2d_{mj} + r_{mj})\\ & +\tfrac{1}{3}G^v_{jm}(d_{mi} - r_{mi}) + \tfrac{1}{3}G^v_{im}(d_{mj} - r_{mj})\\ & +\tfrac{1}{3}G^n_{jm}(d_{mi} + 2r_{mi}) + \tfrac{1}{3}G^n_{im}(d_{mj} + 2r_{mj})\\ & +2G^v_{km}r_{km}r_{ij} - \tfrac{1}{2}(G^v_{ij} + G^v_{ji} + G^n_{ij} + G^n_{ji})\\ & +(G^v_{mk} + G^n_{mk})(Z^f_{ikmj} - Z^r_{ikmj} - Z^d_{ikmj})\\ & -\hat{C}_r f_{pq}[Z^f_{ijpq} - Z^r_{ijpq} + \tfrac{2}{3}\delta_{pq}(r_{ij} - f_{ij}) + \tfrac{1}{3}\delta_{ij}(r_{pq} - f_{pq})]\,.\end{aligned} \tag{39}$$

Closure of (38) and (39) in the irrotational case requires a consistent model for the fully symmetric tensors Z^n_{ijpq}, Z^r_{ijpq}, and Z^f_{ijpq}. We have constructed a model for the energy-weighted fourth moment of any vector t_i in terms of its second moment t_{ij} that allows the successful closure of (38) and (39) while maintaining *full realizability.* The same model can be used for each of the three vectors v_i, n_i, and s_i and has the general form

$$Z^t_{ijpq} = \langle V^2 t_i t_j t_p t_q\rangle/q^2 = C_1\,\mathbf{i}\circ\mathbf{i} + C_2\,\mathbf{i}\circ\mathbf{t}$$
$$+C_3,\mathbf{t}\circ\mathbf{t} + C_4\,\mathbf{i}\circ\mathbf{t^2} + C_5\,\mathbf{t}\circ\mathbf{t^2} + C_6\,\mathbf{t^2}\circ\mathbf{t^2}\,. \tag{40}$$

Here $\mathbf{i}$ and $\mathbf{t}$ stand for δ_{ij} and $t_{ij} = \langle V^2 t_i t_j\rangle/q^2$ respectively. Extended tensor notation is used in (40), where the fully symmetric product of two second-rank tensors $\mathbf{a}$ and $\mathbf{b}$ is denoted by

$$\mathbf{a}\circ\mathbf{b} \equiv a_{ij}b_{pq} + a_{ip}b_{jq} + a_{jp}b_{iq} + a_{iq}b_{jp} + a_{jq}b_{ip} + a_{pq}b_{ij}\,. \tag{41}$$

The coefficients C_1-C_6 are functions of the invariants of t_{ij} and determined by enforcing the trace condition $Z^t_{ijkk} = t_{ij}$, 2D realizability conditions for the case when the vectors t_i lie in a plane, and an important identity (see KR94),

$$\mathbf{Z}^a = \frac{1}{16}\mathbf{i} \circ \mathbf{i} - \frac{1}{8}\mathbf{i} \circ \mathbf{b} + \frac{3}{8}\mathbf{Z}^b , \qquad (42)$$

which applies between the fourth moments of two vectors a_i and b_i whenever a_i is randomly distributed about b_i. KR94 gave an earlier $\mathbf{Z}$ model that does not exactly satisfy (42). The new coefficients, which allow the $\mathbf{Z}$ model to satisfy (42), will be published separately.

Next, we consider the performance of the interacting particle representation and one-point models for two cases of irrotational mean deformation. In addition, the IPRM is evaluated for three cases combining mean rotation and mean strain effects.

4. Evaluation of the IPRM and One-Point Models

In this section, the IPRM based on (25), (26), (27), and (29) and the one-point model based on (27), (29), (38), (39), and (40) are evaluated for several cases of deformation of homogeneous turbulence. The examples considered here show that, even with a relatively simple closure for nonlinearity, both models achieve remarkably accurate predictions. The evaluation of the IPRM for rapid mean deformation (large Sk/ϵ) is reported in detail in KR94, where it is shown that the IPRM reproduces an *exact* emulation of RDT. Therefore, in this section, we report only on flows involving weak mean deformation (small Sk/ϵ), where the nonlinear interactions are important.

4.1. IRROTATIONAL AXISYMMETRIC STRAIN

First we consider the performance of the IPRM for the case of homogeneous, initially isotropic, turbulence subjected to irrotational axisymmetric mean deformation. The mean velocity gradient tensor is given by

$$S_{ij} = \begin{pmatrix} S & 0 & 0 \\ 0 & -S/2 & 0 \\ 0 & 0 & -S/2 \end{pmatrix} \qquad (43)$$

with $S > 0$ for contraction (nozzle flow) and $S < 0$ for expansion (diffuser flow).

4.1.1. *Axisymmetric Expansion*

Results for the case of irrotational axisymmetric expansion flow with an initial $Sq_0^2/\epsilon_0 = 0.82$ are shown in Figure 1. Here the axis of symmetry

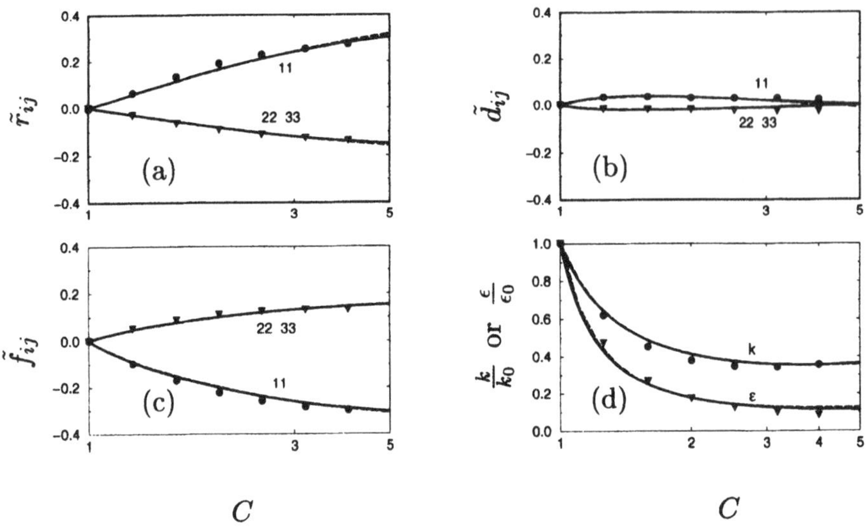

Figure 1. Comparison of the one-point model predictions (dashed lines) with the IPRM results (solid lines) and the 1985 DNS of Lee and Reynolds (symbols) for axisymmetric expansion case EXO ($Sq_0^2/\epsilon_0 = 0.82$). (a)-(c) evolution of the the Reynolds stress, dimensionality, and circulicity anisotropies; 11 component (•), 22 and 33 components (▾). (d) evolution of the normalized turbulent kinetic energy (•) and dissipation rate (▾).

is taken to be x_1. The predictions of the IPRM (solid lines) and those of the one-point model (dashed lines) are compared with the direct numerical simulation (DNS) of Lee and Reynolds (1985), shown as symbols. The evolution histories are plotted against

$$C = \exp\left(\int_0^t |S_{\max}(t')| \, dt' \right),$$

where $S_{\max}$ is the largest principal value of the mean strain tensor. As discussed in Kassinos and Reynolds (1995), the axisymmetric expansion flows exhibit counter-intuitive behavior, where a weaker mean deformation rate produces a level of stress anisotropy $\tilde{r}_{ij}$ that exceeds the one produced under RDT. This effect, which is also supported by the experiments of Choi (1983), is triggered by the different rates of return to isotropy in the $\mathbf{r}$ and $\mathbf{d}$ equations, but it is dynamically controlled by the rapid terms. The net effect is a growth of $\tilde{r}_{ij}$ in expense of $\tilde{d}_{ij}$, which is strongly suppressed. As shown in Figure 1, the predictions of the IPRM and one-point models are almost indistinguishable from each other, and both are able to capture these intriguing effects quite accurately. The predictions of both models for the evolution of the normalized turbulent kinetic energy and dissipation rate are also in good agreement with the DNS.

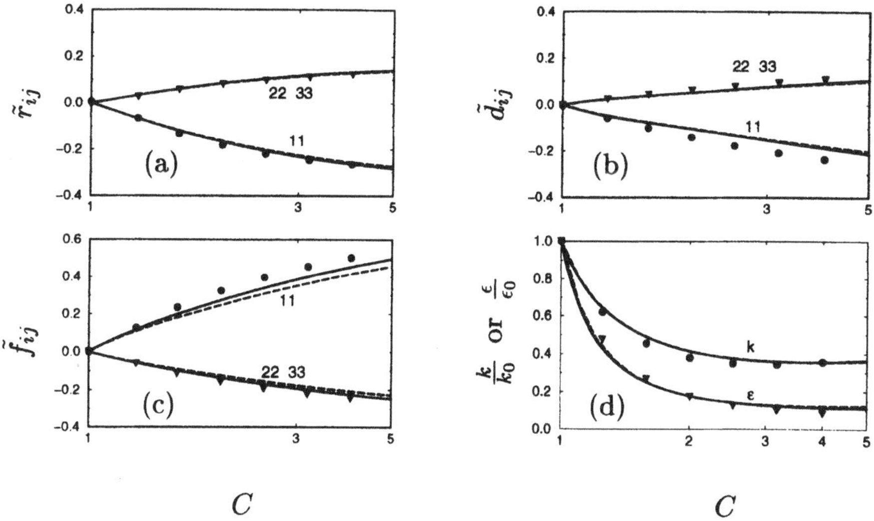

Figure 2. Comparison of the one-point model predictions (dashed lines) with the IPRM results (solid lines) and the 1985 DNS of Lee and Reynolds (symbols) for axisymmetric expansion case AXK ($Sq_0^2/\epsilon_0 = 1.1$). (a)-(c) evolution of the the Reynolds stress, dimensionality, and circulicity anisotropies; 11 component (•), 22 and 33 components (▼). (d) evolution of the normalized turbulent kinetic energy (•) and dissipation rate (▼).

4.1.2. *Axisymmetric Contraction*

A second case of slow irrotational axisymmetric deformation of homogeneous, initially isotropic, turbulence is considered in Figure 2. The initial parameters in this axisymmetric contraction flow ($Sq_0^2/\epsilon_0 = 1.1$) correspond to the slowest run in 1985 simulations of Lee and Reynolds. The anisotropy evolution histories for $\tilde{\mathbf{r}}$, $\tilde{\mathbf{d}}$, and $\tilde{\mathbf{f}}$ predicted by the non-local IPRM (solid lines) and the one-point model (dashed lines) are in satisfactory agreement with the simulation results (symbols). Both models predict decay of the turbulent kinetic energy k and dissipation rate ϵ at the correct rates (see Figure 2d).

4.1.3. *Plane Strain*

A third case of irrotational deformation is considered in Figure 3, where we show results for homogeneous, initially isotropic, turbulence subjected to plane strain. The mean deformation is in the x_2-x_3 plane according to

$$S_{ij} = \begin{pmatrix} 0 & 0 & 0 \\ 0 & -S & 0 \\ 0 & 0 & S \end{pmatrix} . \tag{44}$$

Figures 3a-3c show evolution histories for the three tensor anisotropies for a case of weak irrotational plane strain ($Sq_0^2/\epsilon_0 = 1.0$). Figure 3d shows

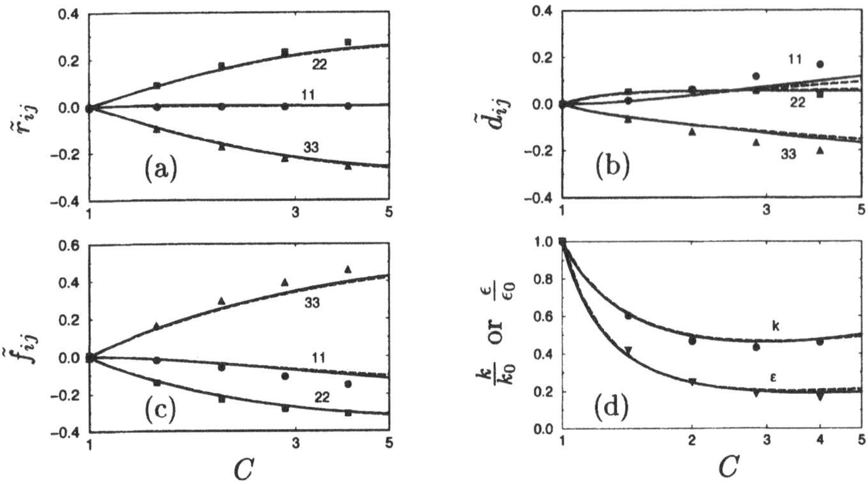

Figure 3. Comparison of the one-point model predictions (dashed lines) with the IPRM results (solid lines) and the 1985 DNS of Lee and Reynolds (symbols) for plane strain case PXA ($Sq_0^2/\epsilon_0 = 1.0$). (a)-(c) evolution of the the Reynolds stress, dimensionality, and circulicity anisotropies; 11 component (●), 22 component (■), and 33 component (▼). (d) evolution of the normalized turbulent kinetic energy (●) and dissipation rate (▼).

the evolution of the normalized turbulent kinetic energy and dissipation rate. Again, the IPRM predictions are shown as solid lines and those of the one-point model as dashed lines. Comparison is made with the 1985 DNS of Lee and Reynolds (symbols). The predictions of the one-point model are practically indistinguishable from those of the IPRM and how both models are in excellent agreement with the DNS results in all cases.

4.2. COMBINATIONS OF MEAN STRAIN AND ROTATION

4.2.1. *Homogeneous Shear*

We first consider the case of homogeneous shear flow with an initial $Sq_0^2/\epsilon_0 = 4.73$, where S is the shear rate. In this case the evaluation is restricted to the IPRM because the formulation of the one-point model was presented only for irrotational deformations.

The IPRM predictions for the components of the normalized Reynolds stress tensor r_{ij} are shown in Figure 4a. The symbols are from the DNS of Rogers and Moin (1986). The agreement between the IPRM predictions and the DNS results is good, but the IPRM somewhat overpredicts r_{11} and underpredicts r_{22}. Figure 4b shows the prediction for the components of the normalized dimensionality tensor d_{ij}, where again the IPRM achieves

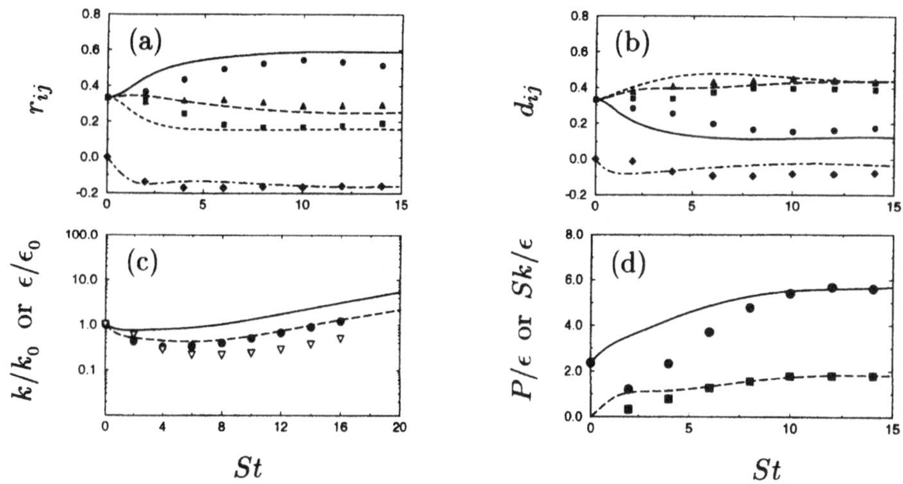

Figure 4. Comparison of IPRM predictions (shown as lines) with the DNS of Rogers and Moin (1986) for homogeneous shear ($Sq_0^2/\epsilon_0 = 4.73$). (a) and (b) evolution of the normalized Reynolds stress and dimensionality tensors; 11 component (——— , ●), 22 component (- - - - , ■), 33 component (- - - - , ▲), 12 component (—·—, ◆). (c) evolution of the normalized turbulent kinetic energy (——— , ●) and dissipation rate (- - - - , ▽). (d) evolution of the dimensionless parameters P/ϵ (- - - - , ■) and Sk/ϵ (——— , ●).

a satisfactory level of accuracy. The evolutions of the normalized turbulent kinetic energy and dissipation rate are shown in Figure 4c. The IPRM does predict the correct growth rates for both k/k_0 and ϵ/ϵ_0, but the agreement between actual values is moderate. Figure 4d shows the evolution of the dimensionless parameters P/ϵ and Sk/ϵ, where P is the production of k. The IPRM predictions (lines) are in good agreement with DNS results (symbols), especially in the period $8 \lesssim St \lesssim 15$ where the DNS was fully developed. The same equilibrium values are predicted for the two dimensionless parameters by both the DNS simulation and the IPRM.

4.2.2. *Homogeneous shear in a rotating frame*

Next we consider the problem of homogeneous shear in a rotating frame. The mean velocity gradient tensor G_{ij}, the frame vorticity Ω_i^f, and frame rotation rate Ω_i are defined by

$$G_{ij} = \begin{pmatrix} 0 & S & 0 \\ 0 & 0 & 0 \\ 0 & 0 & 0 \end{pmatrix}, \qquad 2\Omega_i = \Omega_i^f = (0, 0, \Omega^f)\,. \tag{45}$$

We consider initially isotropic turbulence with $r_{ij} = \frac{1}{3}\delta_{ij}$, $k = k_0$ and $\epsilon = \epsilon_0$. The solution in the case of homogeneous shear in a rotating frame

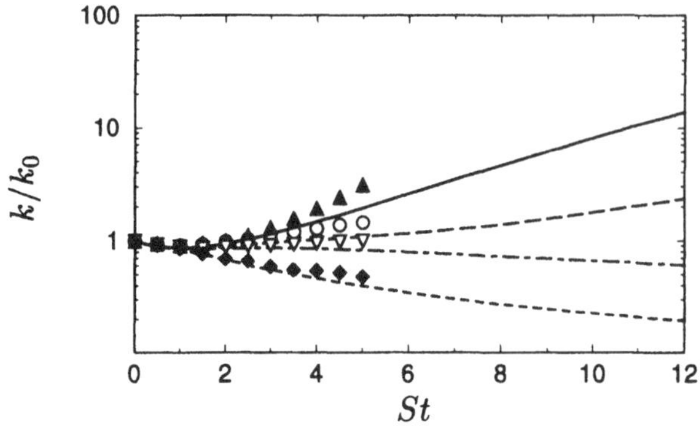

Figure 5. Time evolution of the turbulent kinetic energy in rotating shear flows. IPRM predictions (lines) are compared to the LES of Bradina *et al.* (symbols): $\Omega^f/S = 0$ (---- , ○), $\Omega^f/S = 0.5$ (—— , ▲), $\Omega^f/S = 1.0$ (—·— , ▽), and $\Omega^f/S = -1.0$ (---- , ♦).

depends on the initial conditions only through the dimensionless parameter Sk_0/ϵ_0, and on the frame vorticity through the dimensionless parameter Ω^f/S (Speziale *et al.*, 1991). The value of Ω^f/S determines whether the flow is stable, in which case k and ϵ decay in time, or unstable in which case both k and ϵ grow exponentially in time.

The effect of the ratio Ω^f/S on the time evolution of the normalized kinetic energy k/k_0 is shown in Figure 5. In the absence of DNS or experimental data, we evaluate the model performance using the large-eddy simulations of Bardina *et al.* (1983). Note that the model captures the general trends correctly. For example, it correctly predicts that the highest rate of growth (for both k and ϵ) occurs for $\Omega^f = S/2$, which RDT shows is the most unstable case. It also predicts a weak rate of decay for the case $\Omega^f = S$ and a decay (relaminarization) for $\Omega^f = -S$. The numerical agreement with the LES is reasonable, but the model tends to predict somewhat lower rates of growth, particularly so in the case $\Omega^f = 0.5S$. This problem is also common to all the currently available second-order closures as noted by Speziale *et al.* (1989). However, a detailed comparison of numerical values is probably not meaningful in this case, because the reported LES results came from the filtered field only.

4.2.3. *Elliptic streamline flow*

The elliptic streamline flows combine the effects of mean rotation and plane strain and emulate the conditions encountered in the flow through various sections of turbomachinery. These relatively basic flows provide a challenging test case for turbulence models. For example, direct numerical simulations show exponential growth of the turbulent kinetic energy in elliptic

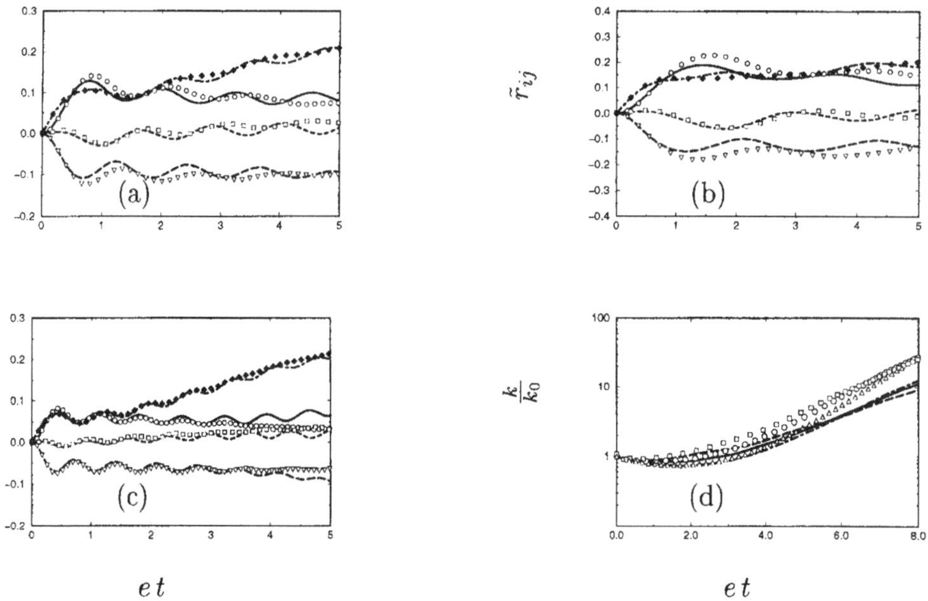

Figure 6. Comparison of IPRM predictions (shown as lines) with the DNS of Blaisdell (shown as symbols) for elliptic streamline flows. Evolution of Reynolds stress anisotropy components for cases (a) $E = 1.4$, (b) $E = 2.0$, and (c) $E = 1.25$: 11 component (——— , ○); 22 component (– – – – , □); 33 component (- - - - , ▽); (—·— , ♦) 13 component . (d) Evolution of the normalized turbulent kinetic energy for cases e3 (——— , ○), e4 (– – – – . □), and e2 (—·— , △).

streamline flows, but standard k-ϵ models (as well as most Reynolds stress models) predict decay of the turbulence. The structure-based model (Kassinos and Reynolds, 1995) does predict an exponential growth, but not yet at the correct rate.

The elliptic streamline flow corresponds to the mean deformation tensor

$$G_{ij} = \begin{pmatrix} 0 & 0 & -\gamma - e \\ 0 & 0 & 0 \\ \gamma - e & 0 & 0 \end{pmatrix} \tag{46}$$

with $0 < |e| < |\gamma|$. Note that the case $e = 0$ corresponds to pure rotation while the case $|e| = |\gamma|$ corresponds to homogeneous shear. The elliptic streamlines in this flow have an aspect ratio given by $E = \sqrt{(\gamma + e)/(\gamma - e)}$. As explained by Blaisdell and Shariff (1994), the important nondimensional parameters for the elliptic streamline flow are (1) the aspect ratio E of the elliptic streamlines and (2) the ratio of the turbulent time scale to the time scale of the mean deformation. The turbulent Reynolds number is also an important parameter, but the IPRM model is based on a high Reynolds number assumption. The IPRM predictions are shown in Figure 5 for three different cases, which correspond to different degrees of streamline ellipticity E (for details see Blaisdell and Shariff, 1994). The notation used here is identical to the one introduced by Blaisdell and Shariff (1994). For example, $S_e^* = ek/\epsilon$ represents the ratio of the turbulent time scale to the mean flow time scale based on the mean strain, and for the cases considered here $S_e^* = 1.69$. We have evaluated the IPRM predictions using preliminary results from the simulations of Blaisdell and found its performance to be very good. The IPRM predicts exponential growth for the turbulent kinetic energy and dissipation rate at the correct rates of growth. The IRPM predictions for the individual components of the Reynolds stress anisotropy tensor were also in very good agreement with the corresponding preliminary DNS results

5. Conclusions

In simple flows with mild mean deformation rates the turbulence has time to come to equilibrium with mean flow and the Reynolds stresses are determined by the strain *rate*. On the other hand, when the mean deformation is very rapid the turbulent structure takes some time to respond, and the Reynolds stresses are determined by the amount of *total* strain.

A good turbulence model should exhibit this viscoelastic character of turbulence, matching the two limiting behaviors and providing a reasonable blend in between. Our goal has been the development of one-point model for engineering use with the proper viscoelastic character. We have shown that to achieve this goal one needs to include structure information in the tensorial base used in the model because non-equilibrium turbulence is inadequately characterized by the turbulent stresses themselves.

The interacting particle representation model (IPRM) is in essence a non-local viscoelastic, structure-based turbulence model. As it was shown here, with a relatively simple model for the nonlinear turbulence-turbulence interactions, the IPRM is able to handle quite successfully a surprisingly wide range of flows. Some of these flows involve paradoxical effects, and the fact that the IPRM is able to reproduce them suggests that perhaps the model captures a significant part of the underlying physics.

A one-point **R-D** model follows directly from the IPRM formulation. The one-point formulation requires additional modeling in order to deal with the non-locality of the rapid-pressure fluctuations, which is treated exactly in the IPRM. Preliminary results shown here indicate that the one-point model has the potential to match the non-local IPRM in accuracy. We are also investigating further extensions to the IPRM that might enable it to become a valuable engineering tool on its own right.

References

Arnold, L., (1974) *Stochastic Differential Equations.* John Wiley and Sons.

Bardina, J, Ferziger, J. H., & Reynolds, W. C. (1983) *Improved turbulence models based on large eddy simulation of homogeneous, incompressible, turbulent flows.* Report TF-19, Thermosciences Division, Department of Mechanical Engineering, Stanford University.

Blaisdell, G. A. & Shariff, K. (1994) Homogeneous turbulence subjected to mean flow with elliptic streamlines. Center for Turbulence Research: *Proceedings of the 1994 Summer Program.*

Blaisdell, G. A. & Shariff, K. (1996) Simulation and modeling of the elliptic streamline flow. Center for Turbulence Research: *Proceedings of the 1996 Summer Program.*

Choi, Kwing-So. (1983) *A study of the return to isotropy of homogeneous turbulence,* Technical Report, Sibley school of Mechanical and Aerospace Engineering, Cornell University, New York.

Durbin, P. A. & Speziale, C. G. (1994) Realizability of second-moment closures via stochastic analysis, *J. Fluid Mech.*, **280**,pp. 395–407.

Jeffreys, H. (1931) *Cartesian tensors,* Cambridge University Press.

Lee, M. J. & Reynolds, W. C. (1985) *Numerical experiments on the structure of homogeneous turbulence".* Report TF-24, Thermosciences Division, Department of Mechanical Engineering, Stanford University.

Mahoney, J. F. (1985) Tensor and Isotropic Tensor Identities, *The Matrix and Tensor Quarterly,* **34(5)**,pp. 85–91.

Mansour, N. N., Shih, T.-H., & Reynolds, W. C. (1991) The effects of rotation on initially anisotropic homogeneous flows, *Phys. Fluids,* **A 3**, pp. 2421–2425.

Kassinos, S. C. and Reynolds, W. C. (1994) *A structure-based model for the rapid distortion of homogeneous turbulence.* Report TF-61, Thermosciences Division, Department of Mechanical Engineering, Stanford University.

Kassinos, S. C. and Reynolds, W. C. (1995) An extended structure-based model based on a stochastic eddy-axis evolution equation. *Annual Research Briefs 1995,* Center for Turbulence Research, NASA Ames/Stanford Univ.

Pope, S. B. (1994) On the relationship between stochastic Lagrangian models of turbulence and second-moment closures, *Phys. Fluids,* **6**, pp. 973–985.

Reynolds, W. C. and Kassinos, S. C. (1995) A one-point model for the evolution of the Reynolds stress and structure tensors in rapidly deformed homogeneous turbulence, *Proc. Roy. Soc. London A,* **451**(1941):87-104.

Rogers, M. M., Moin, P. (1987) The structure of the vorticity field in homogeneous turbulent flows, *J. Fluid. Mech.*, **176**, pp. 33–66.

Speziale, C. G., Sarkar, S., & Gatski, T. B. (1991) Modelling the pressure-strain correlation of turbulence: an invariant dynamical systems approach, *J. Fluid. Mech.*, **227**, pp 245–272.

Speziale, C. G., and Mac Giolla Mhuiris, N (1989) On the prediction of equilibrium states in homogeneous turbulence, *J. Fluid Mech.*, **209**, pp. 591–615.

Speziale, C. G. 1985 Modeling the pressure-velocity correlation of turbulence, *Phys. Fluids,* **28**(8), 69-71.

Speziale, C. G. (1981) Some interesting properties of two-dimensional turbulence, *Phys. Fluids,* **24**(8), 1425-1427.

Tavoularis, S. & Karnik, U. (1989) Further experiments on the evolution of turbulent stresses and scales in uniformly sheared turbulence, *J. Fluid. Mech.*, **204**, pp. 457–478.

Vanslooten, P. R. & Pope, S. B., (1997) Pdf modeling for inhomogeneous turbulence with exact representation of rapid distortions, *Phys. of Fluids,* **9(4)**, pp. 1085–1105.

THE LOW DIMENSIONAL APPROACH TO TURBULENCE

JOHN L. LUMLEY AND PETER BLOSSEY
Sibley School of Mechanical & Aerospace Engineering
Cornell University, Ithaca, New York

Abstract. Our group has developed and applied the low dimensional approach over the past decade, primarily in the setting of the flat-plate turbulent boundary layer. This approach has been able to reproduce much of the dynamics of the turbulence near the wall as well as the response of the boundary layer to various effects such as pressure gradients, streamline curvature, and polymer additives. Recently, the possibility of active control of near-wall turbulence has been explored from the point of view of the models. In this application, the models could be used as interpreters and filters of the necessarily noisy sensor measurements at the wall and as a source of insight into the role of the coherent structures in the generation of drag at the wall. Models of this type offer not only a better understanding of near wall flow, but also the promise of a greater understanding of the dynamics of coherent structures in turbulent shear flows in general.

1. Introduction

Ideas about coherent structures have probably been around since people first stood by rivers and streams, fascinated by the swirls of water moving downstream. In a similar way, flow visualization leads quickly to ideas about the form and dynamics of coherent structures in the turbulent flows, but attempts at determining an analytic form for the coherent structures prove more difficult. Many methods for extracting coherent structures from experimental or numerical data suffer because they are subject to input from the user either through the interpretation of visualizations, the choice of an initial template for pattern recognition, or the choice of threshold levels in conditional sampling techniques. On the other hand, Lumley proposed the use of the Proper Orthogonal, or Karhunen-Loéve, Decomposition for the identification of structures in turbulence (Lumley, 1967). The coherent structures, or empirical eigenfunctions, which emerge from the decom-

M. D. Salas et al. (eds.), Modeling Complex Turbulent Flows, 89–106.

position depend only on the experimental or computational data used to generate the autocorrelation tensor.

Another predecessor to the low dimensional approach to turbulence may be found in the now famous work of (Lorenz, 1963) who attempted to model the variability of weather through a severe truncation of a normal mode projection of Rayleigh-Bénard convection. This work was long unnoticed and was unknown to (Ruelle & Takens, 1971) when they suggested that turbulence might represent an example of a strange attractor. These two examples of the application of dynamical systems theory to fluid dynamics led to much work in the area; however, many of its early applications were in closed flow problems, such as Rayleigh-Bénard or Taylor-Couette flows.

(Aubry *et al.*, 1988) brought together these two approaches by applying the tools of dynamical systems theory to study of the dynamics of coherent structures in the wall region of the turbulent boundary layer. The low dimensional approach has subsequently become quite popular, being picked up by many groups (Ball *et al.*, 1991; Chambers *et al.*, 1988; Deane *et al.*, 1991; Deane & Sirovich, 1991; Glauser & George, 1987a; Glauser & George, 1987b; Glauser *et al.*, 1987; Glauser *et al.*, 1989; Glauser *et al.*, 1992; Glezer *et al.*, 1989; Kirby *et al.*, 1990a; Kirby *et al.*, 1990b; Liu *et al.*, 1994; Moin & Moser, 1989; Sirovich & Deane, 1991). Over the past decade, our group has applied the approach primarily to the flat-plate boundary layer, and this colors our presentation in this paper. In section 2, the dynamics of near wall turbulent flow will be described as background and motivation for the application of low dimensional models in this setting. The construction of the models is laid out in section 3, although a much more complete treatment of this subject is available in (Aubry *et al.*, 1988) and (Holmes *et al.*, 1996). The following two sections treat the application of the models to the study of various physical effects on the boundary layer (e.g. pressure gradients, streamline curvature) and to the control of near-wall turbulence by both active and passive means. The final section points to the possibility of generating empirical eigenfunctions by means of energy stability theory, a method which would make eigenfunction generation much more tractable since it is based on one point statistics whereas the POD relies on the two point correlation tensor.

2. Near Wall Turbulence and the Bursting Process

Many turbulent flows display a combination of organized, or coherent, structures and apparently disorganized, or incoherent, structures. In the wall region of the turbulent boundary layer, coherent structures account for over 80% of the energy in the turbulent fluctuations. (The wall region corresponds to $y^+ = yu_\tau/\nu \leq 40$, where u_τ is the friction velocity and ν the viscosity; the friction velocity is defined as $\mu\partial U/\partial y = \rho u_\tau^2$, with the gradient evaluated at the wall.) We want to develop a low-dimensional model of the flow near the wall that resolves just these most energetic coherent structures. According to observations, these structures are vortices lying parallel to, and close to, the wall, oriented in the streamwise direction. Most often, these vortices appear as unequal pairs, but sometimes are found singly. They are responsible for the streak-like structures seen in flow visualizations and numerical simulations of the boundary layer: slow-moving fluid is lifted away from the wall on the updraft side of the vortex, giving rise to a low-speed streak, while a high-speed streak is produced by the downdraft on the opposite side.

The coherent structures are observed to burst, transferring energy from the large to the small scales and producing turbulent fluctuations. First, the updraft between the vortices strengthens, and the vortices move toward each other. The slow-moving fluid lifted away from the wall by the vortices induces an inflection in the mean velocity profile, which grows in intensity with the updraft and moves father out into the stream. While the updraft is strengthening (at this point, the updraft is referred to as an ejection), the wall shear stress, or drag, is approaching its peak owing to the high-speed fluid being drawn down toward the wall by the downdraft on the opposite side of the vortex. (The high-speed streak is usually spread wider in the spanwise direction than the low-speed streak and hence has a larger impact on the shear stress at the wall.) As the inflection in the mean velocity profile intensifies, it gives rise to a secondary instability and an accompanying burst of Reynolds stress. Through this instability, energy is transferred from the large scales of the turbulence, or the coherent structures, to the smaller scales. As a result of the energy transfer, the vortices are weakened and move apart. Last, high speed fluid comes down from the outer part of the boundary layer, sweeping the wall clean.

The bursting process is responsible for the production of most of the turbulent kinetic energy in the boundary layer; it is also responsible for the maintenance of the turbulent drag on the wall. One way to see the interaction of the bursting process with the drag is through the mean velocity profiles. The inflection points in the mean velocity profile, which give rise to a secondary instability and the generation of turbulent fluctuations, are simply pockets of slow moving fluid which have been ejected away from the wall. These pockets of slow-moving fluid (filled with small scale turbulent fluctuations) continue to propagate into the center of the channel. The propagation of the fluid from the wall into the center of the channel is the mixing of the momentum (deficit) of the wall with the momentum of the stream by the turbulence. Therefore, inhibiting the action of the coherent structures and the bursting process will decrease the mixing of momentum and decrease the drag at the wall. In fact, $T_B u_\tau^2/\nu$ is thought to be approximately constant, so that the drag coefficient is inversely proportional to the inter-burst time, although the data are widely scattered (Lumley & Kubo, 1985). (Here, T_B, u_τ and ν represent the mean time between bursts, the friction velocity, and the kinematic viscosity, respectively.) This relationship coupling the inter-burst time and drag is appealing, but the experimental support that can be found in the literature is somewhat in conflict and disappointing: the data in (Lumley & Kubo, 1985) which is reasonably consistent, represents a careful selection. Recently, however, our own work with the control of bursting in direct simulations of the minimal flow unit[1] has given strong support to this idea. (See section 5.4 for a detailed description of this work.) In the minimal flow unit, we see each burst followed by the ejection of low-momentum fluid into the core of the flow, increasing the drag; suppression of the burst suppresses the ejection, and reduces the drag.

3. The Low Dimensional Approach

The low dimensional models are motivated by a desire to understand the dynamics of the large scales of the turbulent shear flows. This relies on the assumption that the large scales will be relatively independent of the details of the small scales of the motion, although the presence of the small scales must be accounted for. We will identify the large scale structures using the proper orthogonal, or Karhunen-Loève, decomposition (Loève, 1955; Lumley, 1967). Once identified, the structures are projected onto the Navier Stokes equations to yield a coupled, nonlinear system of ordinary differential equations. These equations are closed with a model for the effect of the small scales on the large scales, and numerical simulations and the tools of dynamical systems theory are then applied to study the properties and behavior of the model equations.

3.1. BASIS FUNCTIONS

The empirical eigenfunctions of the proper orthogonal decomposition (POD) were chosen as the basis functions for our low dimensional models, because they are optimal from the viewpoint of convergence. Each eigenfunction of this decomposition is defined as the function which maximizes the mean square correlation between it and an ensemble of realizations of the (ran-

[1](Jiménez & Moin, 1991) introduced the minimal flow unit as a tool for studying near-wall turbulent flow. The minimal flow unit has periodic boundary conditions in the streamwise and spanwise directions and corresponds to the smallest such computational box in which turbulence may be sustained. In a minimal flow unit, each wall layer contains a single set of coherent structures.

dom) velocity field; after the identification of an eigenfunction, the component of each velocity field parallel to the eigenfunction is subtracted. The next eigenfunction maximizes the mean square correlation between it and the component of the velocity fields orthogonal to the previous eigenfunctions. Therefore, the eigenfunctions are mutually orthogonal, and any truncation of the eigenfunction expansion will capture, on average, the maximum kinetic energy among all possible expansions of the same order. In the wall region of the turbulent boundary layer, the first eigenfunction captures 80% of the turbulent kinetic energy on average.[2]

In (statistically) homogeneous directions, the eigenfunctions of the proper orthogonal decomposition are Fourier modes. In the wall layer, the spanwise direction is homogeneous, and the flow is varying slowly in the streamwise direction so that this direction too may, at least locally, be treated as homogeneous. (This approximation is exactly true in fully developed pipe or channel flow. In fact, the eigenfunctions used in (Aubry *et al.*, 1988) were derived from measurements in the wall layer of a pipe flow (Herzog, 1986).) Thus, in the wall layer, the spanwise and streamwise dependence of the eigenfunctions is represented by Fourier modes. Only the wall-normal direction is explicitly treated by the POD.

3.2. GALERKIN PROJECTION

The velocity field is expanded in terms of the eigenfunctions, and the Navier Stokes equations are then projected onto the eigenfunctions to yield a set of coupled, nonlinear ordinary differential equations. The nonlinear terms in the model are quadratic and cubic. The quadratic terms correspond to fluctuation- fluctuation interactions which result from the convective term in the Navier Stokes equations. The cubic terms arise due to a model for the mean velocity profile in which it is expressed as an integral of the Reynolds stress (Tennekes & Lumley, 1972). This mechanism provides an important feedback between the coherent structures, which produce much of the Reynolds stress in the wall region, and their source of energy, the mean velocity profile. As the rolls in the boundary layer strengthen, they eat away at the velocity gradient, thereby weakening their source of energy. This feedback through the cubic term globally stabilizes our model of the boundary layer. The pressure enters the model equations as a boundary term, evaluated at the upper edge of the domain. If the domain of the model were large enough (including, for example, the whole of the channel or extending far away from the wall in the case of a boundary layer), this

[2]Our models for the turbulent boundary layer decompose only the fluctuating velocity field into eigenfunctions. If the full instantaneous velocity field were decomposed, the mean velocity would appear as an eigenfunction (Aubry *et al.*, 1988).

term would disappear. Since our models only extend out to $y^+ = 40$, the pressure will appear as a disturbance in our model accounting for the effect of the outer flow on the near-wall flow.

3.3. MODELING OF SMALL SCALES

The turbulent velocity field in our model is decomposed into three parts: the mean velocity, the resolved (large) scales, and the unresolved (small) scales. The interaction between the mean velocity and the large scales is modeled as described above. However, one must account for the transfer of energy between the resolved and unresolved scales to provide a sink for the energy in the coherent structures. The interaction between the resolved and unresolved scales occurs in two forms: the Leonard stress and the Reynolds stress. The Leonard stress represents the coupling of the resolved and unresolved scales through the nonlinear term in the Navier Stokes equations. This term is neglected from our model based on the assumption of the separation of scales, i.e. if the wavenumber of the resolved scales is much smaller than that of the unresolved scales their interaction will be negligible. However, a gradient transport model is employed to account for the stress of the unresolved scales on the resolved scales, the Reynolds stress. Much like in a large eddy simulation, an eddy viscosity parameterizes the drain of energy on the large scales by the small scale turbulence. This eddy viscosity, or Heisenberg parameter, is treated as a bifurcation parameter in our model, and the behavior of the model is studied for various constant levels of the eddy viscosity. In contrast, a real flow would have varying levels of small scale turbulence and as a consequence varying levels of eddy viscosity, and since the dynamics of the model vary with the level of the eddy viscosity, the real flow will likely display qualitatively different dynamics depending on the intensity of the small scales.[3]

4. Models for the Boundary Layer

Models of varying dimension have been constructed for the wall region of the turbulent boundary layer. The model of (Aubry *et al.*, 1988), containing no streamwise variation (*i.e.*, all modes had zero streamwise wavenumber),

[3]In their study of a low dimensional model for minimal channel flow, (Podvin & Lumley, 1997) explicitly evaluated the energy transfer from the coherent structures to the unresolved scales in a direct simulation. The numerically determined eddy viscosity displayed intermittency in their simulations with spikes in the eddy viscosity matching closely peaks in the shear stress at the wall. (*i.e.*, Energy transfer from the large to small scales — the production of small scale turbulence and Reynolds stress — peaked during the bursting process.) They found that the variation of the eddy viscosity in these intermittent regions led to qualitatively different dynamics in their model of the large scale structures.

was able to reproduce much of the dynamics of the wall region of the turbulent boundary layer. The severe truncation of the model prevents the representation of all of the scales of the flow, but the large scale behavior appears to be well represented by the model. Other models have been constructed with up to 32 modes — 64 degrees of freedom — (Sanghi & Aubry, 1993) which more faithfully reproduce the dynamics of the wall region during the bursting process. Though the larger models more fully mimic the dynamics, even the smallest model with only two active modes reproduces some features of the wall region.

4.1. PROPERTIES OF THE MODELS

For large values of the eddy viscosity, the model behaves as one might expect: all of the turbulent structures die away, leaving only the mean velocity. In phase space, this translates into the global stability of the origin. However, as one decreases the eddy viscosity a ring of fixed points branches off from the origin. These fixed points correspond in physical space to the familiar coherent structures in the wall region: streamwise rolls. For intermediate values of the eddy viscosity, each equilibrium on the ring is a saddle point, and the stable and unstable manifolds of that equilibria are connected with the unstable and stable manifolds, respectively, of the opposite fixed point on the ring. These connections between saddle points are heteroclinic orbits, and trajectories lying close to, but not quite on, the fixed point will spiral away from the fixed point and jump to the opposite fixed point, remaining in its neighborhood for a while before jumping again. Without noise, the period between jumps approaches infinity as the trajectory approaches the (attracting) heteroclinic orbit; however, the pressure signal from the outer part of the boundary layer acts as a disturbance to the system, keeping it from settling down to the fixed points. As a result, the system jumps at irregular intervals. In physical space, the reconstruction of the velocity field corresponding to the jump events qualitatively resembles the bursting process: the intensification of the updraft between the rolls during the jump is followed by the weakening of the rolls and a downdraft and finally, as the trajectory settles down to the opposite fixed point, the re-formation of the rolls. The bursting process is characterized by a transfer in energy from the coherent structures to the small scales of the turbulence; although the severe truncation of the five mode model limits its ability to represent this exchange of energy, the 32 mode model is able to capture the bursting process more accurately since the coherent structures give up their energy to higher wavenumber modes, particularly those with streamwise variation, during the jump.

The probability distribution of the period between jumps matches in shape that observed for the inter-burst time in experiments (Holmes & Stone, 1992; Stone & Holmes, 1989; Stone & Holmes, 1990; Stone & Holmes, 1991). However, the time scales for both burst duration and inter-burst time are unrealistically long, a fact that was not appreciated until recently, due to the inadvertent omission of a factor of $[L_1L_3]^{1/2}$ from the equation in Appendix A of (Aubry *et al.*, 1988). We believe that the discrepancy can be resolved by comparing the reference frames of the model and the experimental data. In the real boundary layer, many sets of structures will convect past a stationary observer. The model, however, focuses on a single structure and follows it through its complete life-cycle. Accounting for the difference in reference frames, the real boundary layer may be thought of as a series of relatively independent coherent structures (evolving according to the model equations) convecting past a stationary observer. The observer sees each structure for a short time before a new structure passes. With this adaptation, the model predicts the correct inter-burst time (Lumley & Podvin, 1996; Podvin *et al.*, 1996).

4.2. RESPONSE TO PHYSICAL EFFECTS

With the knowledge that the models mimic the bursting process, and keeping in mind the connection between inter-burst time and drag, we set out to study the response of our models to slight variations from the original flat-plate equilibrium boundary layer. These studies are related to the problem of 'extra rates of strain' (Bradshaw, 1976), where a turbulent boundary layer displays a substantial response to only a small change in conditions. For example, streamline curvature on the order of a couple of percent can change the drag coefficient by more than twenty percent. This response is a result of the turbulence (small perturbations to a laminar boundary layer yield comparably small responses) and the proper reproduction of this response seems to be a good test for the robustness of our models.

Pressure Gradients In the presence of favorable and unfavorable pressure gradients, the mean time between bursts (and hence, the inverse of the drag coefficient) increases and decreases, respectively, in qualitative accordance with observation (Stone, 1989). That is, in a favorable pressure gradient, the mean time between bursts increases, the drag coefficient drops, and the boundary layer grows more slowly, in both nature and the model. The opposite is true in the case of an unfavorable pressure gradient.

Curved Streamlines (Lumley & Podvin, 1996) approached the problem of adapting the models to flow with curved streamlines. Assuming that the near-wall structures were not significantly affected by the curvature, the flat-wall eigenfunctions were projected onto the Navier Stokes equations in cylindrical polar coordinates. The most significant additional term was a body force in the wall normal momentum equation which has the effect of intensifying or suppressing the coherent structures according to whether the surface is concave or convex. The intensification of the structures on a concave surface results in more frequent bursting and a higher drag coefficient. On the convex wall, the bursting is less frequent, and the boundary layer grows more slowly. (Since the model treats the streamwise direction as homogeneous, it is not able to capture the separated flows associated with strong (convex) surface curvature.)

5. Applications to the Control of Turbulence

Through the bursting process. the coherent structures near the wall are responsible for much of the production of turbulence in the boundary layer; thus interfering with the dynamics of the coherent structures provides a possible mechanism for the control of turbulence. Inhibiting the bursting process in the boundary layer also inhibits the production of turbulence, thereby decreasing momentum transport away from the wall (and, equivalently, turbulent skin friction). Another possible application for the control of turbulence is mixing enhancement. By exciting the action of coherent structures and promoting the production of turbulence, one could make possible more complete mixing of reactants in a combustor (increasing the efficiency of combustion) or faster spreading rates for a jet (lowering the jet noise output).

The possibility of the control of turbulence is approached from three perspectives. First, we look at a passive technique for turbulence control in wall-bounded flows: polymer addition. Empirically, it has been found that the addition of small quantities of polymers can lead to drag reduction of up to 80% in pipe and channel flows. This, combined with an analytical study which showed the equivalence of polymer drag reduction and active control (Aubry *et al.*, 1990), seems to indicate the potential of the active control of turbulence. Next, control is approached from the point of view of the models. Finally, with some insight from the dynamics of the models, we approach the problem of the control of the full fluid system.

5.1. POLYMER DRAG REDUCTION

Since polymers were first found to induce considerable reductions in turbulent skin friction (see (Toms, 1977) for an early history), much study has been devoted to elucidate the mechanism by which the polymers effect this reduction. A theory has been developed to explain the mechanism through which the polymers affect the turbulence (Lumley, 1969; Lumley, 1973;

Lumley & Kubo, 1985). This theory relies on the extension of the polymers in the large strain rate fluctuations which occur outside of the viscous sublayer. The polymers, whose equilibrium state is something like a loosely coiled ball of string, act to increase the viscosity locally once they are extended. Since small eddies are responsible for much of the Reynolds stress in the boundary layer and the additional viscosity from the polymers acts primarily on these small eddies, the momentum transport away from the wall through the Reynolds stress is weakened by the polymers. The boundary layer responds with a thickening of the wall region and the movement of the logarithmic layer away from the wall. Our familiar coherent structures, the rolls and streaks, grow to fill the enlarged wall region. Since the rolls now act across a greater velocity difference, the streamwise velocity fluctuations are enhanced. At the same time, the cross-stream velocity fluctuations are attenuated. The streak spacing in the drag-reduced boundary layer increases with the thickness of the wall region, indicating that the rolls roughly maintain their shape even though their scale has changed.[4] Foreshadowing some of the work described below, (Lumley & Kubo, 1985) suggested that the coherent structures must grow larger (becoming more energetic in the process) in order to remain in equilibrium with the increased losses from the turbulence as a result of the polymers.

Starting from this description of polymer drag-reduced flow, (Aubry *et al.*, 1990) adapted their model of the Newtonian boundary layer (Aubry *et al.*, 1988) to the non-Newtonian (polymer) flow. Assuming that the form of the coherent structures is basically unchanged from the Newtonian boundary layer, they stretched their eigenfunctions in the wall-normal direction to account for the growth of the wall region in the drag-reduced case. Stretching was applied in the spanwise and streamwise directions as well (both to match observations and satisfy continuity). The new eigenfunctions were then projected onto the Navier Stokes equations, yielding a new set of model equations for the drag-reduced flow. These model equations display the same qualitative features as the model of (Aubry *et al.*, 1988): a single fixed point for large values of the eddy viscosity (Heisenberg parameter) and more complex dynamics including heteroclinic orbit for smaller values of the eddy viscosity. However, when the wall region is stretched, analogous behaviors occur for much higher values of the eddy viscosity than in the Newtonian boundary layer. Since experiments indicate that the bursting process persists in polymer drag-reduced flow, we would expect that the eddy viscosity should be larger in a drag-reduced flow than in the Newtonian flow so that the heteroclinic jumps (associated with the

[4]A nice summary of the effect of polymers on the statistics of turbulent channel flow is contained in (Sureshkumar *et al.*, 1997), who performed direct simulations using a visco-elastic model to account for the presence of the polymers.

bursting process) occur in both systems. This indicates that an extra sink of energy is necessary in drag-reduced flow. In a manner consistent with (Lumley & Kubo, 1985), this extra drain can be seen to result from the extension of the polymer molecules, which, once extended, damp the small scales of the turbulence.

5.2. CONTROL OF THE MODELS

Motivated by the identification of bursts in the boundary layer with jumps in the phase space of the model equations, in conjunction with the idea that increasing the time between the bursts will reduce the drag, we developed a strategy for interfering with these jumps. The scheme for delaying the jumps (Coller *et al.*, 1994a; Coller *et al.*, 1994b; Coller *et al.*, 1994c; Coller, 1995) relies on an unexpected interaction: a transverse (shearing) velocity field — such as might be produced by a single vortex or a pair adjacent to the coherent structure — causes the system point in phase space to rotate around the fixed point. As the trajectory settles down to the fixed point along the stable direction, this rotation can be used to direct the system point back to the stable manifold if the trajectory tries to escape along an unstable direction. If the system point tries to escape by moving to the right, a rotation is induced that brings it back to the left, and so forth. (This geometric interpretation corresponds to the two-dimensional case; the multidimensional case is necessarily more complex). The net result of this technique is an average increase of the bursting period (corresponding to a decrease of the drag) of the order of 50% in the best case.

5.3. THE MODEL AS AN INTERPRETER

We envision an implementation of active control with an array of sensors and actuators on the surface under the boundary layer. The sensors provide some information about the flow, most likely measurements of the fluctuating wall shear stress. However, their placement on the surface severely limits the information they are able to provide about the flow and leaves us the considerable task of inferring what is happening in the flow above from the limited information that they are able to supply. A low-dimensional model would be useful here. The model could be used as a Kalman filter to make the most of the partial information available from the sensors (Podvin, 1997). The wall shear stress measurements could be used to estimate the state of the coherent structures above the surface, and the state of the coherent structures would be tracked with the low dimensional model. Measurements at the wall would be continuously fed into the model so that the state of the model would follow, as closely as possible, the state of the boundary layer. Others have based their estimation of the state of the flow above the surface on the presence of the coherent structures in the flow. (Lee *et al.*, 1997) used a neural net to predict the presence of strong up- or downdrafts above the surface, based solely on measurements of spanwise shear stress. Assuming that the coherent structures near the wall evolve (at least locally) in a linear fashion, (Rathnasingham & Breuer, 1997b) placed sensors both upstream and downstream of their actuator and developed a transfer function for the near-wall flow. Their control was applied with the aim of cancelling the disturbances at the downstream sensor.

5.4. CONTROL OF DIRECT SIMULATIONS

Thus far, we have only considered control, both passive and active, in the setting of the model, not in the real flow. (Carlson *et al.*, 1995) have successfully implemented a direct numerical simulation of turbulent channel flow in a minimal flow unit with and without a moving boundary. In our first attempts at control, the moving boundary was used to simulate the actuator, a bump, rising directly into the flow. We had thought that the bump would produce a simple necklace vortex, with the trailing vortices resembling the naturally occurring coherent structures. The simulations, however, suggested that the flow around the bump was rather complicated: the flow splits to pass around the bump, generating a strong vortex pair with an updraft between the vortices; behind the bump, as the flow comes around to rejoin itself, another strong vortex pair of opposite sign is generated underneath the first pair; this stack of two pairs is topped by the necklace vortex, the weakest of the three, that has the same sign as the lowest pair (Carlson & Lumley, 1996b). Though this rather complicated flow may not be so useful from the point of view of control, the bump was able to exert some control in our simulations both locally, by lifting the high speed streak away from the wall, and globally, by deflecting it into the low speed streak, thereby weakening the inflection in the mean velocity profile (Carlson & Lumley, 1996a). One further difficulty was the reversal of drag reduction when the bump was lowered.

The difficulties with the bump led us to explore the possibility of using other actuators. A reasonable point of departure is to determine what actuator velocity field is desired and to move from the velocity field to the actuator. Our ideal actuator velocity field is simply a replica of the coherent structures themselves — the rolls — which we hope to use to inhibit the bursting of the naturally-occurring coherent structures. We feel that raising a sharp-edged flap into the flow will produce this sort of velocity field, with the flap inducing a pair of trailing vortices similar to those from the wing tips of an airplane. It should also be noted that groups at MIT (Rathnasingham & Breuer, 1997a) and Stanford (Jacobson & Reynolds,

1993) have built actuators based on vibrating beams forcing air in and out of a cavity that produce just this sort of velocity field (in a time-averaged sense). Since our pseudospectral codes are restricted to the computation of smooth boundaries and since the small scales induces by the sharp edges of the flap would require much higher resolution than the turbulence alone, we chose to simulate the effect of the flap on the flow through the use of a body force.

As a result, we added a body force to a flat-walled channel flow computation and so induced a pair of counter-rotating vortices, thereby simulating the effect of the flap but not the flap itself. The body force, which acts solely on the wall normal component of velocity, added vorticity of opposite sign to the coherent structures present in the flow. The coherent structures and their associated streaks were weakened. With the weakening of the high speed streak, the drag fell; at the same time, the low-speed streak and the inflection in the mean velocity profile decreased in intensity, delaying the onset of the secondary instability, and the ejection. The streaks thickened in the wall-normal direction (much like in the polymer drag reduced flow) and seemed to stabilize for some time, before the coherent structures strengthened again and caused an ejection and burst. The action of the body force for a duration of 50 wall time units (u_τ/ν) had the effect of suppressing the burst for ten times its duration, resulting in drag reduction of 15% when averaged over 700 wall time units.

Our attempts to control these direct simulations were not optimized, nor were they particularly coordinated with our understanding of what happens in the phase space of the model; nevertheless, both schemes have achieved drag reductions near 20% in the vicinity of the applied control. Of even more interest is a reduction in drag on the wall opposite the applied control in the simulations with the body force. Presumably, this reflects the reduced production of turbulence resulting from the suppression of bursting.

6. Estimation of Eigenfunctions Using Energy Stability Theory

One limitation of the application of the low dimensional approach to turbulence lies in the computation of the empirical eigenfunctions from the proper orthogonal decomposition (POD). The POD requires the two point velocity correlation tensor as input. This tensor is difficult to measure experimentally, especially in the neighborhood of the wall. ((Herzog, 1986) performed his measurements in glycerine to increase resolution near the wall.) Although direct or large eddy simulations may be used to generate the tensor, the cost of producing the large statistical sample necessary for well-converged eigenfunctions may be prohibitive, or one may wish to study flows for which no simulation database exists.

Following a suggestion of (Lumley, 1971), (Poje & Lumley, 1995) approached the problem of determining the POD eigenfunctions for the wall region of the turbulent boundary layer from energy stability theory using only one point statistics (rather than the two point correlations required for the POD). The method relies on the same decomposition of the velocity field: mean, large scales, and small scales, as described above in section 3.1. With the mean velocity and Reynolds stress profiles specified as an input and the effect of the small scales modeled with an eddy viscosity (which is specified and varies in the wall-normal direction), an equation for the evolution of coherent, or large scale, energy is derived. The mode which maximizes the growth rate of coherent energy is chosen as an approximation to the POD eigenfunction (that contains the most kinetic energy on average). In applying the method to the wall region, (Poje & Lumley, 1995) found an anisotropic eddy viscosity to be necessary to properly model the stresses induced by the coherent structures. An isotropic eddy viscosity prescribes a phase-locking of the small scale stresses with respect to the large scales and also precludes any coupling of the streamwise and cross-stream velocities (in a streamwise invariant mode) through the small scale stresses. (The anisotropic eddy viscosity is important here because these equations are being used to determine the form of the large scale structures. An isotropic eddy viscosity was sufficient for our purposes in section 3, because the form of the large scales was already determined by the POD, and only the stresses due to the small scales were accounted for by the gradient transport model.) Once the anisotropic eddy viscosity was implemented, the modes generated by energy stability theory closely matched those from the POD. In addition to prescribing the shape of the modes in the wall-normal direction, (Poje & Lumley, 1995) were also able to predict the peak of the energy spectrum in the spanwise direction, which corresponds to the mean streak spacing near the wall.

Building on their previous work (Lumley & Poje, 1997) applied this method to study flows with density fluctuations, in particular sheared Rayleigh-Bénard convection. Before coming to the energy stability approach, however, the application of the POD to flows with density fluctuations must be addressed. A simple way of avoiding this issue is to apply the POD to the velocity and temperature/density fields separately. In this method, all of the coupling of the fields will come from the equations. The POD, however, is designed to extract the maximum amount of information possible from the two point correlation tensor. In the case of flows with density fluctuations, this tensor contains information about the heat/density fluxes which is not used if the velocity and temperature are treated separately by the POD. A complication which must be faced when applying the POD to the velocity and temperature jointly is the relative weighting of information about the two fields. (Lumley & Poje, 1997) address the issue of weighting by normalizing the velocity components by the rms kinetic energy and the density/temperature fluctuation by its rms value. While this may not be the optimal weighting, it assures each eigenfunction will capture equal amounts of information about the temperature and velocity fields. With only the mean velocity and temperature profile as inputs (and an isotropic eddy viscosity), the energy stability method produced a coherent field which well approximated the second order statistics computed from DNS. In addition, the predicted eigenfunctions followed the response of the full fluid system to changes in the strength of the shear relative to the heating. For weak shear, plume-like structures are displayed in the simulations; similarly, the most energetic modes predicted by energy stability theory have nonzero wavenumber in both the streamwise and spanwise direction. When the shear increases, energy stability theory matches the simulations with the convection cells becoming more like rolls and aligning themselves with the stream.

7. Conclusions

While the determination of approximate POD modes through energy stability theory may promise easier access to the low dimensional approach to turbulence, more generally, the low dimensional approach allows a compact representation of the dynamics of the large scales of turbulent shear flows. What we have attempted to show here, through specific applications in the boundary layer, is the ability of the models to replicate the response of a full fluid system to many physical effects. It is our belief that such models are capable of shedding light on a variety of problems, from jets and shear layers to panel flutter and sonar self-noise, if only large scale information is required.

Acknowledgements

This manuscript describes work carried out over a ten year period, necessarily involving many past and present students and colleagues. JLL thanks them all for their contributions, but particularly Phil Holmes, Gal Berkooz and Sid Leibovich. JLL is supported in part by Contract No. F49620-92-J-0287 jointly funded by the U. S. Air Force Office of Scientific Research (Control and Aerospace Programs), and the U. S. Office of Naval Research, in part by Grant No. F49620-92-J-0038, funded by the U. S. Air Force Office of Scientific Research (Aerospace Program), and in part by the Physical Oceanography Programs of the U. S. National Science Foundation (Contract No. OCE-901 7882) and the U. S. Office of Naval Research (Grant No. N00014-92-J-1547). This material is based in part upon work supported under a National Science Foundation Graduate Research Fellowship awarded to PNB. We would like to thank the Cornell Theory Center for providing computational resources, and in particular Bruce Land who helped with visualizations. The material in this article appeared in a somewhat different form in the *Annual Review of Fluid Mechanics*, vol. 30, 1998.

References

Aubry, N., Holmes, P., Lumley, J. L., and Stone, E., 1988. "The dynamics of coherent structures in the wall region of a turbulent boundary layer," *J. Fluid Mech.* **192**, pp. 115–173.

Aubry, N., Lumley, J. L., and Holmes, P., 1990. "The effect of modeled drag reduction in the wall region," *Theoret. Comput. Fluid Dynamics* **1**, pp. 229–248.

Ball, K. S., Sirovich, L., and Keefe, L. R., 1991. "Dynamical eigenfunction decomposition of turbulent channel flow," *Int. Jour. for Num. Meth. in Fluids* **12**, pp. 585–604.

Bradshaw, P., 1976. "Complex turbulent flows," in *Proceedings of the 14th IUTAM Congress*, ed. W. T. Koiter, Delft, The Netherlands: North Holland, pp. 103–113.

Carlson, H. A., Berkooz, G., and Lumley, J. L., 1995. "Direct numerical simulation of flow in a channel with complex, time-dependent wall geometries: a pseudospectral method," *J. Comput. Phys.* **121**, pp. 155–175.

Carlson, H. A. and Lumley, J. L., 1996a. "Active control in the turbulent boundary layer of a minimal flow unit," *J. Fluid Mech.* **329**, pp. 341–371.

Carlson, H. A. and Lumley, J. L., 1996b. "Flow over an obstacle emerging from the wall of a channel," *AIAA J* **34**(5), pp. 924–931.

Chambers, D. H., Adrian, R. J., Moin, P., Stewart, D., and Sung, H. J., 1988. "Karhunen-Loève expansion of Burgers model of turbulence," *Phys. Fluids* **31**, pp. 2573–2582.

Coller, B. D., 1995. *Suppression of heteroclinic bursts in boundary layer models*, PhD thesis, Cornell University.

Coller, B. D., Holmes, P., and Lumley, J. L., 1994a. "Control of bursting in boundary layer models," *Appl. Mech Rev.* **47**(6), part 2, pp. S139–S143, Mechanics USA 1994, ed. A. S. Kobayashi.

Coller, B. D., Holmes, P., and Lumley, J. L., 1994b. "Controlling noisy heteroclinic cycles," *Physica D* **72**, pp. 135–160.

Coller, B. D., Holmes, P., and Lumley, J. L., 1994c. "Interaction of adjacent bursts in the wall region," *Phys. Fluids* **6**(2), pp. 954–961.

Deane, A. E., Keverkidis, I. G., Karniadakis, G. E., and Orszag, S. A., 1991. "Low-dimensional models for complex flows: Application to grooved channels and circular cyliders," *Phys. Fluids A* **3**(10), pp. 2337–2354.

Deane, A. E. and Sirovich, L., 1991. "A computational study of Raleigh-Bénard convection Part I. Rayleigh number scaling," *J. Fluid Mech.* **222**, pp. 231–250.

Glauser, M., Leib, S. J., and George, W. K., 1987. "Coherent structures in the axisymmetric turbulent jet mixing layer," in *Turbulent Shear Flows 5*, Berlin/New York: Springer-Verlag.

Glauser, M. N. and George, W. K., 1987a. "Orthogonal decomposition of the axisymmetric jet mixing layer including azimuthal dependence," in *Advances in Turbulence*, ed. G. Comte-Bellot and J. Mathieu, Berlin/New York: Springer-Verlag.

Glauser, M. N. and George, W. K., 1987b. "An orthogonal decomposition of the axisymmetric jet mixing layer utilizing cross-wire velocity measurements," in *Proceedings, Sixth Symposium on Turbulent Shear Flows*, Toulouse, France, pp. 10.1.1–10.1.6.

Glauser, M. N., Zheng, X., and Doering, C. R., 1989. "The dynamics of organized structures in the axisymmetric jet mixing layer," in *Turbulence and Coherent Structures*, ed. M. Lesieur and O. Metais, Boston: Kluwer Academic.

Glauser, M., Zheng, X., and George, W. K., 1992. "The streamwise evolution of coherent structures in the axisymmetric jet mixing layer," in *Studies in Turbulence*, ed. T. B. Gatski, S. Sarkar, and C. G. Speziale, New York: Springer-Verlag, pp. 207–222.

Glezer, A., Kadioglu, A. J., and Pearlstein, A. J., 1989. "Development of an extended proper orthogonal decomposition and its application to a time periodically forced plane mixing layer," *Phys. Fluids A* **1**, pp. 1363–73.

Herzog, S., 1986. *The Large Scale Structure in the Near Wall Region of a Turbulent Pipe Flow*, PhD thesis, Cornell Univ.

Holmes, P. J., Berkooz, G., and Lumley, J. L., 1996. *Turbulence, Coherent Structures, Dynamical Systems and Symmetry*, New York/Cambridge: Cambridge University Press.

Holmes, P. J. and Stone, E., 1992. "Heteroclinic cycles, exponential tails and intermittency in turbulence production," in *Studies in Turbulence*, ed. T. B. Gatski, S. Sarkar, and C. G. Speziale, Springer-Verlag, pp. 179–189.

Jacobson, S. A. and Reynolds, W. C., 1998. "Active control of streamwise vortices and streaks in boundary layers," *J. Fluid Mech.* **360**, pp. 179–211.

Jiménez, J. and Moin, P., 1991. "The minimal flow unit in near-wall turbulence, *J. Fluid Mech.* **225**, pp. 213–240.

Kirby, M., Boris, J., and Sirovich, L., 1990a. "An eigenfunction analysis of axisymmetric jet flow," *J. Comput. Phys.* **90**(1), pp. 98–122.

Kirby, M., Boris, J., and Sirovich, L., 1990b. "A proper orthogonal decomposition of a simulated supersonic shear layer," *Int. Jour. for Num. Meth. in Fluids* **10**, pp. 411–428.

Lee, C., Kim, J., Babcock, D., and Goodman, R., 1997. "Application of neural networks to turbulence control for drag reduction," *Phys. Fluids* **9**(6), pp. 1740–1747.

Liu, Z.-C., Adrian, R. J., and Hanratty, T. J., 1994. "Reynolds-number similarity of orthogonal decomposition of the outer layer of turbulent wall flow," Technical Report TAM 748, UILU-ENG-94-6004, University of Illinois, Department of Theoretical and Applied Mechanics.

Loève, M., 1955. *Probability theory*, Van Nostrand.

Lorenz, E. N., 1963. "Deterministic nonperiodic flow," *J. Atmos. Sci.* **20**, pp. 130–141.

Lumley, J. L., 1967. "The structure of inhomogeneous turbulence," in *Atmospheric Turbulence and Wave Propagation*, ed. A. M. Yaglom and V. I. Tatarski, Moscow: Nauka.

Lumley, J. L., 1969. "Drag reduction by additives," *Ann. Rev. Fluid Mech.* **1**, pp. 367–384.

Lumley, J. L., 1971. "Some comments on the energy method," in *Developments in Mechanics 6*, ed. L. Lee and A. Szewczyk, Notre Dame Press.

Lumley, J. L., 1973. "Drag reduction in turbulent flow by polymer additives," *J. Polym. Sci. Macromol. Rev.* **7**, pp. 263–290.

Lumley, J. L. and Kubo, I., 1985. "Turbulent drag reduction by polymer additives: A survey," in *The Influence of Polymer Additives on Velocity and Temperature Fields*, ed. B. Gampert, New York: Springer-Verlag, pp. 3–24.

Lumley, J. L. and Podvin, B., 1996. "Dynamical systems theory and extra rates of strain in turbulent flows," *Journal of Experimental and Thermal Fluid Science* **12**, pp. 180–189, Peter Bradshaw Symposium.

Lumley, J. L. and Poje, A. C., 1997. "Low-dimensional models for flows with density fluctuations," *Phys. Fluids* **9**(7), pp. 2023–2031.

Moin, P. and Moser, R. D., 1989. "Characteristic-eddy decomposition of turbulence in a channel," *J. Fluid Mech.* **200**, pp. 471–509.

Podvin, B., 1997. *Turbulence, information and dynamical systems*, PhD thesis, Cornell University.

Podvin, B., Gibson, J., Berkooz, G., and Lumley, J. L., 1996. "Lagrangian and Eulerian view of the bursting period," *Phys. Fluids* **9**(2), pp. 433–437.

Podvin, B. and Lumley, J. L., 1997. "A low dimensional approach for the minimal flow unit," *J. Fluid Mech.*, Submitted.

Poje, A. C. and Lumley, J. L., 1995. "A model for large scale structures in turbulent shear flows," *J. Fluid Mech.* **285**, pp. 349–369.

Rathnasingham, R. and Breuer, K. S., 1997a. "Coupled fluid-structural characteristics of actuators for flow control," *AIAA J.* **35**(5), pp. 832–37.

Rathnasingham, R. and Breuer, K. S., 1997b. "System identification and control of a turbulent boundary layer," *Phys. Fluids* **9**(7), pp. 1867–69.

Ruelle, D. and Takens, F., 1971. "On the nature of turbulence," *Communs. Math. Phys.*, pp. 137–151.

Sanghi, S. and Aubry, N., 1993. "Mode interaction models for near-wall turbulence," *J. Fluid Mech.* **247**, pp. 455–488.

Sirovich, L. and Deane, A. E., 1991. "A computational study of Raleigh-Bénard convection Part II. Dimension considerations," *J. Fluid Mech.* **222**, pp. 251–265.

Stone, E., 1989. *A Study of Low Dimensional Models for the Wall Region of a Turbulent Layer*, PhD thesis, Cornell University.

Stone, E. and Holmes, P. J., 1989. "Noise induced intermittency in a model of a turbulent boundary layer," *Physica D* **37**, pp. 20–32.

Stone, E. and Holmes, P. J., 1990. "Random perturbations of heteroclinic cycles," *SIAM J. on Appl. Math.* **50**(3), pp. 726–743.

Stone, E. and Holmes, P. J., 1991. "Unstable fixed points, heteroclinic cycles and exponential tails in turbulence production," *Phys. Lett. A* **155**, pp. 29–42.

Sureshkumar, R., Beris, A. N., and Handler, R. A., 1997. "Direct numerical simulation of the turbulent channel flow of a polymer solution," *Phys. Fluids* **9**(3), pp. 743–755.

Tennekes, H. and Lumley, J. L., 1972. *A First Course in Turbulence*, The MIT press.

Toms, B. A., 1977. "On the early experiments on drag reduction by polymers," *Phys. Fluids* **20**(10), pp. S3–S5.

MODELING NON-EQUILIBRIUM TURBULENT FLOWS

CHARLES G. SPEZIALE

Aerospace & Mechanical Engineering Department
Boston University
Boston, Massachusetts

Abstract. Traditional turbulence models invoke a range of equilibrium assumptions that renders them incapable of describing turbulent flows where the departures from equilibrium are large. Even the commonly used second-order closures have an implicit equilibrium assumption for the pivotal pressure-strain correlation that makes them incapable of describing such non-equilibrium turbulent flows. It will be shown that explicit algebraic stress models can be partially extended to non-equilibrium turbulent flows by a Padé approximation. Then, by implementing a relaxation time approximation, second-order closures are obtained where – to the lowest order – the rapid pressure-strain correlation is represented by models that depend nonlinearly on the invariants of the non-dimensional strain rates. However, unlike in many of the more recent second-order closures, linearity is maintained in the Reynolds stress anisotropy tensor consistent with the definition of the rapid pressure-strain correlation. It will be demonstrated by a variety of examples how this leads to an improved performance in non-equilibrium turbulence without compromising the predictions for the near-equilibrium case. A new approach to large-eddy simulations will also be presented that allows subgrid scale stress models to continuously go to Reynolds stress models in the coarse mesh/infinite Reynolds number limit. Furthermore, the modeling of the turbulent dissipation rate will be considered, particularly in regard to the non-equilibrium effects of vortex stretching and anisotropic dissipation. The status of these recent developments and the prospects for future research will be thoroughly discussed.

M. D. Salas et al. (eds.), Modeling Complex Turbulent Flows, 107–137.

1. Introduction

Turbulence models have typically been formulated subject to a variety of equilibrium assumptions employing standard benchmark flows such as the equilibrium turbulent boundary layer and homogeneous shear flow in equilibrium. When these models are then applied to non-equilibrium flow situations, they usually break down entirely. Even existing second-order closure models perform poorly in such circumstances, including the somewhat simplified non-equilibrium test cases of rapidly distorted homogeneous shear flow and plane strain turbulence. In the standard formulation of second-order closures, the pivotal pressure-strain correlation is modeled based on what is tantamount to a homogeneous equilibrium assumption (see Speziale 1996). This fact – along with the use of gradient transport models for the turbulent diffusion – makes these models, at best, capable of describing turbulent flows with moderate departures from equilibrium and mild inhomogeneous effects. In addition, the scale determining equation – which is usually based on the turbulent dissipation rate – commonly employs a standard equilibrium hypothesis whereby it is assumed that the leading order part of the destruction of dissipation exactly counterbalances the production of dissipation by vortex stretching (see Bernard, Thangam and Speziale 1992). In so far as the dissipation rate is concerned, the Kolmogorov assumption of local isotropy is typically invoked which formally requires that the turbulent Reynolds number be extremely high and that the departures from equilibrium not be too large (cf. Durbin and Speziale 1991). In practical flow situations, these assumptions can be violated. Thus, there is a definite need for turbulence models that can be applied to these non-equilibrium flow situations.

In this paper it will be shown how models can be developed that are more suitable for non-equilibrium turbulent flows. Both two-equation models and second-order closures will be considered where the former are obtained from the latter using the homogeneous, equilibrium hypothesis. The two-equation models – which are in the form of nonlinear, anisotropic eddy viscosity models with strain-dependent coefficients – are extended to the non-equilibrium domain via a Padé approximation that regularizes while building in some limited agreement with Rapid Distortion Theory (RDT) (see Speziale and Xu 1996). Their derivation will be thoroughly discussed. Then, second-order closures can be obtained by a standard relaxation time approximation wherein agreement with the Crow (1968) constraint is automatically built in. This approach leads to second-order closures which, to the lowest order, depend on the non-dimensional invariants of both the irrotational and rotational strain rates. However, consistent with the definition of the rapid pressure-strain correlation, tensorial linearity of this term with respect to the Reynolds stress anisotropy tensor – and, hence, the energy spectrum tensor – is maintained. Many recent rapid pressure-strain models have been proposed that are inconsistent since they are nonlinear in the Reynolds stress anisotropy tensor and, hence, nonlinear in the energy spectrum tensor. Using these results, a non-equilibrium extension of the Speziale, Sarkar and Gatski (SSG) (1991) second-order closure model will be developed. It will be shown that homogeneous shear flow and plane strain turbulence are much better described for a wide range of shear and strain rates from the equilibrium regime to the RDT regime. Here, something similar in spirit to the correspondence principle in physics is implemented so that the models are applicable to non-equilibrium flow situations without compromising their performance in the equilibrium turbulent flows that they have been principally formulated to describe.

In addition, the role of non-equilibrium vortex stretching in the dissipation rate transport equation will be discussed along with the effects of anisotropic dissipation. Furthermore, a new time-dependent Reynolds-Averaged Navier-Stokes (RANS) and Large-Eddy Simulation (LES) capability will be presented. The unique feature of this new approach is that subgrid scale models go continuously to Reynolds stress models in the coarse mesh/infinite Reynolds number limit. This allows one to automatically go from DNS to LES to RANS. The implications of these results for the modeling of the non-equilibrium turbulent flows of technological importance will be thoroughly discussed.

2. Theoretical Background

The incompressible turbulent flow of a viscous fluid will be considered which is governed by the Navier-Stokes and continuity equations:

$$\frac{\partial v_i}{\partial t} + v_j \frac{\partial v_i}{\partial x_j} = -\frac{\partial P}{\partial x_i} + \nu \nabla^2 v_i \tag{1}$$

$$\frac{\partial v_i}{\partial x_i} = 0. \tag{2}$$

In (1)–(2), v_i is the velocity vector, P is the modified kinematic pressure, and ν is the kinematic viscosity; the Einstein summation convention applies to repeated indices. For simplicity, we have restricted our attention to conservative body forces that can be derived from a scalar potential which can be absorbed into the pressure. The velocity and pressure are decomposed into mean and fluctuating parts:

$$v_i = \overline{v}_i + u_i, \quad P = \overline{P} + p \tag{3}$$

where an overbar represents a Reynolds average. In a statistically steady turbulence, time averages can be used whereas in a statistically homogeneous turbulence, spatial averages can be implemented. However, in general turbulent flows – which include the subject of this study – flows may be neither statistically steady nor homogeneous so that ensemble averages must be used. The ensemble average is given by

$$\overline{\phi} = \lim_{N\to\infty} \frac{1}{N} \sum_{\alpha=1}^{N} \phi^{(\alpha)}(\mathbf{x}, t) \tag{4}$$

for any flow variable ϕ (for instance, $\phi \equiv \{v_i, P\}$) where an average is taken over N realizations (or experiments) of a given turbulent flow with the same initial and boundary conditions to within random, small perturbations.

The Reynolds-averaged Navier-Stokes and continuity equations are given by

$$\frac{\partial \overline{v}_i}{\partial t} + \overline{v}_j \frac{\partial \overline{v}_i}{\partial x_j} = -\frac{\partial \overline{P}}{\partial x_i} + \nu \nabla^2 \overline{v}_i - \frac{\partial \tau_{ij}}{\partial x_j} \tag{5}$$

$$\frac{\partial \overline{v}_i}{\partial x_i} = 0 \tag{6}$$

where

$$\tau_{ij} \equiv \overline{u_i u_j}$$

is the kinematic form of the Reynolds stress tensor. This can formally be built-up from solutions of the fluctuating momentum equation, obtained by subtracting (5) from (1), which takes the form

$$\frac{\partial u_i}{\partial t} + \overline{v}_k \frac{\partial u_i}{\partial x_k} = -u_k \frac{\partial u_i}{\partial x_k} - u_k \frac{\partial \overline{v}_i}{\partial x_k} - \frac{\partial p}{\partial x_i} + \nu \nabla^2 u_i + \frac{\partial \tau_{ij}}{\partial x_j}. \tag{7}$$

This equation can be written in symbolic operator notation as follows:

$$\mathcal{N} u_i = 0.$$

Then, the transport equation for the Reynolds stress tensor is obtained from the second moment

$$\overline{u_i \mathcal{N} u_j + u_j \mathcal{N} u_i} = 0.$$

This yields the Reynolds stress transport equation which takes the explicit form (cf. Hinze 1975)

$$\begin{aligned} \frac{\partial \tau_{ij}}{\partial t} + \overline{v}_k \frac{\partial \tau_{ij}}{\partial x_k} &= -\tau_{ik} \frac{\partial \overline{v}_j}{\partial x_k} - \tau_{jk} \frac{\partial \overline{v}_i}{\partial x_k} + \Pi_{ij} \\ &\quad -\varepsilon_{ij} - \frac{\partial C_{ijk}}{\partial x_k} + \nu \nabla^2 \tau_{ij} \end{aligned} \tag{8}$$

where

$$\Pi_{ij} \equiv \overline{p\left(\frac{\partial u_i}{\partial x_j} + \frac{\partial u_j}{\partial x_i}\right)}$$

$$\varepsilon_{ij} \equiv 2\nu \overline{\frac{\partial u_i}{\partial x_k}\frac{\partial u_j}{\partial x_k}}$$

$$C_{ijk} \equiv \overline{u_i u_j u_k} + \overline{pu_i}\delta_{jk} + \overline{pu_j}\delta_{ik}$$

are, respectively, the pressure-strain correlation, dissipation rate tensor and third-order turbulent diffusion correlation. The transport equation for the turbulent kinetic energy

$$K = \frac{1}{2}\tau_{ii}$$

is obtained from a contraction of (8) and is given by

$$\frac{\partial K}{\partial t} + \overline{v}_j \frac{\partial K}{\partial x_j} = \mathcal{P} - \varepsilon - \frac{\partial}{\partial x_j}(\frac{1}{2}\overline{u_i u_i u_j} + \overline{pu_j}) + \nu\nabla^2 K \tag{9}$$

where

$$\mathcal{P} \equiv -\tau_{ij}\frac{\partial \overline{v}_i}{\partial x_j}$$

$$\varepsilon \equiv \nu \overline{\frac{\partial u_i}{\partial x_j}\frac{\partial u_i}{\partial x_j}}$$

are the turbulence production and (scalar) turbulent dissipation rate, respectively. A gradient transport model of the form $\frac{1}{2}\overline{u_i u_i u_j} + \overline{pu_j} = -\frac{\nu_T}{\sigma_k}\partial K/\partial x_j$ is used where σ_k is a constant that is usually taken to be one and ν_T is the eddy viscosity that will be discussed in more detail later.

The turbulent dissipation rate is obtained from a modeled version of the exact transport equation which is derived from the second moment

$$2\nu \overline{\frac{\partial u_i}{\partial x_j}\frac{\partial}{\partial x_j}(\mathcal{N}u_i)} = 0.$$

Its exact form is given by (cf. Speziale 1991)

$$\begin{aligned}\frac{\partial \varepsilon}{\partial t}+\overline{v}_i\frac{\partial \varepsilon}{\partial x_i} &= -2\nu\overline{\frac{\partial u_j}{\partial x_i}\frac{\partial u_j}{\partial x_k}\frac{\partial \overline{v}_i}{\partial x_k}}-2\nu\overline{\frac{\partial u_i}{\partial x_j}\frac{\partial u_k}{\partial x_j}\frac{\partial \overline{v}_i}{\partial x_k}}\\ &-2\nu\overline{\frac{\partial u_i}{\partial x_k}\frac{\partial u_i}{\partial x_m}\frac{\partial u_k}{\partial x_m}}-2\nu\overline{u_k\frac{\partial u_i}{\partial x_j}}\frac{\partial^2 \overline{v}_i}{\partial x_k \partial x_j}\\ &-2\nu\frac{\partial}{\partial x_k}\left(\overline{\frac{\partial p}{\partial x_m}\frac{\partial u_k}{\partial x_m}}\right)-\nu\frac{\partial}{\partial x_k}\left(\overline{u_k\frac{\partial u_i}{\partial x_m}\frac{\partial u_i}{\partial x_m}}\right)\\ &-2\nu^2\overline{\frac{\partial^2 u_i}{\partial x_k\partial x_m}\frac{\partial^2 u_i}{\partial x_k\partial x_m}}+\nu\nabla^2\varepsilon.\end{aligned} \tag{10}$$

This equation is usually modeled in the high-Reynolds-number form

$$\frac{\partial \varepsilon}{\partial t}+\overline{\mathbf{v}}\cdot\nabla\varepsilon = C_{\varepsilon 1}\frac{\varepsilon}{K}\mathcal{P}-C_{\varepsilon 2}\frac{\varepsilon^2}{K}+\frac{\partial}{\partial x_i}\left(\frac{\nu_T}{\sigma_\varepsilon}\frac{\partial \varepsilon}{\partial x_i}\right) \tag{11}$$

where $C_{\varepsilon 1}$ and $C_{\varepsilon 2}$ are constants whereas $\nu_T \equiv C_\mu K^2/\varepsilon$ (with C_μ a constant) is the eddy viscosity. Later, we will discuss the modeling of the dissipation rate equation and extend it to non-equilibrium flows.

Explicit algebraic stress models can be formally obtained by assuming that the turbulence is homogeneous and in equilibrium. For homogeneous turbulent flows, which are spatially uniform in a statistical sense, the Reynolds stress transport equation (8) reduces to the form

$$\dot{\tau}_{ij} = -\tau_{ik}\frac{\partial \overline{v}_j}{\partial x_k}-\tau_{jk}\frac{\partial \overline{v}_i}{\partial x_k}+\Pi_{ij}-\varepsilon_{ij} \tag{12}$$

where a superposed dot represents the time derivative and all other terms are defined as before. Homogeneous turbulence achieves an equilibrium state – which is time-independent – in the limit as the time $t \to \infty$. Homogeneous turbulence in equilibrium (as well as local regions of inhomogeneous turbulent flows where there is a production-equals-dissipation equilibrium) satisfy the constraints:

$$\dot{b}_{ij} = 0 \tag{13}$$

$$-\frac{\partial C_{ijk}}{\partial x_k}+\nu\nabla^2\tau_{ij} = 0 \tag{14}$$

where

$$b_{ij} = \frac{\tau_{ij}-\frac{2}{3}K\delta_{ij}}{2K} \tag{15}$$

is the Reynolds stress anisotropy tensor and (14) applies generally to the deviatoric (traceless) part for inhomogeneous turbulent flows in equilibrium. In physical terms, this is an equilibrium for which convective and transport effects can be neglected; it is the basic equilibrium hypothesis used in the derivation of algebraic stress models (ASM's). *However, it is only globally valid for homogeneous turbulent flows that are in equilibrium.* Inhomogeneous turbulent flows typically only have very narrow regions where there is a production-equals-dissipation equilibrium (e.g., the logarithmic region in turbulent channel flow where models based on the homogeneous equilibrium hypothesis do reasonably well with minor adjustments). That is why we have based our analysis on (12) with the assumption of homogeneity (see Gatski and Speziale 1993).

It follows directly from (13) and (15) that

$$\dot{\tau}_{ij} = \frac{\tau_{ij}}{K}\dot{K} \tag{16}$$

and, hence, by making use of the contraction of (12) – which employs (14) – it can be shown that

$$\dot{\tau}_{ij} = (\mathcal{P} - \varepsilon)\frac{\tau_{ij}}{K} \tag{17}$$

where $\mathcal{P} \equiv -\tau_{ij}\partial\overline{v}_i/\partial x_j$ is again the turbulence production and $\varepsilon \equiv \frac{1}{2}\varepsilon_{ii}$ is the (scalar) turbulent dissipation rate. The substitution of (17) into (12) yields the following equilibrium form of the Reynolds stress transport equation:

$$(\mathcal{P} - \varepsilon)\frac{\tau_{ij}}{K} = -\tau_{ik}\frac{\partial\overline{v}_j}{\partial x_k} - \tau_{jk}\frac{\partial\overline{v}_i}{\partial x_k} + \Pi_{ij} - \frac{2}{3}\varepsilon\delta_{ij} \tag{18}$$

where the Kolmogorov assumption of local isotropy (cf. Hinze 1975) given by

$$\varepsilon_{ij} = \frac{2}{3}\varepsilon\delta_{ij} \tag{19}$$

has also been applied. By making use of (15), we can then rearrange (18) into the alternative form in terms of the Reynolds stress anisotropy tensor:

$$(\mathcal{P} - \varepsilon)b_{ij} = -\frac{2}{3}K\overline{S}_{ij} - K(b_{ik}\overline{S}_{jk} + b_{jk}\overline{S}_{ik} - \frac{2}{3}b_{mn}\overline{S}_{mn}\delta_{ij})$$

$$-K(b_{ik}\overline{\omega}_{jk} + b_{jk}\overline{\omega}_{ik}) + \frac{1}{2}\Pi_{ij} \tag{20}$$

where

$$\overline{S}_{ij} = \frac{1}{2}\left(\frac{\partial\overline{v}_i}{\partial x_j} + \frac{\partial\overline{v}_j}{\partial x_i}\right), \quad \overline{\omega}_{ij} = \frac{1}{2}\left(\frac{\partial\overline{v}_i}{\partial x_j} - \frac{\partial\overline{v}_j}{\partial x_i}\right). \tag{21}$$

In virtually all of the commonly used second-order closure models based on (8), Π_{ij} is modeled in the general form (see Reynolds 1987 and Speziale 1991):

$$\Pi_{ij} = \varepsilon \mathcal{A}_{ij}(\mathbf{b}) + K\mathcal{M}_{ijkl}(\mathbf{b})\frac{\partial \overline{v}_k}{\partial x_l}. \tag{22}$$

As argued in Speziale (1996), (22) is only rigorously justifiable for homogeneous turbulent flows that are in equilibrium. So it is consistent to use the homogeneous, equilibrium assumption in the formulation of algebraic stress models. The substitution of (22) into (20) yields a closed system of algebraic equations for the determination of the Reynolds stress anisotropy tensor in terms of the mean velocity gradients. *This constitutes the general form of the traditional algebraic stress models* (see Rodi 1976). These traditional algebraic stress models are *implicit* since the Reynolds stress tensor appears on both sides of the equation. In order to conduct the algebraic stress model approximation, a linear form for Π_{ij} is usually needed.

It has been shown that the most general nonlinear form of the hierarchy of models (22) for two-dimensional mean turbulent flows in equilibrium systematically reduces to the linear tensorial form (see Speziale, Sarkar and Gatski 1991 and Speziale 1996):

$$\begin{aligned}\Pi_{ij} &= -C_1\varepsilon b_{ij} + C_2 K\overline{S}_{ij} + C_3 K\left(b_{ik}\overline{S}_{jk} + b_{jk}\overline{S}_{ik} - \frac{2}{3}b_{mn}\overline{S}_{mn}\delta_{ij}\right)\\ &\quad + C_4 K(b_{ik}\overline{\omega}_{jk} + b_{jk}\overline{\omega}_{ik})\end{aligned} \tag{23}$$

where only the quadratic return term has been neglected — which is typically small since it is directly proportional to $\varepsilon b_{ik}b_{kj}$ — where $C_1 - C_4$ are constants that depend on the specific model chosen. Speziale, Sarkar and Gatski (1991) showed that a range of strained two-dimensional homogeneous turbulent flows near equilibrium can be collapsed using (23) with the constants $C_1 = 6.80$, $C_2 = 0.36$, $C_3 = 1.25$ and $C_4 = 0.40$. Out of equilibrium, the first two coefficients are functions of the ratio of production to dissipation, $\mathcal{P}/\varepsilon$, and the second invariant, II, of b_{ij} which makes the SSG model proposed by Speziale, Sarkar and Gatski (1991) quasi-linear (these coefficients become constants in equilibrium yielding a linear rapid pressure-strain model which is needed for the algebraic stress model approximation). *Hence, by means of the homogeneous, equilibrium hypothesis, it becomes possible to systematically conduct the algebraic stress model approximation for general nonlinear pressure-strain models.*

The direct substitution of (23) into (20) yields the equation:

$$b_{ij} = \frac{1}{2}g\tau[(C_2 - \frac{4}{3})\overline{S}_{ij} + (C_3 - 2)(b_{ik}\overline{S}_{jk} + b_{jk}\overline{S}_{ik}$$

$$-\frac{2}{3}b_{mn}\overline{S}_{mn}\delta_{ij}) + (C_4 - 2)(b_{ik}\overline{\omega}_{jk} + b_{jk}\overline{\omega}_{ik})] \tag{24}$$

where

$$g = \left(\frac{C_1}{2} + \frac{\mathcal{P}}{\varepsilon} - 1\right)^{-1} \tag{25}$$

and $\tau \equiv K/\varepsilon$ is the turbulent time scale. For homogeneous turbulence, the turbulent kinetic energy K and dissipation rate ε are solutions of the transport equations discussed earlier which simplify to

$$\dot{K} = \mathcal{P} - \varepsilon \tag{26}$$

$$\dot{\varepsilon} = C_{\varepsilon 1}\frac{\varepsilon}{K}\mathcal{P} - C_{\varepsilon 2}\frac{\varepsilon^2}{K}. \tag{27}$$

where $C_{\varepsilon 1}$ and $C_{\varepsilon 2}$ are constants, as before, which, most recently, have assumed the values of 1.44 and 1.83, respectively (cf. Speziale, Sarkar and Gatski 1991). In (26)–(27), a superposed dot again represents a time derivative (in homogeneous turbulence, $K = K(t)$ and $\varepsilon = \varepsilon(t)$). Equations (26)–(27) yield the equilibrium solution

$$\frac{\mathcal{P}}{\varepsilon} = \frac{C_{\varepsilon 2} - 1}{C_{\varepsilon 1} - 1}. \tag{28}$$

The homogeneous equilibrium hypothesis, thus, makes use of the commonly adopted dissipation rate transport equation (11).

If we introduce the dimensionless, rescaled variables (see Gatski and Speziale 1993):

$$S_{ij}^* = \frac{1}{2}g\tau(2 - C_3)\overline{S}_{ij}, \quad \omega_{ij}^* = \frac{1}{2}g\tau(2 - C_4)\overline{\omega}_{ij} \tag{29}$$

$$b_{ij}^* = \left(\frac{C_3 - 2}{C_2 - \frac{4}{3}}\right) b_{ij} \tag{30}$$

then (24) reduces to the simpler dimensionless form

$$b_{ij}^* = -S_{ij}^* - \left(b_{ik}^* S_{jk}^* + b_{jk}^* S_{ik}^* - \frac{2}{3}b_{kl}^* S_{kl}^* \delta_{ij}\right) + b_{ik}^*\omega_{kj}^* + b_{jk}^*\omega_{ki}^* \tag{31}$$

which is *linear* in b_{ij} when (28) is made use of.

The explicit solution to (31), for two-dimensional mean turbulent flows by integrity bases methods, renders the general form (see Pope 1975)

$$\mathbf{b} = \sum_{\lambda} G^{(\lambda)}\mathbf{T}^{(\lambda)}$$

where $\mathbf{T}^{(\lambda)}$ are the integrity bases for which there are only three linearly independent ones in two-dimensions. This solution takes the explicit form (see Gatski and Speziale 1993 and Pope 1975):

$$b^*_{ij} = -\frac{3}{3 - 2\eta^2 + 6\xi^2}[S^*_{ij} + (S^*_{ik}\omega^*_{kj} + S^*_{jk}\omega^*_{ki}) \\ -2(S^*_{ik}S^*_{kj} - \frac{1}{3}S^*_{k\ell}S^*_{k\ell}\delta_{ij})] \tag{32}$$

where

$$\eta = (S^*_{ij}S^*_{ij})^{1/2}, \quad \xi = (\omega^*_{ij}\omega^*_{ij})^{1/2}.$$

For equilibrium turbulent flows, η and ξ are typically less than one whereas when $\eta, \xi >> 1$ we have non-equilibrium turbulent flows that are rapidly distorted (where $\mathcal{P}/\varepsilon$ is also much greater than one). In more familiar terms, (32) is equivalent to the form

$$\begin{aligned} \tau_{ij} \;=\; & \frac{2}{3}K\delta_{ij} - \frac{3}{3 - 2\eta^2 + 6\xi^2}\left[\alpha_1\frac{K^2}{\varepsilon}\overline{S}_{ij} + \alpha_2\frac{K^3}{\varepsilon^2}(\overline{S}_{ik}\overline{\omega}_{kj}\right. \\ & \left. +\overline{S}_{jk}\overline{\omega}_{ki}) - \alpha_3\frac{K^3}{\varepsilon^2}\left(\overline{S}_{ik}\overline{S}_{kj} - \frac{1}{3}\overline{S}_{k\ell}\overline{S}_{k\ell}\delta_{ij}\right)\right] \end{aligned} \tag{33}$$

where, α_1, α_2 and α_3 are constants related to the coefficients $C_1 - C_4$ and g. More specifically,

$$\alpha_1 = \left(\frac{4}{3} - C_2\right)g, \quad \alpha_2 = \frac{1}{2}\left(\frac{4}{3} - C_2\right)(2 - C_4)g^2$$

$$\alpha_3 = \left(\frac{4}{3} - C_2\right)(2 - C_3)g^2, \quad g = \left(\frac{1}{2}C_1 + \frac{\mathcal{P}}{\varepsilon} - 1\right)^{-1}$$

$$\eta = \frac{1}{2}\frac{\alpha_3}{\alpha_1}\frac{K}{\varepsilon}(\overline{S}_{ij}\overline{S}_{ij})^{1/2}, \; \xi = \frac{\alpha_2}{\alpha_1}\frac{K}{\varepsilon}(\overline{\omega}_{ij}\overline{\omega}_{ij})^{1/2} \tag{34}$$

and $\mathcal{P}/\varepsilon$ is given by (28). We can write this equation in the alternative form

$$\begin{aligned} \tau_{ij} = & \frac{2}{3}K\delta_{ij} - \alpha^*_1\frac{K^2}{\varepsilon}\overline{S}_{ij} - \alpha^*_2\frac{K^3}{\varepsilon^2}(\overline{S}_{ik}\overline{\omega}_{kj} \\ & +\overline{S}_{jk}\overline{\omega}_{ki}) + \alpha^*_3\frac{K^3}{\varepsilon^2}\left(\overline{S}_{ik}\overline{S}_{kj} - \frac{1}{3}\overline{S}_{kl}\overline{S}_{kl}\delta_{ij}\right) \end{aligned} \tag{35}$$

where

$$\alpha^*_i = \alpha_i\left(\frac{3}{3 - 2\eta^2 + 6\xi^2}\right) \tag{36}$$

for $i = 1, 2, 3$.

Since this model is formally derived for homogeneous turbulent flows in equilibrium, a singularity can arise when it is applied to turbulent flows that are out of equilibrium through the vanishing of the denominator in (36). Hence, the model needs to be regularized (on the other hand, traditional algebraic stress models have multiple solutions and are also in need of a type of regularization; see Speziale 1997). For these explicit models it can easily be done via a Padé approximation that regularizes while building in some limited agreement with Rapid Distortion Theory (RDT). Thus, far from being a deficiency, the regularization is actually a virtue that extends the range of applicability of the model to regions where it would not normally apply. These issues will be discussed in more detail in the next Section.

3. Non-Equilibrium Turbulence Modeling

We will make use of the RDT solution for homogeneous shear flow. Since shear flows are such a basic benchmark flow in engineering applications, it is important to develop a model that performs well over a wide range of shear rates. The best chance of accomplishing this is to build in agreement with the Rapid Distortion Theory (RDT) test case of large shear rates since the equilibrium case of mild shear rates is already well described. In the RDT solution for homogeneous shear flow, we consider an initially isotropic turbulence subjected to a uniform shear – with shear rate S – where $SK_0/\varepsilon_0 >> 1$ given that K_0 and ε_0 are the initial turbulent kinetic energy and dissipation rate (the mean velocity gradient tensor in this case is given by $\partial\overline{v}_i/\partial x_j \equiv S\delta_{1i}\delta_{2j}$). This constitutes a far from equilibrium test case since it is known that, in homogeneous shear flow, an equilibrium value of $SK/\varepsilon \approx 5$ is ultimately achieved (see Tavoularis and Karnik 1989). The RDT solution – which is obtained from solving the linearized Navier-Stokes equations – is formally an excellent approximation to the full nonlinear Navier-Stokes equations for short elapsed times under a rapid distortion (i.e, for a fraction of an eddy turnover time where $\varepsilon_0 t/K_0 < 1$). The long time asymptotic RDT solution (in the limit as $t^* \rightarrow \infty$ where $t^* = St$) has been obtained analytically by Rogers (1991), who corrected some previously published errors. This asymptotic RDT solution is given by

$$\frac{\tau_{12}}{K_0} = -2ln2 \tag{37}$$

$$K^* = (2ln2)t^* \tag{38}$$

$$b_{11} = \frac{2}{3}, \quad b_{22} = b_{33} = -\frac{1}{3} \tag{39}$$

where $K^* \equiv K/K_0$. In obtaining this solution, use has been made of the fact that

$$\dot{K}^* = -\frac{\tau_{12}}{K_0} \tag{40}$$

in the RDT limit of homogeneous shear flow (the dissipation can be neglected). This asymptotic solution is approached fairly fast (i.e., by the time $St \approx 10$). For sufficiently large shear rates $SK_0/\varepsilon_0 >> 1$, which are required for the RDT approximation, we are guaranteed that the required constraint of $\varepsilon_0 t/K_0 < 1$ is still satisfied at that time.

There are two key elements to the RDT solution for homogeneous shear flow:

(1) The turbulent kinetic energy K^* grows linearly so that $\dot{K}^*$ remains bounded and of order one.

(2) The normal Reynolds stress anisotropies approach a one component state where $b_{11} = 2/3$, $b_{22} = -1/3$ and $b_{33} = -1/3$ (and, hence, all the energy is concentrated in the streamwise component). DNS results of Lee, Kim and Moin (1990) indicate that this happens in a fraction of an eddy turnover time.

From (35), it is obvious that

$$\frac{\tau_{12}}{K_0} = -\frac{1}{2}\alpha_1^* \left(\frac{SK}{\varepsilon}\right)\left(\frac{K}{K_0}\right). \tag{41}$$

in homogeneous shear flow. Since, in the RDT solution, $SK/\varepsilon \to \infty$ while K/K_0 remains of order one, it is obvious from (37)–(38) that

$$\alpha_1^* \sim \frac{1}{\eta} \tag{42}$$

where we have made use of the fact that $\eta \propto \xi \propto SK/\varepsilon$ in homogeneous shear flow. It is clear that the equilibrium model – encompassed by (36) – violates this constraint. This problem can be remedied via a Padé approximation whereby (36) is replaced by a regularized expression that is approximately equal to (36) for the near-equilibrium case (where $\eta < 1$) but has the correct asymptotic behavior, given in (42), for $\eta \gg 1$. One such approximation, which is accurate to $O(\eta^4)$, is as follows (see Speziale and Xu 1996):

$$\alpha_1^* = \frac{(1+2\xi^2)(1+6\eta^5)+\frac{5}{3}\eta^2}{(1+2\xi^2)(1+2\xi^2+\eta^2+6\beta_1\eta^6)}\alpha_1 \tag{43}$$

where β_1 is an arbitrary constant. Equation (43) is regular for all values of η and ξ thus removing the singularity in (36). It yields results that are within one percent of (36) for near-equilibrium turbulent flows (where the latter is valid) and has the correct asymptotic behavior of $\alpha_1^* \sim 1/\eta$ for $\eta \gg 1$. Equations (37)–(40) suggest that β_1 is in the range of 5 – 10. The value of

$$\beta_1 \approx 7$$

has been arrived at (see Speziale and Xu 1996).

Although the regularized expression first derived by Gatski and Speziale (1993) by a Taylor expansion is asymptotically consistent for the b_{11}, b_{22} and b_{33} components, it does not yield the correct tendency to a one component state in the RDT limit. This can be remedied by the alternative regularized form, obtained by a Padé approximation that is accurate to $O(\eta^4)$ (see Speziale and Xu 1996):

$$\alpha_i^* = \frac{(1+2\xi^2)(1+\eta^4)+\frac{2}{3}\eta^2}{(1+2\xi^2)(1+2\xi^2+\beta_i\eta^6)}\alpha_i \tag{44}$$

where β_i is an arbitrary constant ($i = 2,3$). Equation (44) represents an excellent approximation to (36) for near-equilibrium turbulent flows where $\eta < 1$. It is regular for all values of η and ξ, has the correct asymptotic behavior for $\eta \gg 1$ and, for values of β_2 and β_3 of approximately 5, predicts an approach to a one component state consistent with RDT of homogeneous shear flow. The specific values of

$$\beta_2 \approx 6, \quad \beta_3 \approx 4$$

have been arrived at by Speziale and Xu (1996). As stated before, the idea here is to develop a model that will perform well for a wide range of shear rates since shear flows form such a cornerstone of engineering calculations.

Second-order closures that are suitable for non-equilibrium turbulent flows can then be obtained by conducting a relaxation time approximation around the non-equilibrium extension of the explicit ASM given by (35) and (43)–(44). The idea of obtaining second-order closures by a relaxation time approximation around an equilibrium algebraic model is probably first attributable to Saffman (1977) (this stood in contrast with the more commonly adopted approach of directly modeling the higher-order correlations that appear in the Reynolds stress transport equation which was popular even before Launder, Reece and Rodi 1975). However, Saffman (1977) implemented this relaxation time approximation around a simple, nonlinear algebraic representation for the Reynolds stress tensor. In contrast to that approach, we have implemented a relaxation time approximation about the non-equilibrium extension of the explicit ASM written in terms of the Reynolds stress anisotropy tensor (in strained homogeneous turbulent flows, it is only the Reynolds stress anisotropy tensor that equilibrates; the Reynolds stresses grow exponentially). Hence, we have proposed the relaxation model

$$\dot{b}_{ij} = -C_R \frac{\varepsilon}{K}(b_{ij} - b_{ij}^{E^*}) \tag{45}$$

where

$$b_{ij}^{E^*} = -\frac{1}{2}\alpha_1^* \frac{K}{\varepsilon}\overline{S}_{ij} - \frac{1}{2}\alpha_2^* \frac{K^2}{\varepsilon^2}(\overline{S}_{ik}\overline{\omega}_{kj} + \overline{S}_{jk}\overline{\omega}_{ki})$$

$$+\frac{1}{2}\alpha_3^* \frac{K^2}{\varepsilon^2}\left(\overline{S}_{ik}\overline{S}_{kj} - \frac{1}{3}\overline{S}_{kl}\overline{S}_{kl}\delta_{ij}\right)$$

is the non-equilibrium extension of (35), written in terms of the Reynolds stress anisotropy tensor, with $\alpha_1^* - \alpha_3^*$ given by their non-equilibrium forms (43)-(44). In (45), C_R is a dimensionless relaxation coefficient. Consistency with the Crow (1968) constraint (which is important to capture the correct early time behavior) requires that:

$$C_R = \frac{8}{15}\left(\frac{3 - 2\eta^2}{3\alpha_1}\right) \tag{46}$$

in an initially isotropic turbulence subjected to a mild strain. Of course, for strongly strained turbulent flows, (46) must be regularized. One preliminary form that has been considered – motivated by (43) – is given by (see Speziale and Xu 1996):

$$C_R = \frac{8}{15\alpha_1}\frac{1 + \eta^2 + 6\beta_1\eta^6}{1 + \frac{5}{3}\eta^2 + 6\eta^5}. \tag{47}$$

When (45) is rearranged into a transport equation for the Reynolds stress tensor, it yields a model that differs from the traditional models in one notable way: the rapid pressure-strain correlation depends linearly on the Reynolds stress anisotropy tensor, but depends nonlinearly on the invariants of the rotational and irrotational strain rates to the lowest order, i.e.,

$$\mathcal{M}_{ijkl} = \mathcal{M}_{ijkl}(\mathbf{b}; \eta, \xi). \tag{48}$$

This is consistent with the rapid pressure-strain correlation which, by definition, is linear in the energy spectrum tensor and, hence, linear in the Reynolds stress tensor (see Speziale 1996). Models for the rapid pressure-strain correlation that are *fully nonlinear* in b_{ij} are fundamentally inconsistent. Later, we will see that this new approach leads to a notable improvement in the description of homogeneous shear flow under a wide range of shear rates.

By utilizing (48), a non-equilibrium extension of the SSG second-order closure model (Speziale, Sarkar and Gatski 1991) can be obtained. The SSG model assumes the quasi-linear form for the pressure-strain correlation:

$$\begin{aligned}\Pi_{ij} &= -(C_1\varepsilon + C_1^*\mathcal{P})b_{ij} + C_2\varepsilon\left(b_{ik}b_{kj} - \frac{1}{3}b_{k\ell}b_{k\ell}\delta_{ij}\right)\\ &\quad +(C_3 - C_3^* II_b^{1/2})K\overline{S}_{ij} + C_4 K\left(b_{ik}\overline{S}_{jk} + b_{jk}\overline{S}_{ik}\right.\\ &\quad \left. -\frac{2}{3}b_{k\ell}\overline{S}_{k\ell}\delta_{ij}\right) + C_5 K(b_{ik}\overline{\omega}_{jk} + b_{jk}\overline{\omega}_{ik})\end{aligned} \tag{49}$$

where

$$C_1 = 3.4,\ C_1^* = 1.80,\ C_2 = 4.2,\ C_3 = \frac{4}{5}$$

$$C_3^* = 1.30,\ C_4 = 1.25,\ C_5 = 0.40,\ II_b = b_{ij}b_{ij}.$$

The Launder, Reece and Rodi (1975) model is recovered as a special case of the SSG model when

$$C_1 = 3.0,\ C_1^* = 0,\ C_2 = 0,\ C_3 = \frac{4}{5},\ C_3^* = 0,$$

$$C_4 = 1.75,\ C_5 = 1.31.$$

Hence, the SSG model can be easily implemented in any computer code that makes use of the Launder, Reece and Rodi model (the same will be true of the non-equilibrium extension of the SSG model). The SSG model can then be extended to non-equilibrium turbulent flows – by making use of (48) – through a heuristic Padé approximation of the coefficients which also builds in agreement with the RDT solution for plane strain as well as homogeneous shear flow. The coefficients in this model take the form (Xu and Speziale 1997):

$$C_1 = 3.4, \quad C_2 = 4.2, \quad C_3 = 0.8$$

$$C_1^* = \frac{1.8 + 0.225\eta^6}{1 + 0.0625\eta^6 + 0.5\xi^8} \quad C_3^* = \frac{1.3 + 8.84\eta^8}{1 + 9.02\eta^8} \tag{50}$$

$$C_4 = \frac{1.25 + 6.33\eta^6}{1 + 1.52\eta^6 + 0.1\xi^7}, \quad C_5 = \frac{0.4 + 0.114\eta^6}{1 + 0.285\eta^6 + 0.5\xi^8}.$$

This yields results indistinguishable from the SSG model for equilibrium turbulent flows where $\eta, \xi < 1$ (for homogeneous shear flow in equilibrium, $\eta \approx 0.4$ and $\xi \approx 0.7$). But in the rapid distortion limit, where $\eta, \xi \to \infty$, considerably different values of the coefficients are obtained. This is consistent with the findings of Reynolds (private communication) who found that a different form for $\mathcal{M}_{ijkl}$ was needed for the rapid distortion limit compared to equilibrium turbulent flows. The way that this can be achieved is to let $\mathcal{M}_{ijkl}$ be a function of η, ξ as in (48). In Section 5, it will be shown how this leads to a much better description of plane strain turbulence and homogeneous shear flow for a wide range of shear and strain rates that includes the rapid distortion limit.

Some comments are needed concerning the modeling of the turbulent diffusion terms. In many applications of the SSG model, the Mellor and Herring (1973) model has been used for the turbulent diffusion terms represented by C_{ijk}. This model is given by

$$C_{ijk} = -\frac{2}{3}C_s\frac{K^2}{\varepsilon}\left(\frac{\partial\tau_{jk}}{\partial x_i} + \frac{\partial\tau_{ik}}{\partial x_j} + \frac{\partial\tau_{ij}}{\partial x_k}\right) \tag{51}$$

with pressure-diffusion effects neglected (C_s is a constant). This is the asymptotically consistent model for small shear rates which are required to apply the gradient transport hypothesis in (51) or any comparable models of its type. For inhomogeneous turbulent flows, a simple gradient transport model is also used for the turbulent diffusion terms in the relaxation model (45). This model thus takes the inhomogeneous form:

$$\frac{Db_{ij}}{Dt} = -C_R\frac{\varepsilon}{K}(b_{ij} - b_{ij}^{E^*}) + \frac{\partial}{\partial x_k}\left(\frac{\nu_T}{\sigma_b}\frac{\partial b_{ij}}{\partial x_k}\right) \tag{52}$$

where $\nu_T \equiv C_\mu K^2/\varepsilon$ is the eddy viscosity and σ_b is a constant.

An entirely new approach to large-eddy simulations is being considered based on these ideas. In large-eddy simulations (LES), the filtered velocity is obtained from the equations of motion:

$$\frac{\partial\overline{v}_i}{\partial t} + \overline{v}_j\frac{\partial\overline{v}_i}{\partial x_j} = -\frac{\partial\overline{P}}{\partial x_i} + \nu\nabla^2\overline{v}_i - \frac{\partial\tau_{ij}^S}{\partial x_j} \tag{53}$$

where τ_{ij}^S is the subgrid scale stress tensor and

$$\overline{v}_i = \int_D G(\mathbf{x} - \mathbf{x}^*, \Delta)v_i(\mathbf{x}^*)d^3x^* \tag{54}$$

constitutes the spatial filter of v_i given that G is a filter function, Δ is the computational mesh size and D represents the fluid domain. This equation is of the same form as the Reynolds-averaged Navier-Stokes equations (5). The major difference is that the subgrid scale stress tensor τ_{ij}^S is a scaled down version of the Reynolds stress tensor that is less dissipative than τ_{ij}. We will make use of this to develop a combined LES and time-dependent RANS capability. In this new approach, great care is being taken to ensure:

(1) The absence of any test filters or double filtered fields in the subgrid scale stress model that can significantly contaminate the large scales and

necessitate the inversion of filtered quantities which must be avoided (this is equivalent to solving a Fredholm integral equation of the first kind which is ill-posed in a mathematical sense).

(2) The incorporation of a systematically derived anisotropic eddy viscosity that is strain-dependent and allows for the direct integration of subgrid scale models to a solid boundary without the need for any *ad hoc* wall damping functions (see Speziale and Abid 1995).

(3) The recovery of a state-of-the-art Reynolds stress model in the coarse mesh/infinite Reynolds number limit so that an LES goes continuously to a RANS computation.

In mathematical terms, this can be accomplished by models of the form

$$\tau_{ij}^S = [1 - exp(-\beta\Delta/L_K)]^n \tau_{ij} \tag{55}$$

for the subgrid scale stress tensor τ_{ij}^S, where τ_{ij} is a Reynolds stress model, Δ is the computational mesh size, L_K is the Kolmogorov length scale, and β and n are constants. In the limit as $\Delta/L_K \to 0$, all relevant scales are resolved and we have a direct simulation where $\tau_{ij} = 0$; as $\Delta/L_K \to \infty$ and the mesh becomes coarse or the Reynolds number becomes extremely large, we recover a Reynolds stress model and a RANS computation. In between these two limits, we have an LES or a VLES (the latter denotes a large-eddy simulation where the preponderance of the turbulent kinetic energy is unresolved). Here, it should be noted that in the simulation of turbulence, the mesh is fine (or coarse) depending on whether Δ is small (or large) compared to the Kolmogorov length scale $L_K \equiv \nu^{3/4}/\varepsilon^{1/4}$. This automatically brings in a dependence on the turbulence Reynolds number ($R_t \equiv K^2/\nu\varepsilon$) since $L_K = R_t^{-3/4} K^{3/2}/\varepsilon$. An estimate of the Kolmogorov length scale is provided by the modeled transport equation for ε discussed earlier in (11) which will be discussed in more detail later. In order to estimate the Kolmogorov length scale to within 10% it is only required that the dissipation rate be estimated to within 50%. Hence, it is quite reasonable to expect that an acceptable estimate of the Kolmogorov length scale can be obtained (this is a crucial element of this proposed approach). The Reynolds stress model in (55) can be represented by the forms discussed in this Section. A two-equation model – which is upgraded to a second-order closure as needed – is recommended for τ_{ij}. It was recently shown by Speziale and Abid (1995) that these two-equation models can be integrated directly to a solid boundary with no wall damping functions; only the singularity in the ε – transport equation needs to be removed.

Some remarks are needed concerning the choice of a filter in this new approach to large-eddy simulations. We want a filter that yields the minimum contamination of the large scales. The reason for this is clear; defiltering must be avoided since it constitutes an ill-posed mathematical problem as stated earlier. The purpose of practical LES is to predict the Reynolds-averaged fields. In order to do so, the filtered velocity must invariably be used to estimate the large-scale part of the instantaneous velocity which then yields the Reynolds-averaged fields through appropriate ensemble averages. Here, the large scales make the dominant contribution to the most pertinent fields such as the turbulent kinetic energy. A minimum contamination of the large scales can be accomplished with, of the order of, a 128^3 computational mesh using a filter with a compact support – such as the box filter – which has a small filter width of, say, two mesh points. Some of the previously conducted coarse grid LES (which has typically had no more than 32^3 mesh points) must be avoided wherein the filter width has, at times, been as much as 25% of the computational domain, significantly contaminating the large scales. Besides, recent increases in computational capacity have begun to make 128^3 computations much more feasible for engineering calculations (a small compromise to 100^3 computations can always be made). In addition, it should be noted that practical LES – in complex geometries – will require the use of finite difference techniques with a compact filter (these should be based on fourth-order accurate finite difference schemes). Some illustrative calculations will be provided in Section 5 to demonstrate the potential of this new approach for practical engineering calculations.

4. Modeling the Turbulent Dissipation Rate

In this Section we will discuss the modeling of the dissipation rate equation. Getting a reasonable estimate of the turbulent dissipation rate is crucial in the new approach to large-eddy simulations proposed in the last Section. Thus we will discuss the modeling of the turbulent dissipation rate along with its extension to non-equilibrium turbulent flows.

The exact transport equation (10) for ε in homogeneous turbulence can be written in the form:

$$\dot{\varepsilon} = -\varepsilon_{ij}\frac{\partial \overline{v}_i}{\partial x_j} - \varepsilon_{ij}^{(c)}\frac{\partial \overline{v}_i}{\partial x_j} + 2\nu\overline{\omega_i\omega_j\frac{\partial u_i}{\partial x_j}} - 2\nu^2\overline{\frac{\partial \omega_i}{\partial x_j}\frac{\partial \omega_i}{\partial x_j}} \tag{56}$$

where

$$\omega = \nabla \times \mathbf{u}$$

$$\varepsilon_{ij}^{(c)} = 2\nu\overline{\frac{\partial u_k}{\partial x_i}\frac{\partial u_k}{\partial x_j}}$$

are, respectively, the fluctuating vorticity and complementary dissipation rate. This equation can be rewritten as

$$\dot{\varepsilon} = -2\varepsilon(d_{ij} + d_{ij}^{(c)})\frac{\partial \overline{v}_i}{\partial x_j} + \frac{7}{3\sqrt{15}} S_K R_t^{1/2} \frac{\varepsilon^2}{K} - 2\nu^2 \overline{\frac{\partial \omega_i}{\partial x_j}\frac{\partial \omega_i}{\partial x_j}} \tag{57}$$

where

$$S_K \equiv \frac{6\sqrt{15}}{7} \frac{\overline{\omega_i \omega_j \frac{\partial u_i}{\partial x_j}}}{(\overline{\omega_k \omega_k})^{3/2}}$$

$$R_t = K^2/\nu\varepsilon$$

$$d_{ij}^{(c)} = \frac{\varepsilon_{ij}^{(c)} - \frac{1}{3}\varepsilon_{kk}^{(c)}\delta_{ij}}{2\varepsilon}$$

are the velocity derivative skewness, turbulence Reynolds number and anisotropy of the complementary dissipation. For isotropic turbulence,

$$2\nu^2 \overline{\frac{\partial \omega_i}{\partial x_j}\frac{\partial \omega_i}{\partial x_j}} \propto \nu^2 \int_0^\infty \kappa^4 E(\kappa, t) d\kappa.$$

The major contributions to this integral occur at high wavenumbers where the energy spectrum $E(\kappa, t) \propto E(\kappa L_K)$ given that $L_K = \nu^{3/4}/\varepsilon^{1/4}$ is the Kolmogorov length scale. This Kolmogorov scaling yields (Bernard, Thangam and Speziale 1992):

$$2\nu^2 \overline{\frac{\partial \omega_i}{\partial x_j}\frac{\partial \omega_i}{\partial x_j}} \propto [R_t^{1/2} + O(1)]\frac{\varepsilon^2}{K} = \frac{7}{3\sqrt{15}} G_K R_t^{1/2} \frac{\varepsilon^2}{K} + C_{\varepsilon 2}\frac{\varepsilon^2}{K}.$$

Hence,

$$\dot{\varepsilon} = -2\varepsilon(d_{ij} + d_{ij}^{(c)})\frac{\partial \overline{v}_i}{\partial x_j} + \frac{7}{3\sqrt{15}}(S_K - G_K) R_t^{1/2} \frac{\varepsilon^2}{K} - C_{\varepsilon 2}\frac{\varepsilon^2}{K}. \tag{58}$$

The standard high-Reynolds-number equilibrium hypothesis is then typically invoked whereby

$$S_K = G_K, \quad C_{\varepsilon 2} = \text{constant}.$$

Then by assuming that

$$d_{ij} \propto b_{ij}, \quad d_{ij}^{(c)} \propto b_{ij},$$

along with a gradient transport model for the turbulent diffusion terms, (11) is arrived at. This is based on the assumption that the large scales imprint their anisotropy on the small scales through the cascade.

In non-equilibrium turbulent flows, S_K and G_K are out of balance. This leads to the dissipation rate transport equation of the form (see Bernard and Speziale 1992)

$$\frac{\partial \varepsilon}{\partial t} + \overline{\mathbf{v}} \cdot \nabla \varepsilon = C_{\varepsilon 1} \frac{\varepsilon}{K} \mathcal{P} - C_{\varepsilon 2} \frac{\varepsilon^2}{K} + C_{\varepsilon 3} R_t^{1/2} \frac{\varepsilon^2}{K} + \frac{\partial}{\partial x_i}\left(\frac{\nu_T}{\sigma_\varepsilon} \frac{\partial \varepsilon}{\partial x_i}\right). \quad (59)$$

The coefficient $C_{\varepsilon 3}$ represents the imbalance in vortex stretching. For turbulent flows in equilibrium, $C_{\varepsilon 3} = 0$. This term leads to models which are better behaved. For instance, the singularity in plane stagnation point turbulent flows is removed as discussed in Abid and Speziale (1996).

In turbulent flows that are out of equilibrium – or where the turbulence Reynolds number is small – anisotropies in the turbulent dissipation rate must be accounted for. Models for d_{ij} and $d_{ij}^{(c)}$ can be obtained from an analysis of the transport equation for the tensor dissipation which, for homogeneous turbulence, is given by (see Durbin and Speziale 1991):

$$\dot{\varepsilon}_{ij} = N_{ij} - \varepsilon_{ik} \frac{\partial \overline{v}_j}{\partial x_k} - \varepsilon_{jk} \frac{\partial \overline{v}_i}{\partial x_k} + A_{ijk\ell}^{(\varepsilon)} \frac{\partial \overline{v}_k}{\partial x_\ell} \quad (60)$$

where

$$\begin{aligned} N_{ij} &= 2\nu \overline{\left(\frac{\partial u_i}{\partial x_j} + \frac{\partial u_j}{\partial x_i}\right) \frac{\partial u_\ell}{\partial x_k} \frac{\partial u_k}{\partial x_\ell}} - 2\nu \overline{\frac{\partial u_i}{\partial x_k} \frac{\partial u_\ell}{\partial x_k} \frac{\partial u_j}{\partial x_\ell}} \\ &\quad -2\nu \overline{\frac{\partial u_j}{\partial x_k} \frac{\partial u_\ell}{\partial x_k} \frac{\partial u_i}{\partial x_\ell}} - 4\nu^2 \overline{\frac{\partial^2 u_i}{\partial x_k \partial x_m} \frac{\partial^2 u_j}{\partial x_k \partial x_m}} \end{aligned}$$

$$A_{ijk\ell}^{(\varepsilon)} = 4\nu \left(\overline{\frac{\partial u_j}{\partial x_i} \frac{\partial u_\ell}{\partial x_k}} + \overline{\frac{\partial u_i}{\partial x_j} \frac{\partial u_\ell}{\partial x_k}} - \overline{\frac{\partial u_i}{\partial x_\ell} \frac{\partial u_j}{\partial x_k}} \right)$$

In physical terms,

$N_{ij} \equiv$ Production by Vortex Stretching – Destruction by Viscous Diffusion

$A_{ijk\ell}^{(\varepsilon)} \equiv$ Structure and Redistribution Term.

Speziale and Gatski (1997) have proposed the following models:

$$\begin{aligned} A^{(\varepsilon)}_{ijk\ell}\frac{\partial \overline{v}_k}{\partial x_\ell} &= \frac{16}{15}\varepsilon\overline{S}_{ij} + \left(\frac{30}{11}\alpha_3 + \frac{20}{11}\right)\varepsilon\left(d_{ik}\overline{S}_{jk}\right. \\ &\quad \left. + d_{jk}\overline{S}_{ik} - \frac{2}{3}d_{k\ell}\overline{S}_{k\ell}\delta_{ij}\right) \\ &\quad - \left(\frac{14}{11}\alpha_3 - \frac{20}{11}\right)\varepsilon\left(d_{ik}\overline{W}_{jk} + d_{jk}\overline{W}_{ik}\right) \\ &\quad - \left(\frac{14}{11}\alpha_3 - \frac{16}{33}\right)\varepsilon d_{k\ell}\overline{S}_{k\ell}\delta_{ij} \end{aligned} \tag{61}$$

$$N_{ij} = \frac{2}{3}N\delta_{ij} + {}_DN_{ij} \tag{62}$$

$$N = \frac{\varepsilon}{K}\mathcal{P} - C_{\varepsilon 2}\frac{\varepsilon^2}{K} \tag{63}$$

$${}_DN_{ij} = -C_{\varepsilon 5}\frac{\varepsilon}{K}\left(\varepsilon_{ij} - \frac{2}{3}\varepsilon\delta_{ij}\right) \tag{64}$$

based on an expansion technique that makes use of tensor invariance, symmetry properties and the fact that d_{ij} is small so that only linear terms need to be maintained.

By assuming that the dissipation equilibrates on a much faster time scale than the Reynolds stress tensor, the standard equilibrium hypothesis is invoked whereby:

$$\dot{d}_{ij} \doteq 0. \tag{65}$$

This leads to an algebraic system of equations analogous to that obtained in the algebraic stress approximation discussed earlier. For two-dimensional mean turbulent flows, the exact solution is:

$$\begin{aligned} d_{ij} &= -2C_{\mu\varepsilon}\left[\overline{S}^*_{ij} + \left(\frac{\frac{7}{11}\alpha_3 + \frac{1}{11}}{C_{\varepsilon 5} + \mathcal{P}/\varepsilon - 1}\right)\left(\overline{S}^*_{ik}\overline{\omega}^*_{jk} + \overline{S}^*_{jk}\overline{\omega}^*_{ik}\right)\right. \\ &\quad \left. + \left(\frac{\frac{30}{11}\alpha_3 - \frac{2}{11}}{C_{\varepsilon 5} + \mathcal{P}/\varepsilon - 1}\right)\left(\overline{S}^*_{ik}\overline{S}^*_{jk} - \frac{1}{3}\overline{S}^*_{mn}\overline{S}^*_{mn}\delta_{ij}\right)\right] \end{aligned} \tag{66}$$

where

$$\begin{aligned} C_{\mu\varepsilon} &= \frac{1}{15(C_{\varepsilon 5} + \mathcal{P}/\varepsilon - 1)}\left[1 + 2\overline{\omega}^*_{ij}\overline{\omega}^*_{ij}\left(\frac{\frac{7}{11}\alpha_3 + \frac{1}{11}}{C_{\varepsilon 5} + \mathcal{P}/\varepsilon - 1}\right)^2\right. \\ &\quad \left. - \frac{2}{3}\left(\frac{\frac{15}{11}\alpha_3 - \frac{1}{11}}{C_{\varepsilon 5} + \mathcal{P}/\varepsilon - 1}\right)\overline{S}^*_{ij}\overline{S}^*_{ij}\right]^{-1} \end{aligned} \tag{67}$$

$$\overline{S}_{ij}^* = \overline{S}_{ij}\frac{K}{\varepsilon}, \quad \overline{\omega}_{ij}^* = \overline{\omega}_{ij}\frac{K}{\varepsilon}$$

(Speziale and Gatski 1997). The substitution of these algebraic equations into the contraction of the ε_{ij} transport equation yields the scalar dissipation rate equation

$$\dot{\varepsilon} = C_{\varepsilon 1}^* \frac{\varepsilon}{K}\mathcal{P} - C_{\varepsilon 2}\frac{\varepsilon^2}{K} \tag{68}$$

where

$$C_{\varepsilon 1}^* = 1 + \frac{2(1+\alpha)}{15C_\mu}\left[\frac{C_{\varepsilon 5} + C_\mu\eta^2 - 1}{(C_{\varepsilon 5} + C_\mu\eta^2 - 1)^2 + \beta_1^2\xi^2 - \frac{1}{3}\beta_2^2\eta^2}\right]$$

$$\eta = (2\overline{S}_{ij}^*\overline{S}_{ij}^*)^{1/2}, \quad \xi = (2\overline{\omega}_{ij}^*\overline{\omega}_{ij}^*)^{1/2}$$

$$\alpha = \frac{3}{4}\left(\frac{14}{11}\alpha_3 - \frac{16}{33}\right), \quad \beta_1 = \frac{7}{11}\alpha_3 + \frac{1}{11}$$

$$\beta_2 = \frac{15}{11}\alpha_3 - \frac{1}{11}, \quad C_{\varepsilon 5} \approx 5, \quad \alpha_3 \approx 0.6.$$

It should be noted that the contraction of (60) yields $d_{ij}^{(c)}\partial\overline{v}_i/\partial x_j = \alpha d_{ij}\partial\overline{v}_i/\partial x_j$ which was used to obtain this result. The constants α_3 and $C_{\varepsilon 5}$ were evaluated using DNS results for homogeneous shear flow (Rogers, Moin and Reynolds 1986). The constant $C_{\varepsilon 2}$ can be evaluated by an appeal to isotropic turbulence. This dissipation rate model – like all comparable such models – predicts that the turbulent kinetic energy decays according to the standard power law in isotropic turbulence (cf. Speziale 1991):

$$K \sim t^{-\frac{1}{(C_{\varepsilon 2}-1)}}.$$

Based on this, we have taken $C_{\varepsilon 2} = 1.83$ which yields an exponent of approximately 1.2 in agreement with the most cited experimental data (see Comte-Bellot and Corrsin 1971).

For two-dimensional turbulent shear flows that are in equilibrium,

$$C_{\varepsilon 1}^* \approx 1.4$$

which is remarkably close to the traditionally chosen constant value of $C_{\varepsilon 1} = 1.44$. For more general two-dimensional homogeneous turbulent flows that are near equilibrium, this model takes the form:

$$\dot{\varepsilon} = C_{\varepsilon 1}^* C_\mu \frac{\varepsilon^2}{K}\eta^2 - C_{\varepsilon 2}\frac{\varepsilon^2}{K} \tag{69}$$

where

$$C_{\varepsilon 1}^* = C_{\varepsilon 1}^*(\eta,\xi), \quad C_\mu = C_\mu(\eta,\xi).$$

In contrast to this result, the new dissipation rate model of Lumley (1992) takes the form

$$\dot{\varepsilon} = C_1 \frac{\varepsilon^2}{K}\eta - C_{\varepsilon 2}\frac{\varepsilon^2}{K} \tag{70}$$

for near equilibrium turbulent flows where C_1 is a *constant.* The model (70) obviously contains less physics than the model (68) derived herein since it *does not* depend on rotational strains. It has long been recognized that the dissipation rate is dramatically altered by rotations. The results presented in this paper clearly show that this effect can be systematically incorporated by accounting for anisotropic dissipation. To the best knowledge of the author, this constitutes the first rigorous introduction of rotational effects into the scalar dissipation rate equation. Previous attempts to account for rotational effects (see Raj 1975; Hanjalic and Launder 1980; and Bardina, Ferziger and Rogallo 1985) were largely *ad hoc.*

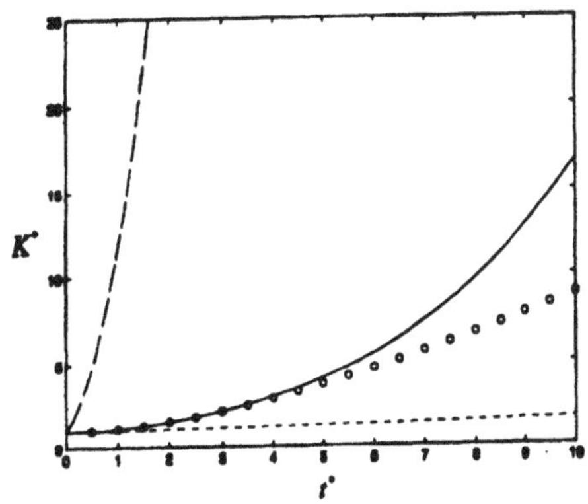

Figure 1. Time evolution of the turbulent kinetic energy in homogeneous shear flow: Comparison of the model predictions for the far from equilibrium initial condition $SK_0/\varepsilon_0 = 50$ with Rapid Distortion Theory. (—) SSG Model; (- -) $K - \varepsilon$ Model; (- - -) Explicit ASM of Gatski and Speziale (1993); (o) RDT.

5. Discussion of Results

We will first consider the far from equilibrium test case in homogeneous shear flow where an initially isotropic turbulence is subjected to the strong shear of $SK_0/\varepsilon_0 = 50$. For this strongly sheared case, RDT constitutes a good approximation for early times. In Figure 1, the time evolution of the turbulent kinetic energy predicted by several models is compared with the RDT solution (see Rogers 1991). It is clear from these results that none of the models are able to predict the correct trend (DNS results have tended to indicate that, for this case, RDT is a good approximation until, at least,

$t^* \equiv St \approx 12$). The interesting point here is that the SSG second-order closure predicts too large a growth rate whereas the explicit ASM of Gatski and Speziale (1993) based on the SSG model yields a growth rate that is far too low. Here, the former problem arises from the fact that pressure-strain models of the form (22) do not apply to turbulent flows that are far from equilibrium; the latter problem is due to the fact that the regularization procedure initially used by Gatski and Speziale (1993) does not apply to turbulent flows that are strongly strained (the eddy viscosity $\nu_T \sim 1/\eta^2$ instead of like $1/\eta$ which explains the low growth rate in this case). On the other hand, the standard $K-\varepsilon$ model renders $\nu_T \sim O(1)$ which explains its enormous growth rate (the standard $K-\varepsilon$ model erroneously predicts that $\dot{K}^* \to \infty$ as $\eta \to \infty$; $K^* \equiv K/K_0$).

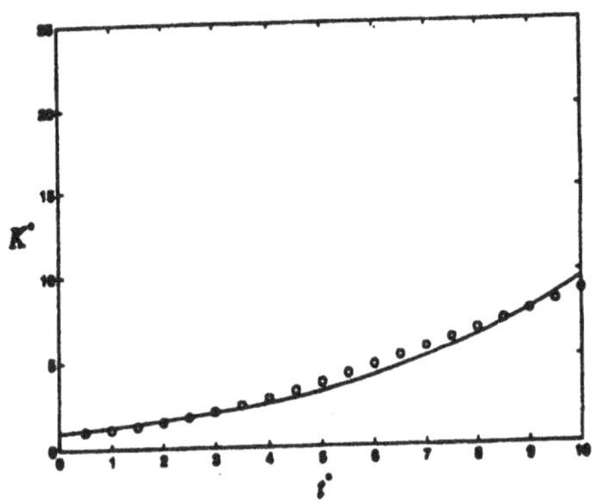

Figure 2. Comparison of the predictions of the (—) new non-equilibrium extension of the explicit ASM for $SK_0/\varepsilon_0 = 50$ with (o) Rapid Distortion Theory for the time evolution of the turbulent kinetic energy in homogeneous shear flow.

Results will now be presented for the non-equilibrium extension of the two-equation model discussed in Section 3. These results correspond to the choice of constants

$$\beta_1 = 7.0, \quad \beta_2 = 6.3, \quad \beta_3 = 4.0$$

in the regularized coefficients $\alpha_1^* - \alpha_3^*$ given in (43)–(44). The predictions of the new explicit ASM for the time evolution of the turbulent kinetic energy are compared in Figure 2 with the RDT solution for the same far from equilibrium test case in homogeneous shear flow where $SK_0/\varepsilon_0 = 50$. With this new non-equilibrium extension, the results are remarkably improved. It is not even necessary to introduce the relaxation time approximation (from (47), the relaxation coefficient is large for this case, rendering its effect small on K^*). In Figure 3, the time evolution of the normal components of the Reynolds stress anisotropy tensor obtained from the new relaxation

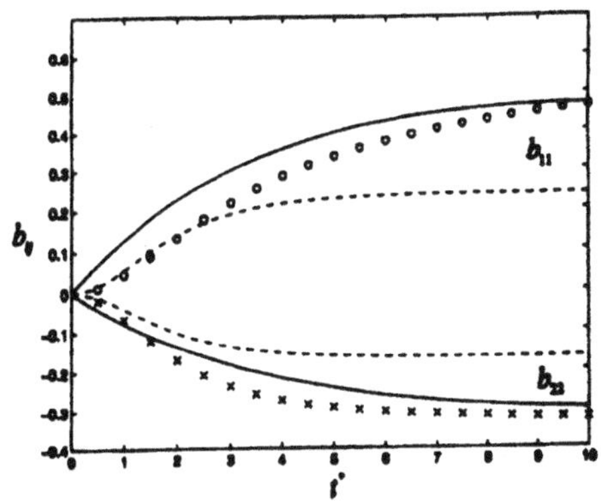

Figure 3. Time evolution of the Reynolds stress anisotropies in homogeneous shear flow: Comparison of the model predictions for $SK_0/\varepsilon_0 = 50$ with Rapid Distortion Theory. (- - -) SSG Model; (—) new relaxation model; (o, ×) RDT.

model (45) (for $SK_0/\varepsilon_0 = 50$) are compared with RDT as well as with the predictions of the SSG second-order closure. Here again, the new model yields a substantial improvement, rendering results that are more properly in line with an approach to a one-component state predicted by RDT. The ability to predict shear flows over a wide range of shear rates should lead to an enhanced time-dependent RANS capability in engineering applications and that is one of our ultimate goals.

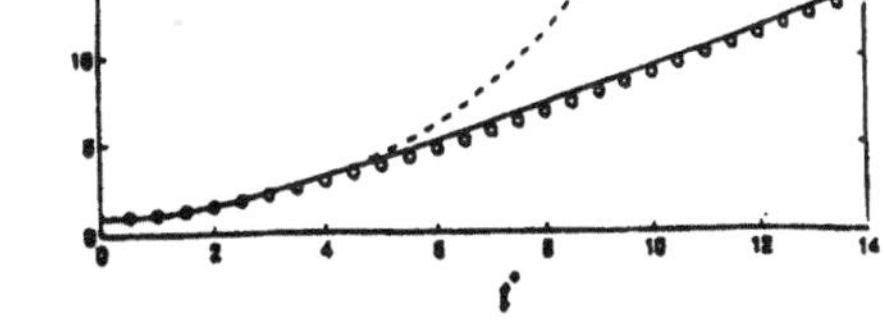

Figure 4. Comparison of the predictions of the (—) new non-equilibrium extension of the SSG model and the (- - -) standard SSG model for $SK_0/\varepsilon_0 = 100$ with (o) Rapid Distortion Theory for the time evolution of the turbulent kinetic energy in homogeneous shear flow.

The performance of the non-equilibrium extension of the SSG model given in (50) will now be discussed. First, the highly non-equilibrium test

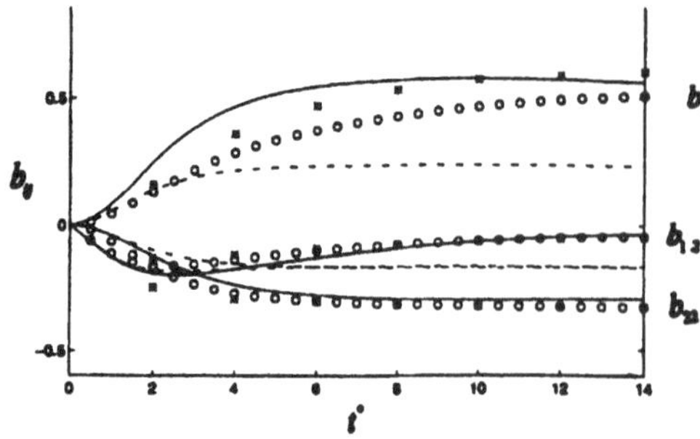

Figure 5. Time evolution of the Reynolds stress anisotropies in homogeneous shear flow: Comparison of the model predictions for $SK_0/\varepsilon_0 = 100$ with Rapid Distortion Theory and DNS. (- - -) Standard SSG Model; (—) new non-equilibrium extension of the SSG model; (o) RDT, (*) DNS of Lee, Kim, Moin (1990).

case of homogeneous shear flow — subjected to a non-dimensional shear rate of $SK_0/\varepsilon_0 = 100$ — will be considered. As can be seen from Figure 4, much better predictions for the growth rate of the turbulent kinetic energy are obtained. The non-equilibrium SSG model yields results that are in line with RDT which has been shown to be an excellent approximation until, at least, $St \approx 12$. The standard SSG model yields a growth rate that is far too large as can be seen in Figure 4. The predictions for Reynolds stress anisotropies are provided in Figure 5. In close approximate agreement with RDT, an approach to a one-component state where b_{12} tends to zero is predicted. In contrast to these results, the standard SSG model — which performs well in near-equilibrium turbulent flows as documented in a variety of test cases by Speziale, Sarkar and Gatski (1991) — yields poor predictions in this case as shown in Figure 5. The case of plane strain — for a large non-dimensional strain rate of $\Gamma K_0/\varepsilon_0 = 100$ — is next considered. The growth rate of the turbulent kinetic energy is displayed in Figure 6 where it is compared with RDT and the direct simulations of Lee and Reynolds (1985) ($t^* \equiv \Gamma t$). It can be seen that discernibly improved predictions to the standard SSG model are obtained which predicts too large a growth rate. The associated Reynolds stress anisotropies are displayed in Figure 7. Again, the non-equilibrium extension of the SSG model yields substantially improved predictions compared to the standard model as shown in Figure 7 without compromising its near-equilibrium predictions which remain virtually identical to those shown in Speziale, Sarkar and Gatski (1991).

Some preliminary LES results will be presented for the developing turbulent boundary layer — integrated through transition — based on the combined time-dependent RANS and LES approach discussed in Section

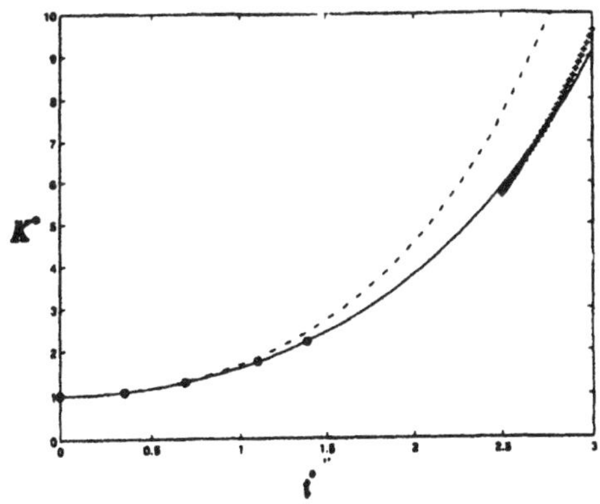

Figure 6. Comparison of the predictions of the (—) new non-equilibrium extension of the SSG model and the (- - -) standard SSG model for $\Gamma K_0/\varepsilon_0 = 100$ with the (o) DNS of Lee and Reynolds (1985) and (×) RDT for the time evolution of the turbulent kinetic energy in plane strain turbulence.

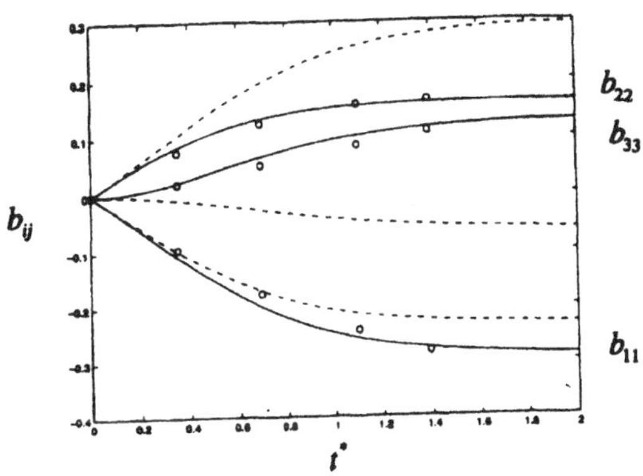

Figure 7. Time evolution of the Reynolds stress anisotropies in plane strain turbulence: Comparison of the model predictions for $\Gamma K_0/\varepsilon_0 = 100$ with DNS. (- - -) Standard SSG Model; (—) new non-equilibrium extension of the SSG model; (o) DNS of Lee and Reynolds (1985).

3. These computations were conducted by H. Fasel and his group at the University of Arizona using an empirically based ramp function — that depends explicitly on the momentum thickness Reynolds number and the mesh size with a simple eddy viscosity model — as a preliminary test of the ideas embodied in (55). In Figure 8, the spanwise vorticity obtained from the LES is shown which compares favorably with the corresponding large-scale results obtained from DNS. It is clear that the subgrid scale model allows the LES to pick up the pertinent flow structures and to be integrated

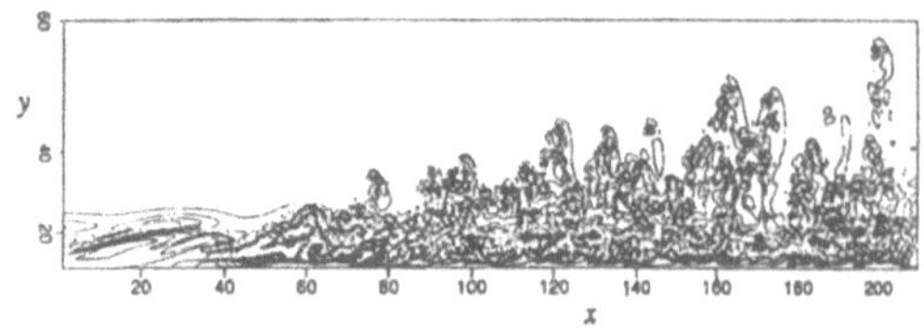

Figure 8. Plot of spanwise vorticity in the developing turbulent boundary layer obtained from LES (computations done by H. Fasel and co-workers at the University of Arizona).

through transition (laminar – turbulent flow). The ramp function, which forms a central part of (55), allows the eddy viscosity to gradually turn on as the flow becomes turbulent. In this regard, the corresponding eddy viscosity is displayed in Figure 9. These preliminary results are extremely encouraging and demonstrate the potential of this new approach for LES and time-dependent RANS.

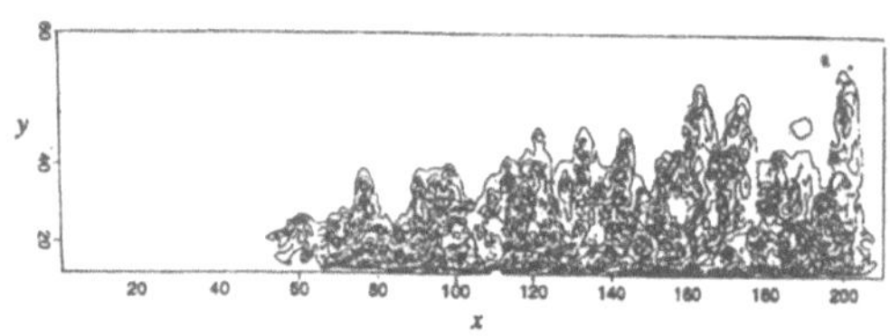

Figure 9. Plot of the eddy viscosity in the developing turbulent boundary layer obtained from LES (computations done by H. Fasel and co-workers at the University of Arizona).

A brief application of the anisotropic dissipation rate model discussed in Section 4 will finally be presented. We have considered the spatially evolving flat plate turbulent wake that was measured recently by Marasli, Champagne and Wygnanski (1991) (see Speziale and Gatski 1997). This is at a Reynolds number $Re_\theta \approx 1000$ based on the momentum thickness. The measurements were taken far enough downstream to ensure complete self-similarity. In Figure 10, results for the turbulent kinetic energy profiles in the far-field wake are displayed that were computed by Cimbala (1995). In this figure, we compare the predicted results for the SSG model – using anisotropic as well as isotropic dissipation – with the far field wake measurements of Marasli, Champagne and Wygnanski (1991). It is clear from these results that the inclusion of the new anisotropic dissipation rate model leads to a considerable improvement. These results are obtained by simply including a variable $C_{\varepsilon 1}$ in what is otherwise the standard dissipation equation as discussed in Section 4.

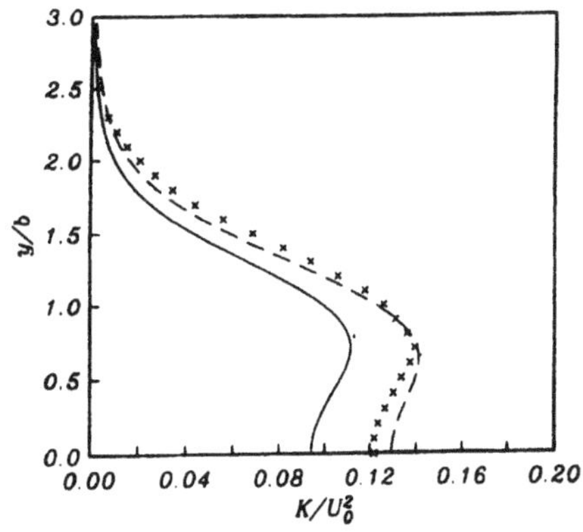

Figure 10. Comparison of the SSG model predictions with the experiments of Marasli *et al.* (1991) for the turbulent kinetic energy profiles in the turbulent plane wake (calculations by J. M. Cimbala 1995).

6. Conclusion

A fundamentally new approach to the modeling of non-equilibrium turbulent flows has been presented. It was shown how explicit algebraic stress models can be extended to non-equilibrium turbulent flows via a Padé approximation and relaxation time approximation. This suggests that the rapid pressure-strain correlation depends on the non-dimensional invariants of the rotational and irrotational strain rates with linearity maintained in b_{ij} as required by its definition. It has led to the development of a non-equilibrium extension of the SSG model. The non-equilibrium extension of the SSG model has been demonstrated to be capable of predicting homogeneous shear flow and plane strain turbulence over a wide range of shear and strain rates from the near-equilibrium regime to the rapid distortion regime. In addition, a new approach to large-eddy simulations has been proposed that allows subgrid scale stress models to go continuously to Reynolds stress models in the coarse mesh/infinite Reynolds limit. It is believed that two-equation models — based on an explicit ASM with these non-equilibrium corrections — are far superior to the Smagorinsky model and can make a significant impact on large-eddy simulations. Here, it is only required that the turbulent dissipation rate be reasonably estimated in non-equilibrium flow situations. In this regard, the modeling of the turbulent dissipation rate in non-equilibrium flows was considered. Arguments were presented that the effects of vortex stretching and anisotropic dissipation must be accounted for in non-equilibrium flows, particularly at lower turbulence Reynolds numbers.

While more work is certainly needed, significant progress has nonetheless been made on the modeling of non-equilibrium turbulent flows. Both in Reynolds stress models for engineering calculations and subgrid scale models for large-eddy simulations. In regard to the latter, the elusive dream of having a DNS go continuously and automatically to an LES and then to a RANS calculation as the mesh becomes coarser may soon be realized. Furthermore, subgrid scale models that build in considerably more turbulence physics than the Smagorinsky model are now at hand. The prospects are bright that turbulence models may start to make a major impact on the calculation of the complex turbulent flows of technological importance.

7. Acknowledgments

This work was supported by the Office of Naval Research under Grant N00014-94-1-0088 (*ARI on Nonequilibrium Turbulence*, Dr. L. P. Purtell, Program Officer). My appreciation goes to Dr. X. H. Xu (Boston University) for conducting the calculations in Figs. 1 – 7 and to Prof. H. Fasel (University of Arizona) for conducting the calculations in Figs. 8 – 9.

References

Abid, R. and Speziale, C.G. (1996) The freestream matching condition for stagnation point turbulent flows: An alternative formulation, *ASME J. Appl. Mech.* **63**, pp. 95-100.

Bardina, J., Ferziger, J.H., and Rogallo, R.S. (1985) Effect of rotation on isotropic turbulence: Computation and modeling, *J. Fluid Mech.* **154**, pp. 321-336.

Bernard, P.S. and Speziale, C.G. (1992) Bounded energy states in homogeneous turbulent shear flow – An alternative view, *ASME J. Fluids Eng.* **114**, pp. 29-39.

Bernard, P.S., Thangam, S., and Speziale, C.G. (1992) The role of vortex stretching in turbulence modeling, in *Instability, Transition and Turbulence*, M.Y. Hussaini, A. Kumar, and C.L. Streett, eds., pp. 563-574, Springer-Verlag, New York.

Cimbala, J.M. (1995) Direct numerical simulations and modeling of a spatially evolving turbulent wake, *Proceedings of the Tenth Symposium on Turbulent Shear Flows* **1**, pp. 6.25 – 6.30, The Pennsylvania State University, State College, PA.

Comte-Bellot, G. and Corrsin, S. (1971) Simple Eulerian time correlation of full- and narrow-band velocity signals in grid-generated isotropic turbulence, *J. Fluid Mech.* **48**, pp. 273-337.

Crow, S. C. (1968) Viscoelastic properties of fine-grained incompressible turbulence, *J. Fluid Mech.* **33**, pp. 1-20.

Durbin, P.A. and Speziale, C.G. (1991) Local anisotropy in strained turbulence at high Reynolds numbers, *ASME J. Fluids Eng.* **113**, pp. 707-709.

Gatski, T.B. and Speziale, C.G. (1993) On explicit algebraic stress models for complex turbulent flows, *J. Fluid Mech.* **254**, pp. 59-78.

Hanjalic, K. and Launder, B.E. (1980) Sensitizing the dissipation equation to irrotational strains, *ASME J. Fluids Eng.* **102**, pp. 34-40.

Hinze, J.O. (1975) *Turbulence*, 2nd ed., McGraw-Hill, New York.

Launder, B.E., Reece, G.J., and Rodi, W. (1975) Progress in the development of a Reynolds stress turbulence closure, *J. Fluid Mech.* **68**, pp. 537-566.

Lee, M.J. and Reynolds, W.C. (1985) Numerical experiments on the structure of homogeneous turbulence, *Stanford University Technical Report No. TF-24*.

Lee, M.J., Kim, J., and Moin, P. (1990) Structure of turbulence at high shear rate, *J. Fluid Mech.* **216**, pp. 561-583.

Lumley, J.L. (1992) Some comments on turbulence, *Phys. Fluids A* **4**, pp. 203-211.

Marasli, B., Champagne, F., and Wygnanski, I. (1991) On linear evolution of unstable disturbances in a plane turbulent wake, *Phys. Fluids A* **3**, pp. 665-674.

Mellor, G.L. and Herring, H.J. (1973) A survey of mean turbulent field closure models, *AIAA J.* **11**, pp. 590-599.

Pope, S.B. (1975) A more general effective viscosity hypothesis, *J. Fluid Mech.* **72**, pp. 331-340.

Raj, R. (1975) Form of the turbulence dissipation equation as applied to curved and rotating turbulent flows, *Phys. Fluids* **18**, pp. 1241-1244.

Reynolds, W.C. (1987) Fundamentals of turbulence for turbulence modeling and simulation, in *Lecture Notes for Von Kármán Institute, AGARD Lect. Ser. No. 86*, pp. 1-66, NATO, New York.

Reynolds, W.C. (1990) The potential and limitations of direct and large-eddy simulations, *Lecture Notes in Physics* **357**, pp. 313-343.

Rodi, W. (1976) A new algebraic relation for calculating the Reynolds stresses, *ZAMM* **56**, pp. T219-221.

Rogers, M.M. (1991) The structure of a passive scalar field with a uniform mean gradient in rapidly sheared homogeneous turbulent flow, *Phys. Fluids A* **3**, pp. 144-154.

Rogers, M.M., Moin, P., and Reynolds, W.C. (1986) The structure and modeling of the hydrodynamic and passive scalar fields in homogeneous turbulent shear flow, *Stanford University Technical Report No. TF-25.*

Saffman, P.G. (1977) Results of a two-equation model for turbulent flows and development of a relaxation stress model for application to straining and rotating flows, *Proc. Project SQUID Workshop on Turbulence in Internal Flows*, S. Murthy, ed., pp. 191-231, Hemisphere.

Speziale, C.G. (1991) Analytical methods for the development of Reynolds-stress closures in turbulence, *Ann. Rev. Fluid Mech.* **23**, pp. 107-157.

Speziale, C.G. (1996) Modeling of Turbulent Transport Equations, in *Simulation and Modeling of Turbulent Flows*, T.B. Gatski, M.Y. Hussaini, and J.L. Lumley, eds., pp. 185-242, Oxford Univ. Press, New York.

Speziale, C.G. (1997) Comparison of explicit and traditional algebraic stress models of turbulence, *AIAA J.* **35**, pp. 1506-1509.

Speziale, C.G. and Abid, R. (1995) Near-wall integration of Reynolds stress turbulence closures with no wall damping, *AIAA J.* **33**, pp. 1974-1977.

Speziale, C.G. and Gatski, T.B. (1997) Analysis and modeling of anisotropies in the dissipation rate of turbulence, *J. Fluid Mech.* **344**, pp. 155-180.

Speziale, C.G. and Xu, X.H. (1996) Towards the development of second-order closure models for non-equilibrium turbulent flows, *Int. J. Heat & Fluid Flow* **17**, pp. 238-244.

Speziale, C.G., Sarkar, S., and Gatski, T.B. (1991) Modeling the pressure-strain correlation of turbulence: An invariant dynamical systems approach, *J. Fluid Mech.* **227**, pp. 245-272.

Tavoularis, S. and Karnik, U. (1989) Further experiments on the evolution of turbulent stresses and scales in uniformly sheared turbulence, *J. Fluid Mech.* **204**, pp. 457-478.

Xu, X.H. and Speziale, C.G. (1997) A non-equilibrium extension of the SSG model, to be published.

DEVELOPMENT OF ALGEBRAIC REYNOLDS STRESS MODEL FOR NON-EQUILIBRIUM TURBULENCE

SHARATH S. GIRIMAJI
Institute for Computer Applications in Science and Engineering
NASA Langley Research Center, Hampton, Virginia

Abstract. For over 20 years, the Rodi (1976) weak-equilibrium assumption has been invoked to derive algebraic Reynolds stress model from its transport equation. In this paper, we describe a more general procedure for formal development of an algebraic Reynolds stress model in non-equilibrium turbulence. In this approach, the departure from equilibrium of anisotropy (b_{ij}) is parameterized in terms of the departure from equilibrium of relative strain rate ($\omega = \varepsilon/SK$, where S is some characteristic strain rate). An algebraic solution of the Reynolds stress anisotropy transport equation in non-equilibrium turbulence is derived analytically for the case of two-dimensional mean flows employing quasilinear pressure-strain correlation model. The present model is compared with the full Reynolds stress closure model against direct numerical simulation (DNS) data and rapid distortion theory (RDT) resulting in good agreement.

1. Introduction

Despite recent advances in Reynolds stress closure methods (RSCM), large-eddy simulations (LES) and direct numerical simulations (DNS), the two-equation turbulence model is the workhorse of practical turbulent flow computations. The more sophisticated two-equation models employ non-linear algebraic constitutive relationships between turbulent Reynolds stress and the mean velocity gradients. Nonetheless, the current algebraic models suffer from important remediable deficiencies and are yet to realize their full potential. One of the major shortcomings of the current algebraic models, which serves as a motivation for this paper, is their invalidity in non-equilibrium turbulence. Recently, there has been another important motivation for developing more accurate non-equilibrium algebraic models.

M. D. Salas et al. (eds.), Modeling Complex Turbulent Flows, 139–160.

Unsteady Reynolds-Averaged Navier-Stokes (URANS) computations and Very Large-Eddy Simulations (VLES) are emerging as viable approaches for computing complex turbulent flows. These calculations can be described as coarse LES where the spectral cut-off is in the energy containing scales of motion rather than in the inertial range, as in LES. Simple eddy-viscosity (Smagorinsky) type models which are perhaps adequate in LES are no longer acceptable. Eddy-viscosity models are valid only when the length and time scales of the resolved motion are much larger than those of unresolved motion and the latter can be assumed to be in quasi-equilibrium with the former. In the context of VLES, the scales of the resolved and unresolved motion are not very different and the subgrid stresses are not in equilibrium with the imposed resolved scales of motion. There is, therefore, a need for a sophisticated non-equilibrium algebraic model for the subgrid scale stresses.

Algebraic models based on Reynolds stress closure equation. Decades of turbulence modeling experience and knowledge is incumbent in the RSCM model. But for its large computational expense, RSCM is the best option for computing practical flows. Therefore, an inexpensive algebraic surrogate to the RSCM is extremely desirable. Rodi (1976) was the first to derive an algebraic Reynolds stress model from its transport equation in the weak equilibrium limit of turbulence characterized by

$$\frac{db_{ij}}{dt^*} \approx 0, \tag{1}$$

where d/dt^* represents substantial derivative following a fluid particle in dimensional time. This methodology has since been further advanced by, among others, Pope (1976), Taulbee (1992), Gatski and Speziale (1993) and Girimaji (1996). The algebraic model of Girimaji (1996) is the exact analytic weak-equilibrium solution of the Reynolds stress evolution equation in two-dimensional mean flows for quasilinear pressure-strain correlation models. There have been attempts to extend the applicability of this methodology to non-equilibrium turbulence, Taulbee (1992) and Speziale and Xu (1996). Neither of these two models, are however, formal approximate solutions to the Reynolds stress evolution equations in non-equilibrium turbulence.

Objective of present work. The objective of this paper is to examine the Reynolds stress evolution equations in non-equilibrium turbulence to determine if the memory and history effects dominate the evolution rendering algebraic modeling unrealistic, or a unique characterization, of the nature of a constitutive relationship, between Reynolds stress and local mean deformation variables is possible. If such an algebraic constitutive relationship exists, we will then attempt to obtain a model for that behavior with the highest degree of fidelity possible to the Reynolds stress evolution equation. Our method is equivalent to seeking an algebraic approximation to the numerical solution of the Reynolds stress closure equations.

Two important *caveats* to note are (i) the algebraic models inherit all the deficiencies of the parent pressure-strain correlation model; and (ii) the turbulent transport effect is either neglected or modeled such that the anisotropy evolution is unaffected by transport.

(The work presented here is a part of a larger body of work on reduced dimensional modeling of Reynolds stress equations, Girimaji, 1998a.)

2. Reynolds stress anisotropy transport equation

The exact Reynolds stress transport equation in an arbitrary non-inertial reference frame undergoing rotation with angular velocity Ω_i is given by

$$\frac{\partial \overline{u_i u_j}}{\partial t^*} + U_k \overline{u_i u_j}_{,k} + 2\Omega_m (e_{mkj}\overline{u_i u_k} + e_{mki}\overline{u_j u_k}) = P_{ij} - \varepsilon_{ij} + \phi_{ij} + \mathcal{D}_{ij}, \quad (2)$$

where e_{ijk} is the alternating tensor. The terms, respectively, are the time rate of change, advection, Coriolis acceleration, production (P_{ij}), dissipation (ε_{ij}), pressure-strain correlation (ϕ_{ij}) and turbulent transport ($\mathcal{D}_{ij}$) of Reynolds stress:

$$\begin{aligned} P_{ij} &= -\overline{u_i u_k}\frac{\partial U_j}{\partial x_k} - \overline{u_j u_k}\frac{\partial U_i}{\partial x_k}; \quad \varepsilon_{ij} = 2\nu \overline{\frac{\partial u_i}{\partial x_k}\frac{\partial u_j}{\partial x_k}} \\ \mathcal{D}_{ij} &= \frac{\partial}{\partial x_l}[-\overline{p u_i}\delta_{jl} - \overline{p u_j}\delta_{il} + \nu\frac{\partial \overline{u_i u_j}}{\partial x_l} - \overline{u_i u_j u_l}]. \end{aligned} \quad (3)$$

The production and dissipation rate of turbulent kinetic energy are, respectively, $P = \frac{1}{2}P_{ii}$ and $\varepsilon = \frac{1}{2}\varepsilon_{ii}$. The dissipation rate tensor can be considered isotropic in high Reynolds number flows:

$$\varepsilon_{ij} = \frac{2}{3}\varepsilon\delta_{ij}. \quad (4)$$

We restrict ourselves to nearly homogeneous turbulence with negligible turbulent transport ($\mathcal{D}_{ij} \approx 0$).

Most of the currently used pressure-strain correlation models can be expressed in the following quasi-linear form:

$$\begin{aligned} \phi_{ij} &= -(C_1^0\varepsilon + C_1^1 P)b_{ij} + C_2 K S_{ij} + C_4 K(b_{ik}W_{jk}^{**} + b_{jk}W_{ik}^{**}) \\ &\quad + C_3 K(b_{ik}S_{jk}^* + b_{jk}S_{ik}^* - \frac{2}{3}b_{mn}S_{mn}^*\delta_{ij}), \end{aligned} \quad (5)$$

where the C's are model coefficients and we use the following definitions

$$\begin{aligned} S^*_{ij} &= \frac{1}{2}\left(\frac{\partial U_i}{\partial x_j} + \frac{\partial U_j}{\partial x_i}\right); \quad \overline{\omega_{ij}} = \frac{1}{2}\left(\frac{\partial U_i}{\partial x_j} - \frac{\partial U_j}{\partial x_i}\right); \\ b_{ij} &= \frac{\overline{u_i u_j}}{2K} - \frac{1}{3}\delta_{ij}; \quad W^{**}_{ij} = \overline{\omega_{ij}} + e_{mji}\Omega_m. \end{aligned} \tag{6}$$

We use the pressure-strain model of Girimaji (1998b) which is sensitive to non-equilibrium turbulence and has the right limiting behavior in the rapid distortion limit.

The modeled anisotropy evolution equation can be derived from the Reynolds stress equation:

$$\begin{aligned} \frac{db_{ij}}{dt^*} &= -b_{ij}(L_1^0 \frac{\varepsilon}{K} - L_1^1 b_{mn} S^*_{mn}) \\ &\quad + L_2 S^*_{ij} + L_3(b_{ik}S^*_{jk} + b_{jk}S^*_{ik} - \frac{2}{3} b_{lm} S^*_{lm}\delta_{ij}) \\ &\quad + L_4(b_{ik}W^*_{jk} + b_{jk}W^*_{ik}). \end{aligned} \tag{7}$$

In the above equation W_{ij} represents the total normalized vorticity given by

$$W^*_{ij} = \overline{\omega_{ij}} + \frac{C_4 - 4}{C_4 - 2} e_{mji}\Omega_m. \tag{8}$$

and the pressure-strain correlation models are redefined as:

$$L_1^0 \equiv \frac{C_1^0}{2} - 2; \;\; L_1^1 \equiv C_1^1 + 2; \;\; L_2 \equiv \frac{C_2}{2} - \frac{2}{3}; \;\; L_3 \equiv \frac{C_3}{2} - 1; \;\; L_4 \equiv \frac{C_4}{2} - 1. \tag{9}$$

The turbulent kinetic energy evolves according to

$$\frac{dK}{dt^*} = P - \varepsilon, \tag{10}$$

and the modeled evolution equation of dissipation is

$$\frac{d\varepsilon}{dt^*} = C_{e1}\frac{\varepsilon}{K}P - C_{e2}\frac{\varepsilon^2}{K}. \tag{11}$$

We define the following norm of the mean deformation rate tensor

$$\eta = S^*_{ij}S^*_{ij} + W^*_{ij}W^*_{ij}, \tag{12}$$

which is used to non-dimensionalize the equations:

$$\begin{aligned} S_{ij} = S^*_{ij}/\sqrt{\eta}; &\qquad W_{ij} = W^*_{ij}/\sqrt{\eta}; \\ dt = \sqrt{\eta}dt^*; &\qquad \omega = \varepsilon/(\sqrt{\eta}K), \end{aligned} \tag{13}$$

where ω is the ratio of the turbulence to mean flow strain rates and is called the relative strain rate. In the above equations, asterisk is used to represent dimensional quantities and the corresponding non-dimensional quantity is written without the asterisk.

In normalized time, the anisotropy transport equation is:

$$\begin{aligned} \frac{db_{ij}}{dt} &= -b_{ij}(L_1^0\omega - L_1^1 b_{mn}S_{mn}) + L_2 S_{ij} + L_3(S_{ik}b_{kj} + b_{ik}S_{kj} \\ &\quad -\frac{2}{3}b_{mn}S_{mn}\delta_{ij}) + L_4(W_{ik}b_{kj} - b_{ik}W_{kj}). \end{aligned} \tag{14}$$

The evolution equation of ω is easily obtained from those of the turbulent kinetic energy and dissipation:

$$\frac{d\omega}{dt} = -2\omega(C_{e1} - 1)b_{mn}S_{mn} - (C_{e2} - 1)\omega^2. \tag{15}$$

The first term on the right hand side of equation (15) represents the production of the relative strain rate whereas the second terms represents the destruction.

In homogeneous turbulence, evolution equations (14) and (15) represent an autonomous non-linear dynamical system of equations. These equations form the foundation from which the algebraic Reynolds stress models are to be derived. The objective now is to determine the conditions under which the above equations permit an algebraic approximation for b_{ij}.

Weak-equilibrium limit. Consider the case when the timescale of evolution of the relative strain (ω) is much larger than that of anisotropy. Then, in the timescale of ω, the anisotropy will evolve rapidly from any arbitrary initial condition and attain the so-called weak-equilibrium state (first introduced by Rodi, 1976) described by:

$$\omega \approx \text{constant}; \quad \frac{db_{ij}}{dt} \approx 0. \tag{16}$$

The anisotropy value for a given ω is obtained by setting

$$\begin{aligned} \frac{db_{ij}}{dt} &= -b_{ij}(L_1^0\omega - L_1^1 b_{mn}S_{mn}) + L_2 S_{ij} + L_3(S_{ik}b_{kj} + b_{ik}S_{kj} \\ &\quad -\frac{2}{3}b_{mn}S_{mn}\delta_{ij}) + L_4(W_{ik}b_{kj} - b_{ik}W_{kj}) \\ &= 0. \end{aligned} \tag{17}$$

The conditions under which the weak-equilibrium approximation is valid are not known. In turbulence far from equilibrium, there is no reason to expect the timescale of relative strain to be very different from that of anisotropy.

Non-equilibrium flows. In order to develop a more appropriate algebraic modeling procedure which is formally valid away from the equilibrium state, we now examine the behavior of the coupled set of ordinary differential equations (14 and 15). The inherent non-linearity of the coupled equations renders an analytical study of the equations difficult. Therefore, we resort to a numerical study of the system of equations for a select set of initial conditions and parameters (strain and rotation rate tensors).

2.1. NUMERICAL EXAMINATION OF THE EQUATIONS

The equations (14) and (15) are integrated from specified initial conditions in two prototypical flows: plane shear and strain. The normalized strain and rotation rates for the two cases are given respectively by

$$S_{ij} = \frac{1}{2}\begin{pmatrix} 0 & 1 & 0 \\ 1 & 0 & 0 \\ 0 & 0 & 0 \end{pmatrix}; \quad W_{ij} = \frac{1}{2}\begin{pmatrix} 0 & -1 & 0 \\ 1 & 0 & 0 \\ 0 & 0 & 0 \end{pmatrix}; \tag{18}$$

$$S_{ij} = \frac{1}{2}\begin{pmatrix} 1 & 0 & 0 \\ 0 & -1 & 0 \\ 0 & 0 & 0 \end{pmatrix}; \quad W_{ij} = \frac{1}{2}\begin{pmatrix} 0 & 0 & 0 \\ 0 & 0 & 0 \\ 0 & 0 & 0 \end{pmatrix}. \tag{19}$$

Each case is investigated for the initial condition of $b_{ij}(t=0) = 0$ and a variety of values for the relative strain rate. The exact initial conditions are given in the figure captions. A low initial value of ω ($\ll 1$) corresponds to rapid distortion and a large value to mild distortion. For the results presented in this section, the pressure-strain correlation model of Girimaji (1997) is used:

$$\begin{aligned}
C_1^0 &= 3.4;\ C_1^1 = 1.8;\ C_2^* = 0.36;\ C_3 = 1.25;\ C_4^* = 0.4. \\
C_{e1} &= 1.44;\ C_{e2} = 1.88. \\
C_2 &= \begin{cases} C_2^* & \text{for } \frac{\omega}{\omega_0} \geq 1.; \\ 0.8 - (0.8 - C_2^*)(\omega/\omega^0)^{0.25} & \text{for } \frac{\omega}{\omega_0} < 1. \end{cases} \\
C_4 &= \begin{cases} C_4^* & \text{for } \eta_1 \geq 0.5; \\ 2.0 - (2.0 - C_4^*)[\frac{\eta_1}{1-\eta_1}]^{.75} & \text{for } \eta_1 < 0.5 \end{cases}
\end{aligned} \tag{20}$$

As mentioned before, this pressure-strain correlation model is valid in non-equilibrium turbulence and is ideally suited for developing non-equilibrium algebraic models.

The results of the numerical integration are examined in the b_{ij} - ω phase space. The results of the homogeneous shear case is presented in Fig.1. The phase portrait of b_{12} and ω is plotted in Fig. 1a and that of b_{11} and ω in Fig. 1b. The plane strain results are plotted in Fig.2 (b_{11} and b_{22} *vs.* ω). In both the cases, the b_{ij} - ω phase portrait exhibits three distinct stages of evolution: the early transient stage, the intermediate slow-manifold stage and the equilibrium state.

Early transient stage. In the rapid distortion cases, the early transient stage is characterized by fast evolution of b_{ij} with little change in ω. Since the initial value of ω is very small, we have

$$\frac{d\omega}{dt} = -2\omega(C_{e1}-1)b_{mn}S_{mn} - (C_{e2}-1)\omega^2 \approx 0. \tag{21}$$

Since $b_{ij}(t=0)=0$, the early evolution of anisotropy is given by

$$b_{ij}(t) \approx L_2 t S_{ij}. \tag{22}$$

At slightly later times, all the production and pressure-strain distribution terms are important and the evolution is well described by the rapid distortion theory.

In the mild distortion cases, the relative strain rate is farther removed from its equilibrium value than is the anisotropy. As a result, b_{ij} remains nearly constant at its initial value whereas ω undergoes a rapid decay due to the dominance of its destruction term. The ω evolution can be approximated by

$$\omega(t) = \frac{\omega(0)}{1+\omega(0)(C_{e2}-1)t}, \tag{23}$$

where $\omega(0)$ is the initial value of the strain rate.

The transient stage typically lasts about one S^*t^* time unit. During this stage, the values taken by the relative strain rate and anisotropy are dominated by memory and history effects. Therefore, an algebraic model for Reynolds stress in this regime of turbulence seems difficult. One may be left with no other option but solve for Reynolds stresses in this stage of turbulence using the differential evolution equations. Speziale and Xu (1996) also reach such a conclusion and the propose an exponential relaxation of Reynolds stress anisotropy to its equilibrium value. We will not consider the algebraic modeling of this regime of non-equilibrium turbulence further in this paper.

Equilibrium turbulence. For all the initial conditions (rapid and mild distortions) considered here, each of the flows asymptotes to its respective fixed point described by

$$\frac{db_{ij}}{dt} = 0; \;\; and \;\; \frac{d\omega}{dt} = 0. \tag{24}$$

The model fixed points are in good agreement with those determined from direct numerical simulations and laboratory experiments (see Speziale *et al.*, 1993). Indeed, most pressure-strain correlation models are calibrated to yield good agreement at the equilibrium state of these two prototypical flows.

Slow-manifold turbulence. After the initial-condition dependent transient state, en route to ultimate convergence to the fixed point, the anisotropy and relative strain rate pass through the slow manifold stage. The slow manifold, also sometimes called sliding mode, as its name indicates, is a reduced dimension space on which the system variables evolve slowly toward their fixed point values. In the entire phase space of variables, the evolution rate is the smallest on the slow manifold. Independent of the initial condition, the solution to the system is first drawn to the slow manifold and along this manifold to the fixed point attractor. The slow manifold stage lasts several S^*t^* time units, much longer than the early transient stage. During this stage of evolution, the anisotropy is an unique function of the relative strain rate, independent of the initial condition or elapse time. As a result, algebraic characterization of anisotropy in terms of relative strain rate appears quite feasible.

It is important to understand the difference between the equilibrium state and the slow manifold stage of turbulence. The fixed point (or equilibrium state) is a point of dimension zero in the anisotropy-relative strain phase space. At the fixed point, the evolution rates of all the system variables are zero. On the other hand, in the slow manifold state, the turbulence parameters occupy a reduced (but non-zero) dimensional space. The slow-manifold stage is a non-equilibrium state of turbulence since the system parameters are still undergoing evolution towards the fixed point.

The object of the remainder of the paper is to develop algebraic models for the equilibrium and slow-manifold stage of non-equilibrium turbulence. In the interest of tractable analytical development, we restrict our consideration to two-dimensional mean flows. However, we do expect the model developed to be useful in three-dimensional flows.

3. Equilibrium turbulence modeling

When evolution equations (14) and (15) are subjected to the equilibrium conditions given in (24), the resulting set of algebraic equations can be written as

$$\begin{aligned} 0 &= -B_{ij}(L_1^0\Omega - L_1^1 B_{mn}S_{mn}) + L_2 S_{ij} + L_3(S_{ik}B_{kj} + B_{ik}S_{kj} \\ &\quad -\frac{2}{3}B_{mn}S_{mn}\delta_{ij}) + L_4(W_{ik}B_{kj} - B_{ik}W_{kj}), \end{aligned} \tag{25}$$

and

$$0 = -2\Omega C_{e1}^{*} B_{mn} S_{mn} - C_{e2}^{*} \Omega^2. \tag{26}$$

The equilibrium values of anisotropy and strain-rate ratio are B_{ij} and Ω and we now seek to obtain explicit expressions for them. From equation (26) we have:

$$\Omega = \frac{-2C_{e1}^{*} B_{mn} S_{mn}}{C_{e2}^{*}}, \tag{27}$$

which leads to the following autonomous equation for the equilibrium anisotropy:

$$\begin{aligned} -B_{ij}(2L_1^0 \frac{C_{e1}^{*}}{C_{e2}^{*}} + L_1^1) B_{mn} S_{mn} \quad = \quad & L_2 S_{ij} + L_4 (W_{ik} B_{kj} - B_{ik} W_{kj}) \\ & + L_3 (S_{ik} B_{kj} + B_{ik} S_{kj} \\ & - \frac{2}{3} B_{mn} S_{mn} \delta_{ij}). \end{aligned} \tag{28}$$

Representation theory will now be used to solve the anisotropy equation. The most general, physically permissible tensor representation for the anisotropy in terms of the strain and rotation rate in the case of two-dimensional mean flow is given by (Girimaji 1996)

$$B_{ij} = G_1 S_{ij} + G_2 (S_{ik} W_{kj} - W_{ik} S_{kj}) + G_3 (S_{ik} S_{kj} - \frac{1}{3} \eta_1 \delta_{ij}), \tag{29}$$

where,

$$\eta_1 = S_{ij} S_{ij}; \quad \text{and} \quad \eta_2 = W_{ij} W_{ij} \quad \text{so that} \quad \eta_1 + \eta_2 = 1. \tag{30}$$

In the above equations G_1 - G_3 are scalar functions of the invariants of strain and rotation rate tensors. For two dimensional mean flows, we have the following identities:

$$\begin{aligned} S_{ik} S_{kj} &= \frac{1}{2} \eta_1 \delta_{ij}^{(2)}; & W_{ik} W_{kj} &= -\frac{1}{2} \eta_2 \delta_{ij}^{(2)}; \\ S_{ik} S_{kl} S_{lj} &= \frac{1}{2} \eta_1 S_{ij}; & S_{ik} W_{kl} S_{lj} &= -\frac{1}{2} \eta_1 W_{ij}; \\ W_{ik} S_{kl} W_{lj} &= \frac{1}{2} \eta_2 S_{ij}; & S_{mn} B_{mn} &= G_1 \eta_1, \end{aligned} \tag{31}$$

where $\delta_{ij}^{(2)}$ and δ_{ij} are two and three dimensional delta functions respectively. We retain terms up to quadratic power (in strain and rotation rate) in their original form and invoke their two-dimensional property only when these terms appear in cubic and higher power terms. For example, $S_{ik} S_{kj}$

is retained as such when it appears by itself: whereas, when it appears as a part of a cubic or higher power term we invoke $S_{ik}S_{kj} = \frac{1}{2}\eta_1\delta_{ij}^{(2)}$ to write $S_{ik}S_{kj}S_{jl} = \frac{1}{2}\eta_1 S_{il}$. In invoking the two-dimensional property for reducing only cubic and higher power terms, it is hoped that the three-dimensional effect is approximately accounted for up to the quadratic term. Using these rules, we write

$$
\begin{aligned}
S_{ik}B_{kj} &+ B_{ik}S_{kj} - \frac{2}{3}B_{mn}S_{mn}\delta_{ij} = \frac{1}{3}\eta_1 G_3 S_{ij} + 2G_1(S_{ik}S_{kj} - \frac{1}{3}\eta_1\delta_{ij}), \\
W_{ik}B_{kj} &- B_{ik}W_{kj} = -G_1(S_{ik}W_{kj} - W_{ik}S_{kj}) + 2\eta_2 G_2 S_{ij}. \qquad (32)
\end{aligned}
$$

The results from the representation theory are substituted into the equilibrium anisotropy equation (28) leading to

$$
\begin{aligned}
&-G_1\eta_1 L^*[G_1 S_{ij} + G_2(S_{ik}W_{kj} - W_{ik}S_{kj}) + \\
&G_3(S_{ik}S_{kj} - \frac{1}{3}\eta_1\delta_{ij})] \\
&= L_2 S_{ij} + L_3[\frac{1}{3}\eta_1 G_3 S_{ij} + 2G_1(S_{ik}S_{kj} - \frac{1}{3}\eta_1\delta_{ij})] \\
&+L_4[S_{ik}W_{kj} - W_{ik}S_{kj}) + 2\eta_2 G_2 S_{ij}], \qquad (33)
\end{aligned}
$$

where L^* is a model constant given by

$$L^* = 2L_1^0\frac{C_{e1} - 1}{C_{e2} - 1} + L_1^1. \qquad (34)$$

The coefficients of each tensor on either side of equation (33) have to be equal. Comparing the coefficients of S_{ij} we get

$$-L^*\eta_1 G_1^2 = L_2 + \frac{1}{3}L_3 G_3\eta_1 2L_4\eta_2 G_2. \qquad (35)$$

The coefficients of $(S_{ik}W_{kj} - W_{ik}S_{kj})$ and $(S_{ik}S_{kj} - \frac{1}{3}\eta_1\delta_{ij})$ yield, in that order,

$$G_2 = \frac{L_4}{L^*\eta_1}; \text{ and } G_3 = -\frac{2L_3}{L^*\eta_1}. \qquad (36)$$

The autonomous equation for G_1 is obtained by substituting equation (36) in equation (35):

$$-L^*\eta_1 G_1^2 = L_2 - \frac{2}{3}\frac{L_3^2}{L^*} + 2\frac{L_4^2}{L^*}\frac{\eta_2}{\eta_1}, \qquad (37)$$

from which we obtain

$$G_1^2 = \frac{1}{\eta_1}[-\frac{L_2}{L^*} + \frac{2}{3}\frac{L_3^2}{L^{*2}} - 2\frac{L_4^2}{L^{*2}}\frac{(1-\eta_1)}{\eta_1}]. \qquad (38)$$

The next step is to make the appropriate choice for the sign of G_1. Girimaji (1997b) demonstrates that the negative value corresponds to a stable fixed point (attractor) and positive value to an unstable fixed point (repeller). Therefore, the physically correct model for G_1 must be

$$G_1 = -\frac{1}{\sqrt{\eta_1}}\sqrt{-\frac{L_2}{L^*} + \frac{2}{3}\frac{L_3^2}{L^{*2}} - 2\frac{L_4^2}{L^{*2}}\frac{(1-\eta_1)}{\eta_1}} \tag{39}$$

Equations (39), (36) and (29) describe the equilibrium structure of the anisotropy tensor. Further

$$B_{lm}S_{lm} = G_1\eta_1 = -\sqrt{\eta_1}\sqrt{-\frac{L_2}{L^*} + \frac{2}{3}\frac{L_3^2}{L^{*2}} - 2\frac{L_4^2}{L^{*2}}\frac{1-\eta_1}{\eta_1}}, \tag{40}$$

which leads to the following expression for the relative strain rate at equilibrium:

$$\Omega = -2\frac{C_{e1}-1}{C_{e2}-1}B_{lm}S_{lm} = 2\frac{C_{e1}-1}{C_{e2}-1}\sqrt{\eta_1}\sqrt{-\frac{L_2}{L^*} + \frac{2}{3}\frac{L_3^2}{L^{*2}} - 2\frac{L_4^2}{L^{*2}}\frac{1-\eta_1}{\eta_1}}. \tag{41}$$

Equations (39), (36) and (41) completely describe the self-similarity state of equilibrium turbulence.

One-equation turbulence model. The preceding equilibrium analysis establishes a formal link between the Reynolds stress closure equation and one-equation level of turbulence models. Consider the following one-equation turbulence model:

$$\begin{aligned} \overline{u_i u_j} &= 2K(B_{ij} + \frac{1}{3}\delta_{ij}), \\ \varepsilon &= K\Omega\sqrt{\eta}, \end{aligned} \tag{42}$$

where B_{ij} and Ω are functions only of the mean velocity gradients and K is obtained from solving an evolution equation. In close vicinity of the fixed point, the above model will perform as well as the parent RSCM. Such a one-equation model will in fact be better than many of the two-equation models, especially those based on the Boussinesq approximation. Unlike the standard K-ε model, this one-equation model, due to its strong fidelity to RSCM near equilibrium, is sensitive to system rotation and streamline curvature. Even in the case of simple homogeneous shear, the equilibrium anisotropy predicted by this model will be superior to the standard K-ε model. Formal development of one-equation models from the Reynolds stress closure equation using this strategy is currently underway.

4. Slow-manifold turbulence modeling

The results of Section 2 clearly demonstrate that, in the slow-manifold regime of non-equilibrium turbulence, the Reynolds stress anisotropy can be uniquely determined knowing the relative strain rate (and, of course, mean strain and rotation rates). We now attempt to parameterize anisotropy algebraically in terms of relative strain rate. The anisotropy evolution equation with respect to relative strain rate is obtained by dividing equation (14) by equation (15):

$$\begin{aligned}
[2\omega(C_{e1}-1)b_{mn}S_{mn} &+ (C_{e2}-1)\omega^2]\frac{db_{ij}}{d\omega} \\
&= b_{ij}(L_1^0\omega - L_1^1 b_{mn}S_{mn}) - L_2 S_{ij} \\
&\quad - L_3(S_{ik}b_{kj} + b_{ik}S_{kj} - \frac{2}{3}b_{mn}S_{mn}\delta_{ij}) \\
&\quad - L_4(W_{ik}b_{kj} - b_{ik}W_{kj}).
\end{aligned} \tag{43}$$

The behavior of b_{ij} and ω in their phase space can be monitored by merely integrating the above equation from its initial condition. In fact, if we know any one point on a particular b_{ij}-ω phase portrait, the remainder of the portrait can, in principle, be reconstructed using the above equation.

We propose to construct the slow-manifold stage of the portrait by starting from the equilibrium values of anisotropy and relative strain rate:

$$b_{ij}(\omega) - B_{ij} = \int_{\Omega}^{\omega(t)} \frac{db_{ij}}{d\omega} d\omega. \tag{44}$$

Had the expression for $\frac{db_{ij}}{d\omega}$ been simple enough, one could perhaps have performed the integration analytically leading to an algebraic expression for $b_{ij}(\omega)$. Since that is not the case, we need to make one further simplification to obtain an algebraic model.

New algebraic modeling approximation. The mean value theorem in the vicinity of the equilibrium state leads to,

$$\frac{db_{ij}}{d\omega} \approx \alpha \frac{b_{ij}(\omega) - B_{ij}}{\omega(t) - \Omega}, \tag{45}$$

where α is a relaxation factor. This phase-space approximation corresponds in physical space to

$$\frac{db_{ij}}{dt} \approx \alpha \frac{b_{ij}(\omega) - B_{ij}}{\omega(t) - \Omega}[\frac{d\omega}{dt}]. \tag{46}$$

(For the sake of simplicity, we will take $\alpha = 1$.) The above equation corresponds to the new algebraic approximation for non-equilibrium turbulence. In contrast to the weak-equilibrium simplification (equation 1), in this case the rate of change of anisotropy is non-zero when the relative strain rate is evolving in time.

Decompose the anisotropy and relative strain rate as follows:

$$\begin{aligned} b_{ij} &= B_{ij} + \beta_{ij} \\ \omega &= \Omega + \theta, \end{aligned} \tag{47}$$

where β_{ij} and θ are respectively the deviations of anisotropy and turbulent strain rate from their equilibrium values. For a turbulence field evolving from a given initial state, the equilibrium values B_{ij} and Ω are not functions of time and are obtained as shown previously. The evolution equation of β_{ij} is derived from subtracting equation (25) from equation (14):

$$\begin{aligned} \frac{d\beta_{ij}}{dt} &= -\beta_{ij}[L_1^0(\Omega+\theta) - L_1^1(B_{mn}+\beta_{mn})S_{mn}] + L_2' S_{ij} \\ &+L_3(S_{ik}\beta_{kj} + \beta_{ik}S_{kj} - \frac{2}{3}S_{mn}\beta_{mn}\delta_{ij}) \\ &+L_4(W_{ik}\beta_{kj} - \beta_{ik}W_{kj}) - B_{ij}[L_1^0\theta - L_1^1\beta_{mn}S_{mn}], \end{aligned} \tag{48}$$

where L_2' represents the difference between the current value of L_2 and its equilibrium value:

$$L_2' = L_2(b_{ij},\omega) - L_2(B_{ij},\Omega). \tag{49}$$

Note that only L_2 is a function of the current state of turbulence and all other model coefficients are constant given the mean strain and rotation rates.

The evolution equation for θ is

$$\frac{d\theta}{dt} = -2\theta(C_{e1}-1)(B_{mn}+\beta_{mn})S_{mn} - 2\Omega(C_{e1}-1)\beta_{mn}S_{mn} - (C_{e2}-1)\theta(2\Omega+\theta). \tag{50}$$

The evolution equation of the anisotropy tensor as a function of the relative strain rate is obtained by dividing equation (48) by equation (50):

$$\begin{aligned} (\phi_1 + \phi_2\beta_{mn}S_{mn})\frac{d\beta_{ij}}{d\theta} &= \beta_{ij}[\phi_3 - L_1^1\beta_{mn}S_{mn}] - L_2' S_{ij} \\ &-L_3(S_{ik}\beta_{kj} + \beta_{ik}S_{kj} - \frac{2}{3}S_{mn}\beta_{mn}\delta_{ij}) \\ &-L_4(W_{ik}\beta_{kj} - \beta_{ik}W_{kj}) + B_{ij}[L_1^0\theta - L_1^1\beta_{mn}S_{mn}] \end{aligned} \tag{51}$$

where

$$\begin{aligned}
\phi_1 &= 2\theta(C_{e1}-1)B_{mn}S_{mn} + (C_{e2}-1)\theta(2\Omega+\theta); \\
\phi_2 &= 2(\Omega+\theta)(C_{e1}-1); \\
\phi_3 &= L_1^0(\Omega+\theta) - L_1^1 B_{mn}S_{mn}.
\end{aligned} \tag{52}$$

Note that ϕ_1 - ϕ_3 are independent of β_{ij}. We now attempt to solve equation (51) for β_{ij} in terms of θ. This is an autonomous equation β_{ij} requiring only the specification of an 'initial' condition in order to obtain a solution. We use the equilibrium state of turbulence as the natural initial condition:

$$\beta_{ij} = 0 \ \text{ at } \ \theta = 0. \tag{53}$$

We again invoke representation theory to solve for β_{ij}. From equation (51), we first note that β_{ij} can be a tensor function only of S_{ij} and W_{ij} and not of their derivatives or integrals, for the strain and rotation rate tensors are not functions of time or of θ. From the results in Section 2 it is clear that the anisotropy on the slow-manifold is not a function of the initial state of turbulence. Again, for the sake of simplicity, restricting consideration two-dimensional mean flows, the general representation for b_{ij} is

$$\beta_{ij} = J_1 S_{ij} + J_2(S_{ik}W_{kj} - W_{ik}S_{kj}) + J_3(S_{ik}S_{kj} - \frac{1}{3}\eta_1\delta_{ij}). \tag{54}$$

The coefficients J_1 - J_3 are now scalar functions of the invariants of S_{ij} and W_{ij} and θ. The derivative of β_{ij} is

$$\frac{d\beta_{ij}}{d\theta} = \frac{dJ_1}{d\theta}S_{ij} + \frac{dJ_2}{d\theta}(S_{ik}W_{kj} - W_{ik}S_{kj}) + \frac{dJ_3}{d\theta}(S_{ik}S_{kj} - \frac{1}{3}\eta_1\delta_{ij}). \tag{55}$$

Using the results from the previous section we have

$$\begin{aligned}
\beta_{mn}S_{mn} &= J_1\eta_1 \\
S_{ik}\beta_{kj} + \beta_{ik}S_{kj} &- \frac{2}{3}\beta_{mn}S_{mn}\delta_{ij} = \frac{1}{3}\eta_1 J_3 S_{ij} + 2J_1(S_{ik}S_{kj} - \frac{1}{3}\eta_1\delta_{ij}), \\
W_{ik}\beta_{kj} &- \beta_{ik}W_{kj} = -J_1(S_{ik}W_{kj} - W_{ik}S_{kj}) + 2\eta_2 J_2 S_{ij}.
\end{aligned} \tag{56}$$

The anisotropy evolution equation can now be written using representation theory as (invoking equations 55 and 56)

$$
\begin{aligned}
&(\phi_1 + J_1\eta_1\phi_2)[\frac{dJ_1}{d\theta}S_{ij} + \frac{dJ_2}{d\theta}(S_{ik}W_{kj} - W_{ik}S_{kj}) \\
&+\frac{dJ_3}{d\theta}(S_{ik}S_{kj} - \frac{1}{3}\eta_1\delta_{ij})] \\
= \;& (\phi_3 - L_1^1 J_1\eta_1)[J_1 S_{ij} + J_2(S_{ik}W_{kj} - W_{ik}S_{kj}) + J_3(S_{ik}S_{kj} - \frac{1}{3}\eta_1\delta_{ij})] \\
&+(L_1^0\theta - L_1^1 J_1\eta_1)[G_1 S_{ij} + G_2(S_{ik}W_{kj} - W_{ik}S_{kj}) \\
&+G_3(S_{ik}S_{kj} - \frac{1}{3}\eta_1\delta_{ij})]L_3[\frac{1}{3}\eta_1 J_3 S_{ij} + 2J_1(S_{ik}S_{kj} - \frac{1}{3}\eta_1\delta_{ij})] \\
&-L_4[-J_1(S_{ik}W_{kj} - W_{ik}S_{kj}) + 2\eta_2 J_2 S_{ij}] - L_2' S_{ij} \qquad (57)
\end{aligned}
$$

Again comparing the coefficients of various tensors on either side of the equation leads to the evolution equations for the scalar functions J_1 - J_3:

$$
\begin{aligned}
(\phi_1 + J_1\eta_1\phi_2)\frac{dJ_1}{d\theta} &= J_1(\phi_3 - L_1^1 J_1\eta_1) + G_1(L_1^0\theta - L_1^1 J_1\eta_1) \qquad (58) \\
&\quad -\frac{1}{3}L_3\eta_1 J_3 - 2L_4(1-\eta_1)J_2 - L_2' \\
(\phi_1 + J_1\eta_1\phi_2)\frac{dJ_2}{d\theta} &= J_2(\phi_3 - L_1^1 J_1\eta_1) + G_2(L_1^0\theta - L_1^1 J_1\eta_1) + L_4 J_1 \\
(\phi_1 + J_1\eta_1\phi_2)\frac{dJ_3}{d\theta} &= J_3(\phi_3 - L_1^1 J_1\eta_1) + G_3(L_1^0\theta - L_1^1 J_1\eta_1) - 2L_3 J_1.
\end{aligned}
$$

It does not appear to be easy to solve these equations analytically for all values of θ.

Invoking the new algebraic modeling assumption (equation 45) we have

$$
\frac{d\beta_{ij}}{d\theta} \approx \frac{\beta_{ij}}{\theta}; \quad \frac{dJ_1}{d\theta} \approx \frac{J_1}{\theta}; \quad \frac{dJ_2}{d\theta} \approx \frac{J_2}{\theta}; \quad \frac{dJ_3}{d\theta} \approx \frac{J_3}{\theta}. \qquad (59)
$$

Substituting this simplification in equation (58) we get the following algebraic equations for J's:

$$
\begin{aligned}
(\phi_1 + J_1\eta_1\phi_2)\frac{J_1}{\theta} &= J_1(\phi_3 - L_1^1 J_1\eta_1) + G_1(L_1^0\theta - L_1^1 J_1\eta_1) \qquad (60) \\
&\quad -\frac{1}{3}L_3\eta_1 J_3 - 2L_4(1-\eta_1)J_2 - L_2' \\
(\phi_1 + J_1\eta_1\phi_2)\frac{J_2}{\theta} &= J_2(\phi_3 - L_1^1 J_1\eta_1) + G_2(L_1^0\theta - L_1^1 J_1\eta_1) + L_4 J_1 \\
(\phi_1 + J_1\eta_1\phi_2)\frac{J_3}{\theta} &= J_3(\phi_3 - L_1^1 J_1\eta_1) + G_3(L_1^0\theta - L_1^1 J_1\eta_1) - 2L_3 J_1.
\end{aligned}
$$

Coefficients J_2 and J_3 can be expressed in terms of J_1:

$$
J_2 = \theta\frac{G_2 L_1^0\theta - J_1(L_1^1\eta_1 G_2 - L_4)}{\phi_1 - \phi_3\theta + J_1\eta_1(\phi_2 + L_1^1\theta)}, \qquad (61)
$$

and,

$$J_3 = \theta \frac{G_3 L_1^0 \theta - J_1(L_1^1 \eta_1 G_3 + 2L_3)}{\phi_1 - \phi_3\theta + J_1\eta_1(\phi_2 + L_1^1\theta)}. \tag{62}$$

Substitution of these expressions into the equation for J_1 yields

$$\begin{aligned} J_1[\phi_1 - \phi_3\theta \quad + \quad & J_1\eta_1(\phi_2 + L_1^1\theta)]^2 = -L_2'\theta[\phi_1 - \phi_3\theta + J_1\eta_1(\phi_2 + L_1^1\theta)] \\ & -\frac{\eta_1}{3} L_3\theta^2[G_3 L_1^0\theta - J_1(G_3 L_1^1\eta_1 + 2L_3)] \\ & -2\eta_2 L_4\theta^2[G_2 L_1^0\theta - J_1(G_2 L_1^1\eta_1 - L_4)] \\ & +G_1(L_1^0\theta^2 - J_1\theta\eta_1 L_1^1)[\phi_1 - \phi_3\theta + J_1\eta_1(\phi_2 + L_1^1\theta)]. \end{aligned} \tag{63}$$

After algebraic manipulations, this equation can be written in the standard cubic equation form:

$$J_1^3 + pJ_1^2 + qJ_1 + r = 0, \tag{64}$$

where

$$\begin{aligned} p &= \frac{2(\phi_1 - \phi_3\theta) + \eta_1 G_1 L_1^1\theta}{\eta_1(\phi_2 + L_1^1\theta)} \\ q &= [(\phi_1 - \phi_3\theta)^2 - G_1\eta_1 L_1^0\theta^2(\phi_2 + L_1^1\theta) + G_1\eta_1 L_1^1\theta(\phi_1 - \phi_3\theta) \\ & \quad -2L_4\eta_2\theta^2(G_2 L_1^1\eta_1 - L_4) - \frac{\eta_1}{3} L_3\theta^2(G_3 L_1^1\eta_1 + 2L_3) \\ & \quad +L_2'\theta\eta_1(\phi_2 + L_1^1\theta)]/[\eta_1^2(\phi_2 + L_1^1\theta)^2] \\ r &= [-L_1^0\theta^2\{G_1(\phi_1 - \phi_3\theta) - 2G_2 L_4\eta_2\theta \\ & \quad -G_3 L_3 \frac{\eta_1}{3}\theta\} + L_2'\theta(\phi_1 - \phi_3\theta)]/[\eta_1^2(\phi_2 + L_1^1\theta)^2]. \end{aligned} \tag{65}$$

We need to solve this equation to determine J_1.

A cubic equation – depending upon the parameters p, q, and r – has either one real root and two complex roots or all three real roots. Since J_1 has to be real, the choice is obvious when only one root is real. When the parameters are such that all roots are real, the choice is more difficult. Girimaji (1997a) has an extensive discussion on the selection criteria that can be used to single out the physically most appropriate root. The most important selection criterion is the requirement of continuity of J_1 in the function space. By requiring that J_1 be a smooth (differentiable) function of p, q and r we can narrow down the choice to one root. We adopt the same selection procedure here and present only the final expression for J_1.

$$J_1 = \begin{cases} -\frac{p}{3} + (-\frac{b}{2} + \sqrt{D})^{\frac{1}{3}} + (-\frac{b}{2} - \sqrt{D})^{\frac{1}{3}}, & \text{for } D > 0; \\ -\frac{p}{3} + 2\sqrt{\frac{-a}{3}} cos(\frac{\theta}{3}), & \text{for } D < 0 \text{ and } b < 0; \\ -\frac{p}{3} + 2\sqrt{\frac{-a}{3}} cos(\frac{\theta}{3} + \frac{2\pi}{3}), & \text{for } D < 0 \text{ and } b > 0. \end{cases} \tag{66}$$

The various quantities in the above equation are given by

$$\begin{aligned} a &\equiv (q - \frac{p^2}{3}); \quad b \equiv \frac{1}{27}(2p^3 - 9pq + 27r); \\ D &= \frac{b^2}{4} + \frac{a^3}{27}; \quad cos(\theta) = \frac{-b/2}{\sqrt{-a^3/27}}. \end{aligned} \tag{67}$$

The slow-manifold anisotropy model is

$$\begin{aligned} b_{ij} &= (G_1 + J_1)S_{ij} + (G_2 + J_2)[S_{ik}W_{kj} - W_{ik}S_{kj}] \\ &+ (G_3 + J_3)[S_{ik}S_{kj} - \frac{1}{3}\eta_1\delta_{ij}], \end{aligned} \tag{68}$$

where J_1, J_2 and J_3 are given by equations (66), (61) and (62) respectively. This non-equilibrium algebraic model is formally valid after the initial transient stage, even in rapidly distorted turbulence.

5. Results and discussion

In this section, the non-equilibrium model is compared against DNS data and RSCM. First, we examine the accuracy of the mean-value theorem simplification of equation (45). Any difference between the slow-manifold of the RSCM and the non-equilibrium algebraic model must stem from this approximation. The model calculation is plotted in figures 1 and 2. The agreement between the model and the slow-manifold of the RSCM is excellent: in fact, the trajectories are indistinguishable.

Next, we evaluate the model in rapidly distorted turbulence against real DNS data and rapid distortion theory (RDT) results. This would constitute a stringent test of the algebraic model, as the rapid distortion limit represents the extreme non-equilibrium state. Kinetic energy evolution in rapidly distorted ($SK_0/\varepsilon_0 = 100$) homogeneous plane shear flow is considered in Figure 3. The algebraic model is compared against RDT solution. Calculations from the full Reynolds stress closure models of SSG and Girimaji (1997b) are also shown. The SSG model is formally valid only near equilibrium turbulence and, expectedly, performs poorly. The RSCM of Girimaji (1997b) which is expressly designed for non-equilibrium flows captures the kinetic energy growth very well. The algebraic model, which is based on the Girimaji pressure-strain model, also performs very well. The comparison in the rapidly distorted ($\Gamma K_0/\varepsilon_0$) plane-strain case is performed in Figure 4. Again, the performance of the algebraic model is comparable to that of RSCM (Girimaji) while being better than that of RSCM (SSG).

Overall, the present non-equilibrium model performance in rapidly distorted flows is quite good. Its performance is slightly inferior to the RSCM of Girimaji (1997) and better than that of RSCM (SSG). The non-equilibrium algebraic model is formally valid after the initial transient stage, even in rapidly distorted turbulence. The transient stage must not be confused with rapid distortion. The transient stage is encountered in mildly distorted flows as well. Mild and rapidly distorted turbulence state are well described by the slow-manifold after approximately one St time unit.

6. Conclusion

In this paper, we first demonstrate that turbulence evolution (according to modeled Reynolds stress evolution equation) consists of three parts: the initial transient stage, the slow-manifold stage and the equilibrium state. Algebraic modeling, in the traditional sense, is not possible in the initial-transient stage as the history effects dominate the evolution. Algebraic modeling, however, is possible in the slow-manifold and the equilibrium states. A new algebraic modeling assumption valid in slow-manifold turbulence is presented in equation (45). Using this approximation, an algebraic expression for the solution of the Reynolds stress model is found in the slow-manifold stage of non-equilibrium turbulence. The model is compared against DNS data, RDT solution and RSCM calculations resulting in good agreement.

7. Acknowledgements

This research was supported by the National Aeronautics and Space Administration under NASA Contract No. NAS1-19480. The data for figures 3 and 4 was provided by Professor C. G. Speziale of Boston University.

References

Gatski, T. B. and Speziale, C. G. (1993). "On explicit algebraic stress models for complex turbulent flows," *J. Fluid Mech.* **254**, pp. 59–78.

Gibson, M. M. and Launder, B. E. (1978). "Ground effects of pressure fluctuations in the atmospheric boundary layer," *J. Fluid Mech.* **86**, pp. 491–511.

Girimaji, S. S. (1996). "Fully explicit and self-consistent algebraic Reynolds stress model," *Theo. and Comp. Fluid Dyn.* **8**, pp. 387–402.

Girimaji, S. S. (1998a). "Low-dimensional (algebraic) modeling of Reynolds stresses in non-equilibrium turbulence," to be submitted.

Girimaji, S. S. (1998b). "Pressure-strain correlation modeling of complex turbulent flows," Submitted to the *J. Fluid Mech.*

Launder, B. E., Reece, G. J. and Rodi, W. (1975). "Progress in the Development of Reynolds Stress Turbulence Closure," *J. Fluid Mech.* **68**, pp. 537–566.

Pope, S. B. (1975). "A More General Effective-Viscosity Hypothesis," *J. Fluid Mech.* **72**, pp. 331–340.

Speziale, C. G., Sarkar, S. and Gatski, T. B. (1991). "Modeling the pressure-strain correlation of turbulence: An invariant dynamical system approach," *J. Fluid Mech.* **227**, pp. 245–272.

Speziale, C. G. and Xu, X. H. (1996). "Towards the development of second-order closure models for non-equilibrium turbulent flows," *Int. J. Heat and Fluid Flow* **17**, pp. 245–272.

Taulbee, D. B. (1992). "An improved algebraic stress model and corresponding nonlinear stress model," *Phys. Fluids A* **4**, pp. 2555–2561.

Rodi, W. (1976). "A New Algebraic Relation for Calculating Reynolds Stress," *ZAMM* **56**, pp. T219–T221.

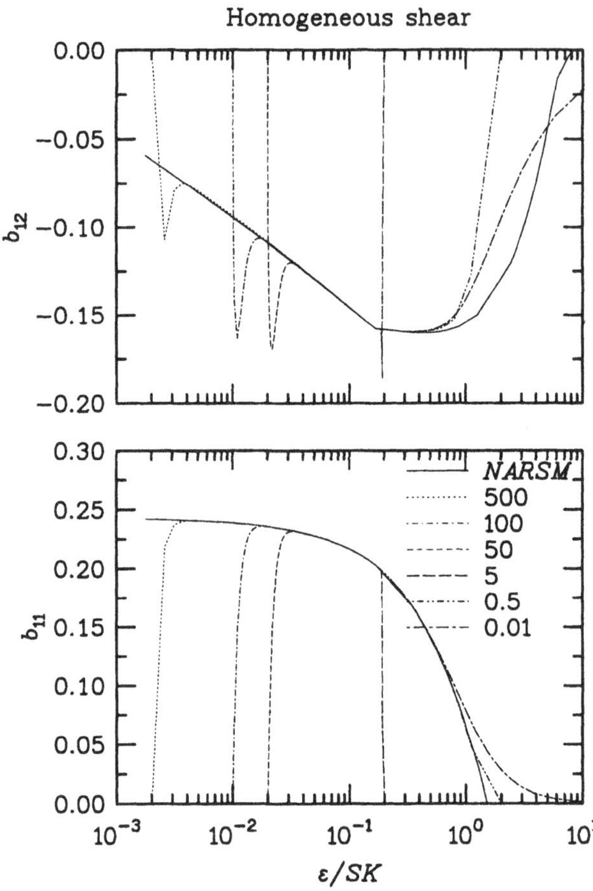

Figure 1. The $b_{ij} - \omega$ phase-space diagram for homogeneous turbulence. Trajectories of full Reynolds-stress calculations for various SK_0/ε_0 values are shown. The slow manifold computed by the non-equilibrium algebraic Reynolds stress model (NARSM) is also shown.

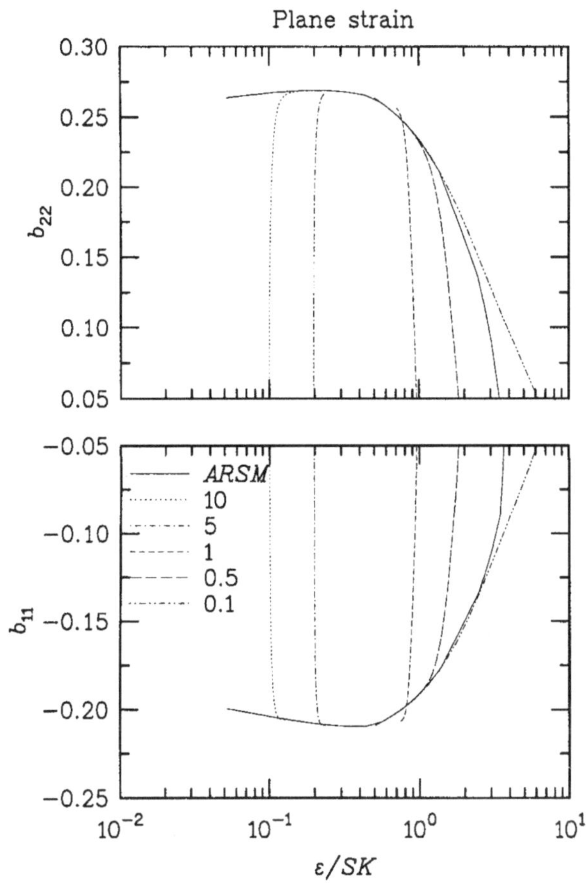

Figure 2. The $b_{ij} - \omega$ phase-space diagram for plane strain. Trajectories of full Reynolds-stress calculations for various SK_0/ε_0 values are shown. The slow manifold computed by the non-equilibrium algebraic Reynolds stress model (NARSM) is also shown.

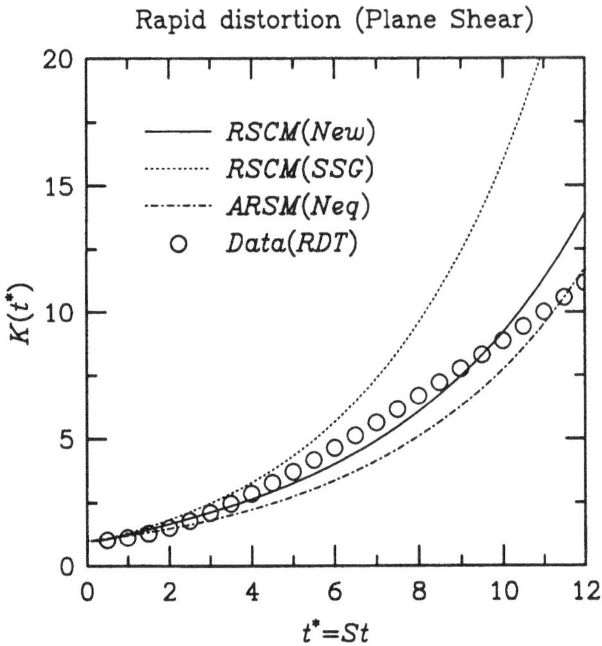

Figure 3. Evolution of kinetic energy in rapidly distorted homogeneous shear turbulence: $SK_0/\varepsilon_0 = 100$.

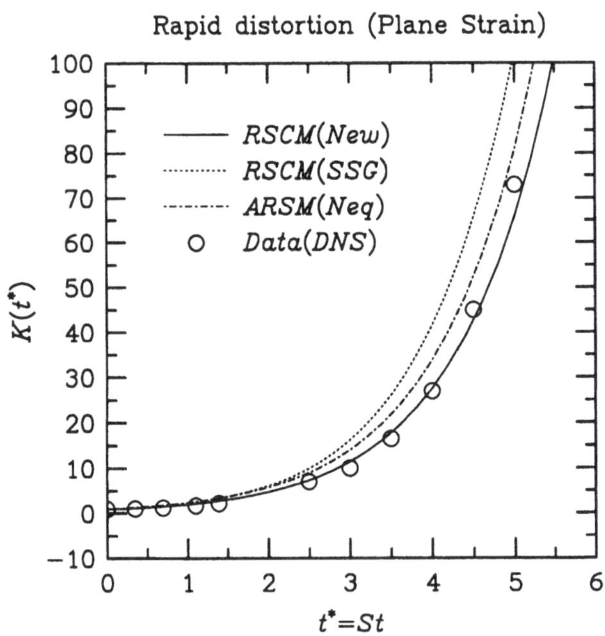

Figure 4. Evolution of kinetic energy in rapidly distorted plane strain turbulence: $\Gamma K_0/\varepsilon_0 = 100$.

TOWARD A VORTEX METHOD SIMULATION OF NON-EQUILIBRIUM TURBULENT FLOWS

PETER S. BERNARD
Department of Mechanical Engineering
University of Maryland
College Park, Maryland

1. Introduction

Many of the most important physical processes associated with turbulent motion are best characterized in terms of the dynamics of the vorticity field. Two obvious examples include the action of vortex stretching in sending energy to small dissipative scales, and the self-replication mechanism by which quasi-streamwise vortices are produced adjacent to solid boundaries. The latter process maintains the Reynolds shear stress by forcing fluid ejections and sweeps (Bernard *et al.*, 1993).

It is not evident that traditional RANS modeling (Speziale, 1991), which in its simplest form approximates the Reynolds stress correlation, is the optimal approach to take when accounting for flow processes primarily involving the dynamics of the vorticity field. At best it is an indirect approach, so it is natural to consider alternatives having substantially closer connections to processes involving vortex motions. This viewpoint has been a primary impetus for the development of vortex methods (Puckett, 1993) for flow simulation, wherein a flow field is represented by a collection of convecting and interacting vortical elements with the Biot-Savart law providing a basis for recovering the velocity field. One may reasonably hope, under the right circumstances, that such direct models of vortex motions have an advantage in turbulent flow prediction.

At least three important technical improvements to vortex methods occurring over the last decade have added to the likelihood of their successful application to turbulent flow simulation. In the first place, the potentially debilitating effect of the singularity in the Biot-Savart law – which can generate excessively large velocities when two vortices get within close range – has been eliminated by smoothing techniques which desingularize the

M. D. Salas et al. (eds.), Modeling Complex Turbulent Flows, 161–181.

Biot-Savart kernel without preventing convergence of the numerical scheme (Puckett, 1993). Such methods are now routinely used in applications of the vortex method.

A second concern is the prediction of viscous vorticity diffusion; a significant aspect of turbulent flow simulation adjacent to solid boundaries. While this has been modeled with some success via random walk techniques (Chorin, 1973), these are ultimately unsatisfactory for turbulent flow simulation since they can introduce large, non-physical flow perturbations. Recently, however, a variety of deterministic schemes for computing viscous vorticity diffusion over a field of vortex elements have been devised (Bernard, 1995; Cottet & Mas-Gallic, 1990; Fishelov, 1990; Ogami & Akamatsu, 1990; Russo, 1993; Shankar & van Dommelen, 1996). Free of the pseudo-turbulent energy generated by randomly jumping vortices, they are better positioned to accommodate the randomness inherent in real turbulent motion. Moreover, in laminar flows, deterministic schemes are capable of giving accurate noise-free predictions (Bernard, 1995).

Finally, though the nominal cost of the Biot-Savart velocity field calculation for N vortices is $O(N^2)$ – a cost which, if necessary, would severely limit the practical resolution of vortex simulations – recently developed fast multipole methods (Greengard & Rokhlin, 1987; Strickland & Amos, 1992; Strickland & Bray, 1994; Winckelmans *et al.*, 1995; Winckelmans *et al.*, 1996) require just $O(N)$ effort. These open up the door to calculations with very large numbers of elements. For example, even with the use of parallel algorithms, a calculation with $O(10^5)$ vortices is barely feasible using $O(N^2)$ summation. In contrast, a three-dimensional $O(N)$ fast multipole scheme can accommodate at least $O(10^6)$ vortices – if not many more (Winckelmans *et al.*, 1995; Winckelmans *et al.*, 1996). With this degree of resolution, the modeling of complex turbulent flows becomes feasible.

Though smooth, fast, deterministic vortex methods may have the technical capability, in so far as speed and resolution is concerned, to model turbulent flow, it still remains to show through concrete examples that they can produce legitimate representations of physical turbulent flow fields. In particular, though vortex methods have been applied in the past to complex three-dimensional high Reynolds number flows, these generally have complicating factors, e.g. complex geometry, transient conditions and so on, which make it difficult to assess the degree to which turbulence *per se* is being well represented. The present work is directed at determining what specific characteristics a vortex method needs to have in order to successfully model turbulence and to illustrate this capability through specific example. In view of the well documented statistical and structural properties of channel flow obtained from physical experiments and direct numerical simulations (Bernard *et al.*, 1993; Kim *et al.*, 1987), this case will be the focal point of the present work. A brief discussion of a calculation of a prolate spheroid flow will also be given to illustrate the generalization of the present method to complex flows.

2. Special Considerations for Turbulence Modeling

By way of illustrating previous applications of vortex methods to complex 3D flows, Fig. 1 shows the instantaneous computed vortex field associated with flow past a hemispherical model of a four ribbon parachute (Strickland, 1994). Axisymmetric conditions are imposed so the computational elements are vortex rings. The calculation starts with the hemisphere at rest; subsequently it is ramped up to a constant velocity and then, after some time, brought to rest. The view in Fig. 1 is of the vortex wake generated during constant motion, just prior to deceleration. Air is seen to travel between the ribbons and through the hole at the axis of symmetry.

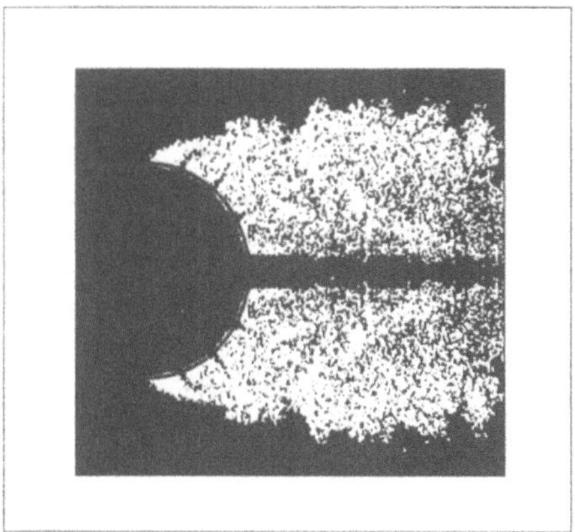

Figure 1. Vortex method simulation of flow past a hemispherical parachute model (Figure supplied courtesy of Sandia National Laboratories).

The complex and apparently chaotic distribution of vortices in the wake suggests that the flow is turbulent in some sense. A video presentation of these results (Strickland, 1997) shows that the vortices indeed move in complex trajectories. The extent to which this random field, as well as others produced in related calculations, is physical is difficult to ascertain due to the complex configuration and paucity of experimental data. By concentrating on a channel flow simulation, as the present work does, it is possible to sidestep these uncertainties.

Unlike the axisymmetric flow shown in Fig. 1, general 3D calculations using vortex filaments incur the full effects of vortex stretching, including a substantial flux of energy to small dissipative scales. Such effects are clear, for example, in 3D computations of the evolution of a spherical vortex sheet and the time developing 3D wake behind an accelerated airfoil with elliptical loading (Winckelmans *et al.*, 1995; Winckelmans *et al.*, 1996). Another, more direct, example is provided by an evolving turbulent vortex filament (Chorin82, 1982) which undergoes relentless stretching and folding as energy cascades to small scales.

It is clear from these examples that numerical implementations of the vortex method are destined to be impractical if they attempt to resolve all scales as they appear in the calculation. In particular, given an opportunity, the comparatively uninteresting small scale features of the flow will consume an ever increasing fraction of the computational resources. Thus, to be useful for turbulent flow modeling, vortex methods must prevent or inhibit the formation of small scale vortices.

A recipe for scale limitation is an important outcome of a statistical analysis of vortex motion by Chorin (Chorin, 1994). In fact, justification may be found for the particular strategy illustrated in Fig. 2 wherein folded vortex segments are eliminated from the calculation as they form, by deleting the hairpin and reattaching the ends (Chorin, 1993). Chorin shows that such hairpin removal effectively prevents the development of fine scale features of the turbulence, without disrupting the normal energy cascade as characterized by Kolmogorov's inertial range law. To be able to partake of this computational remedy, however, it is necessary that tubes or filaments be the principal computational element of the vortex method, as will be the case here.

Figure 2. Hairpin removal via Chorin's algorithm.

Adjacent to boundaries, vortex methods applied to turbulent flow simulations need to capture viscous vorticity diffusion from the surface – the source of new vorticity in the flow field – as well as mimic the production of the quasi-streamwise vortices maintaining the Reynolds shear stress. For high Reynolds numbers the viscous region containing high vorticity is thin and most efficiently represented through an array of high aspect ratio vortex sheets. These are also well suited for capturing the primarily wall-normal mean vorticity diffusion. Vortex tubes can be used to directly represent both coherent and non-coherent structures in the outer flow. For the special case of channel flow, an appropriate arrangement of vortical elements is as depicted in Fig. 3.

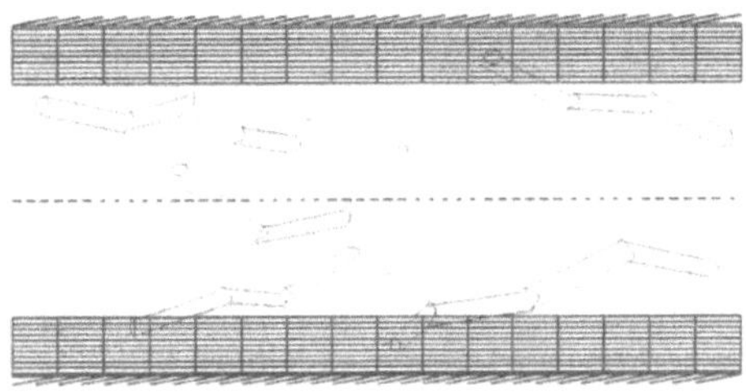

Figure 3. Vortex sheet and tube arrangement for channel flow.

Note that Fig. 3 suggests that no attempt is made to prohibit tubes from wandering into the sheet region, since, apart from their coupling through the velocity field, the dynamical evolution of sheets and tubes can be treated separately, as will be discussed below. The major concern here is rather that an algorithm be employed which satisfies the dynamical requirements underlying the quasi-streamwise vortex regeneration process. A workable method in this direction is described below in Sec. 5.

3. Vortex Sheet and Filament Scheme

The present scheme advances vortex sheets and tubes through time according to the following algorithm. For the sheets, the terms in the 3D vorticity transport equation are solved according to the prescription:

$$\underbrace{\frac{\partial \boldsymbol{\Omega}}{\partial t} \quad + \quad (\nabla \boldsymbol{\Omega})\mathbf{u}}_{convect\ sheets\ \&\ interpolate} \quad = \quad \underbrace{(\nabla \mathbf{u})\boldsymbol{\Omega} + \frac{1}{R_e}\nabla^2 \boldsymbol{\Omega},}_{finite\ differences} \qquad (1)$$

where R_e is the Reynolds number. In this, convection of the sheets is immediately followed by their interpolation back into the mesh, as illustrated in Fig. 4. The end result is to have the sheet vorticity field available on the uniform mesh at the start of every time step so that finite differences can be efficiently used to estimate the stretching and wall-normal diffusion terms. Note that without the use of the mesh structure, evaluation of the stretching term would require multiple velocity evaluations for each sheet, while a more computationally intensive deterministic scheme such as Fishelov's method (Bernard, 1995; Fishelov, 1990), would be necessary to model diffusion.

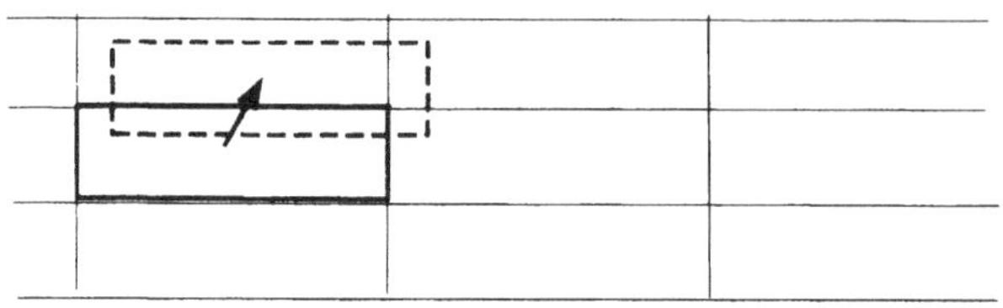

Figure 4. Sheet convection and interpolation.

Time marching of the filaments is accomplished according to the algorithm:

$$\underbrace{\frac{\partial \boldsymbol{\Omega}}{\partial t} + (\nabla \boldsymbol{\Omega})\mathbf{u} = (\nabla \mathbf{u})\boldsymbol{\Omega}}_{\textit{convect filament endpoints}} + \underbrace{\frac{1}{R_e}\nabla^2 \boldsymbol{\Omega}}_{\textit{decay model}} , \tag{2}$$

part of which is illustrated in Fig. 5. Apart from the treatment of the diffusion term, which is discussed in more detail below, this is a standard approach for filament calculations (Chorin, 1993). The diffusion term may be justifiably ignored in external flows where the viscous decay of vortices can be assumed to take place downstream of the region of interest. For a channel flow with periodic conditions, however, vortex filament decay becomes an important issue, since without it, the vortex population can be expected to grow without bound, i.e. viscous decay of filaments is likely to play an important role in maintaining an equilibrium number of vortices.

4. Velocity Field

Following standard practice (Puckett, 1993), the velocity field is computed as a numerical approximation to

$$\mathbf{u}(\mathbf{x}, t) = \int_{\Re^3} K_\eta(\mathbf{x} - \mathbf{x}')\boldsymbol{\Omega}(\mathbf{x}', t)d\mathbf{x}', \tag{3}$$

where K_η is the smoothed Biot-Savart kernel. In the case of sheets,

$$K_\eta = \begin{cases} K & |\,\mathbf{x}\,| \geq \eta \\ K\left(\frac{5}{2} - \frac{3}{2}\left(\frac{\mathbf{x}}{\eta}\right)^2\right)\frac{|\mathbf{x}|^3}{\eta^3} & |\,\mathbf{x}\,| < \eta \end{cases}$$

Figure 5. Filament convection and stretching.

where

$$K(x,y,z) = -\frac{1}{4\pi \mid \mathbf{x} \mid^3} \begin{pmatrix} 0 & -z & y \\ z & 0 & -x \\ -y & x & 0 \end{pmatrix}, \qquad (4)$$

and (3) may be simplified to

$$\int_{\Re^3} K_\eta(\mathbf{x}-\mathbf{x}')\boldsymbol{\Omega}(\mathbf{x}',t)d\mathbf{x}' = \mathbf{A}(\mathbf{x}-\mathbf{x}_i) \times \boldsymbol{\Omega}_i \qquad (5)$$

if it is assumed that the vorticity is constant over the sheets. In the far field for a rectangular Cartesian coordinate system

$$\begin{aligned} A_1 &= -\frac{h_j}{8\pi}\ln\left(\frac{r_{11}-Z_1}{r_{11}+Z_1}\frac{r_{21}+Z_1}{r_{21}-Z_1}\frac{r_{22}-Z_2}{r_{22}+Z_2}\frac{r_{12}+Z_2}{r_{12}-Z_2}\right) \\ A_2 &= -\frac{h_j}{4\pi}\left(\tan^{-1}\frac{X_2Z_2}{Yr_{22}} - \tan^{-1}\frac{X_2Z_1}{Yr_{21}} - \tan^{-1}\frac{X_1Z_2}{Yr_{12}} + \tan^{-1}\frac{X_1Z_1}{Yr_{11}}\right) \\ A_3 &= -\frac{h_j}{8\pi}\ln\left(\frac{r_{11}-X_1}{r_{11}+X_1}\frac{r_{21}+X_1}{r_{21}-X_1}\frac{r_{22}-X_2}{r_{22}+X_2}\frac{r_{12}+X_2}{r_{12}-X_2}\right) \end{aligned}$$

where $r_{mn}^2 = X_m^2 + Y^2 + Z_n^2$, $m,n = 1,2$, $X_1 = (x_i - x_j - l_j)/\eta$, $X_2 = (x_i - x_j + l_j)/\eta$, $Z_1 = (z_i - z_j - d_j)/\eta$, $Z_2 = (z_i - z_j + d_j)/\eta$, and $Y = (y_i - y_j)/\eta$. Within the distance η of the sheets, local versions of these formulas may be obtained analytically.

In the case of tubes, the common approximation

$$\int_{\Re^3} K_\eta(\mathbf{x}-\mathbf{x}')\boldsymbol{\Omega}(\mathbf{x}',t)d\mathbf{x}' = -\frac{\Gamma_i}{4\pi}\frac{\mathbf{r}_i \times \mathbf{s}_i}{\mid \mathbf{r}_i \mid^3}\,\phi(r/\sigma) \tag{6}$$

may be made where

$$\phi(r) = \left(1 - \left(1 - \frac{3}{2}r^3\right)\right) e^{-r^3}$$

is a high-order smoothing function. Additionally, when the velocity due to a sheet at x_i is sought at points $\mathbf{x}$ satisfying $\mid \mathbf{x} - \mathbf{x}_i \mid \;\gtrsim .5$, (5) and (6) are fully equivalent if the sheet is viewed as being a tube with

$$\mathbf{s} = \Delta s \frac{\boldsymbol{\Omega}}{\mid \boldsymbol{\Omega} \mid} \tag{7}$$

and

$$\Gamma \Delta s = \mid \boldsymbol{\Omega} \mid \; V_s \tag{8}$$

where V_s is the volume of the sheet and Δs is arbitrary. The relative simplicity of the tube formulas in comparison to (5) yields a significant gain in efficiency in computing these far field interactions.

In summary, the velocity is computed from

$$\mathbf{u}(\mathbf{x},t) = \sum_{i=1}^{N_{sheets}} \mathbf{A}(\mathbf{x}-\mathbf{x}_i) \times \boldsymbol{\Omega}_i - \frac{1}{4\pi}\sum_{i=1}^{N_{segments}} \frac{\mathbf{r}_i \times \mathbf{s}_i}{\mid \mathbf{r}_i \mid^3}\Gamma_i\,\phi(r/\sigma) \tag{9}$$

together with a potential flow that forces the non-penetration condition to be satisfied at solid boundaries. The latter field is readily computed from standard techniques using boundary source elements at the locations of the wall sheets (Hess & Smith, 1967; Hess, 1990).

5. Vortex Creation

The appearance of new vortex filaments in the calculation has both a numerical and physical connotation. In the case of the former, the natural movement of vorticity away from the sheet region must be provided for by the creation of new filaments. On the other hand, the wall region may be regarded as the foundry out of which new coherent vortical structures appear to both maintain the Reynolds stress and effect the energy cascade to dissipative scales. A successful vortex method must be sensitive to the physical process by which new structures appear, yet not so unconstrained as to attempt to capture all flow details, and thus make the numerical expense of the simulation prohibitive.

One approach toward accommodating these requirements is to limit the creation of new filaments to just those flow events where there is significant ejection of vorticity from the near wall region. This is generally consistent with the quasi-streamwise self-replication model shown in Fig. 6, which is a conceptualization of the ways in which a streamwise oriented vortex can affect the flow near a boundary. The large parent vortex here causes fluid upwelling that redirects spanwise vortex lines into the wall-normal direction, thus creating the pattern of + and − wall-normal vorticity shown in the figure. Forward shearing of the perturbed vorticity creates streamwise vorticity both counter-rotating and co-rotating to the parent. The latter may enhance the circulation of the parent to which it is adjacent. In view of the prodigious amounts of spanwise vorticity produced near the boundary, vortex reorientation such as described here, is likely to be a significant dynamical aspect of the flow.

The parent vortex also causes crossflow over the surface which generates counter-rotating streamwise vorticity at the wall. This collects on the upwash side of the vortex and may potentially be ejected away from the boundary. It may be speculated that this augments the creation of new vortices which are counter-rotating to the parent.

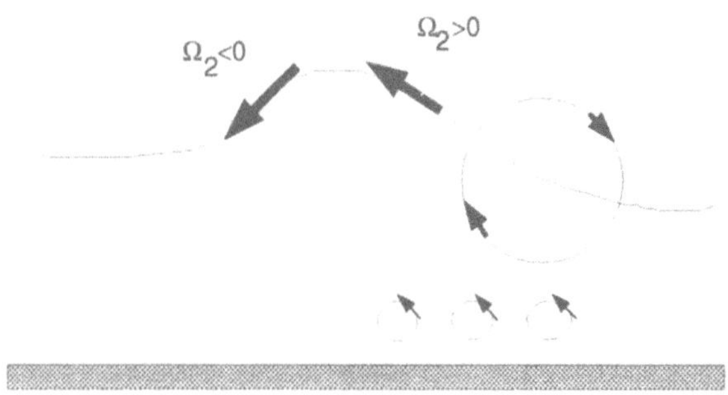

Figure 6. Conceptualization of the self-replication process. Flow is into the page.

During ejection motions, the top layer of sheets in the mesh covering a boundary convect beyond the sheet region. This provides a numerical means for implementing the physical vortex creation model: when sufficient vorticity is ejected outwards, it is formed into a new filament with strength derived from (7) - (8). Further efficiency is gained by grouping together new filaments created at the same location over several consecutive time steps. Discarding the remainder of the outwardly migrating vorticity may be justified if it contributes only to the random background vorticity and is not dynamically essential to the turbulent flow field. Depending on the threshold for creating new structures, this may or may not be satisfied in practice.

6. Vortex Destruction

Vortices in a turbulent flow environment undergo continuous and strong interactions with other nearby vortices. At the same time, the cumulative effect of viscous diffusion erodes the strength of vortex cores. The end result, presumably for all structures, is to render them unrecognizable as coherent entities, i.e. within a finite time, individual vortices become lost to the background field. Capturing this dynamical property of turbulence in the context of a vortex method poses a number of problems, since, once created, vortex filaments tend to persist indefinitely in one form or another unless consciously evicted from the flow. For well resolved grid based schemes, there is no analogous problem.

Several strategies may be employed to encourage physically appropriate vortex interactions in the simulation. As was pursued by Chorin (Chorin, 1993), two approximately anti-parallel vortex segments in close proximity can be canceled and the remainder reattached in mimicry of the vortex reconnection process. Together with hairpin removal, this helps prevent runaway growth in the filament population.

To fully eliminate vortices from the flow, two approaches have been followed here, since it is not yet clear which is the optimal one to take. In the first, vortices containing more than a fixed number of segments are deleted whenever they appear after filament subdivisions. Invariably, such vortices have been in the computation a long time. Moreover, they are usually distant from the boundaries and without evident coherency, i.e. they tend not to have a favored orientation.

An alternative approach for pruning aged vortices considers the accumulated effect of vortex stretching versus viscous spreading on each individual tube. The well known Gaussian core solution (Leonard, 1975) establishes an equilibrium radius for tubes, r_e, for which the effects of vortex stretching are in equilibrium with diffusion. For a tube whose length increases from Δs to $\Delta s'$ in time Δt it may be shown that

$$r_e = 2\sqrt{\frac{\Delta t}{R_e(\Delta s'/\Delta s - 1)}}. \tag{10}$$

Generally, the radius of a vortex is either above or below the r_e value associated with the local stretching experienced by the vortex at the given Reynolds number. If $r > r_e$, the vortex tends to get thinner, if $r < r_e$, the vortex tends to widen. While it may be possible to model this process (Rossi, 1995), it is unsuited to the present purposes, since it fails to provide for termination of old structures. The latter objective can be fulfilled, however, by providing for a loss of circulation from the tubes whenever $r < r_e$. The amount is determined from the extent to which vorticity would diffuse beyond r due to the imbalance of vortex stretching and diffusion. The necessary relation is

$$\Gamma^{n+1} = \Gamma^n \left(1 - \frac{4\Delta t}{R_e}\left(\frac{1}{r^2} - \frac{1}{r_e^2}\right)\right), \tag{11}$$

where Γ^n is the circulation at time n. If $r > r_e$, the circulation may be left the same, i.e.

$$\Gamma^{n+1} = \Gamma^n, \tag{12}$$

and if the tube faces a net contraction, i.e. $\Delta s' < \Delta s$, then the maximum circulation loss can be assessed, namely

$$\Gamma^{n+1} = \Gamma^n \left(1 - \frac{4\Delta t}{R_e}\frac{1}{r^2}\right). \tag{13}$$

For tubes composed of multiple segments, the average of Δs and $\Delta s'$ over all segments can be used.

As the scheme is presently constituted, all vortices have the identical radius. Since high shear near the walls leads to high stretching rates and hence small r_e, the influence of (11) or (13) in reducing Γ should be most pronounced away from the wall where vortices are less organized and experience slower rates of stretching. Whenever the circulation of a vortex drops below a threshold, it is dropped from the calculation, thus providing a second means for eliminating vortices.

Preliminary computations, such as those described below, suggest that the two vortex destruction mechanisms function acceptably well at a practical level, though it remains to fully explore the extent to which they provide the correct dynamical behavior. In particular, further study should reveal whether one or the other of the methods used separately is to be preferred.

7. Numerical Results

A channel flow has been simulated through several thousand time steps covering a time period over which a given vortex convects through the periodic flow domain approximately 15 times. The streamwise and spanwise period are taken to be 2.5 while the channel is of unit height. The Reynolds number is nominally set to $R_\tau = U_\tau H/\nu = 250$, where U_τ is the friction velocity and H is the channel width. In wall units, the channel is $625 \times 250 \times 625$. The initial grid of sheets on each wall is $16 \times 10 \times 16$ for a total of 5120. The sheets have dimension $0.135 \times 0.016 \times 0.135$ so that they cover only the regions $0 \leq y \leq .144$ and $.856 \leq y \leq 1$.

For the calculations reported here, the $O(N^2)$ velocity summation formula was parallelized on an IBM SP-2 computer. Though this approach

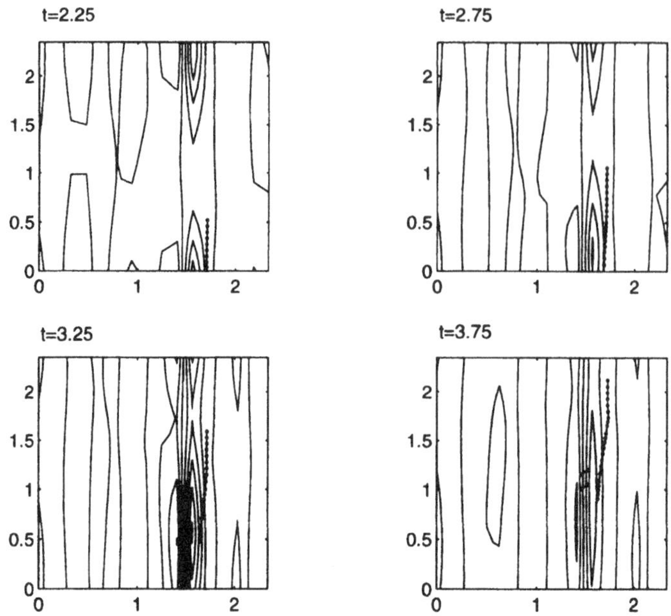

Figure 7. Simulated vortex self-replication.

yields speedup inversely proportional to the number of processors, the number of vortex segments that can be practically simulated is still limited to $\approx 20,000$, since periodicity requires that at least nine images of the computational field be included in finding the velocity at a given point.

The initial vorticity distribution in the sheets is taken to be that corresponding to the mean vorticity profile in a channel flow simulation, together with a single tilted quasi-streamwise vortex on each channel wall. Early times of the simulation show the breakdown of the flow into turbulence. At first, new vortices appear near the initial ones. Later, they appear at other spanwise positions. Eventually, large sections of the domain are covered by vortices. The particular calculation discussed here is not in a final equilibrium – partly due to previous adjustments of the numerical scheme which have not had sufficient time to equilibrate. Nonetheless, the evolution of the field is sufficiently well behaved to allow for some general conclusions to be drawn about the capabilities of the method in modeling turbulent flow.

Before considering the fully turbulent state, it is useful to consider the capabilities of the vortex self-replication model discussed in Section 5. For this purpose, the sequence of events transpiring from an initial state consist-

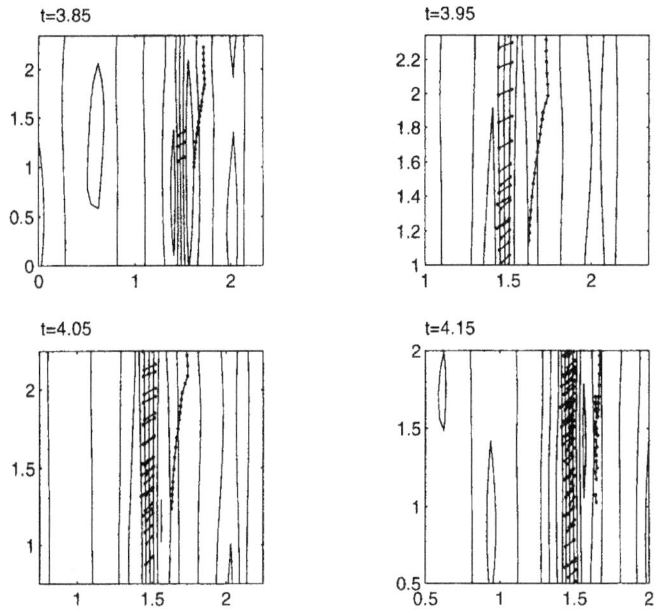

Figure 8. Simulated vortex self-replication.

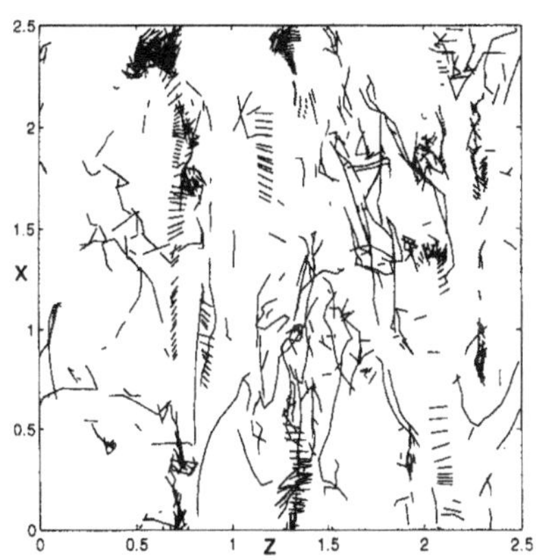

Figure 9. Near-wall computed vortex structure.

ing of a single vortex are shown in Figs. 7 and 8. The view is looking down on the lower channel wall. The contours of wall normal (Ω_2) vorticity at the highest grid plane, are also included. The sequence of images in Fig. 7 show

how the convecting quasi-streamwise structure (indicated by the connected sequence of dots) causes the development of significant + and − Ω_2 regions. Each time the amplitude of vorticity being ejected beyond the sheet region passes a threshold, the numerical scheme inserts a new vortex into the flow as seen at time $t = 3.75$. The new vortices are initially in the $\Omega_2 < 0$ region (see Fig. 6), but later, as in the last image in Fig. 8, some have appeared where $\Omega_2 > 0$ as well. Figure 8 shows that as the parent vortex generates a continuous series of new vortices, the newest ones – at least initially – are created downstream of the previous ones in coordination with the rapid convection and stretching of the downstream part of the parent. Since the new structures are on the ejection side of the parent, however, they move downstream at least as rapidly as the parent vortex. Because of periodicity, the trail of new vortices eventually extends along the whole streamwise period. After appearing in the flow, the new vortices undergo stretching and reorientation very much as conceptualized in Fig. 6. In particular, by $t = 4.05$ the first of the new vortices have been divided into two segments as a result of stretching. At $t = 4.15$, many others have followed suit. Viewed from the side, the new structures tend to be above the upstream ends of the parent, consistent with other observations (Bernard *et al.*, 1993; Brooke, 1993; Miyake *et al.*, 1997).

The vortex creation mechanisms viewed in isolation in Figs. 7 and 8 are at work in the long time calculations, as well, as shown in Fig. 9. This is a view of the lower channel wall for $0 \leq y^+ \leq 30$. A number of long quasi-streamwise vortices are evident: these are purely a product of the calculation over many hundreds of time steps, i.e. the initial quasi-streamwise vortices used in perturbing the flow have long since aged and been deleted from the flow.

The groups of short parallel aligned vortices in Fig. 9 at several spatially intermittent regions are the newest vortices. Their character is very similar to that observed in Fig. 8 under more controlled circumstances. It may also be observed in Fig. 9 that the new vortices generally appear in close proximity to quasi-streamwise vortices.

Additional evidence for the physicality of the computed vortex structure is found in a plot of the streamwise velocity fluctuation on a plane close to the boundary given in Fig. 10. Two regions of low u, elongated in the streamwise direction are evident. Such streaky structure is a hallmark of turbulent boundary layers.

A view of the complete vortex element field from the side is given in Fig. 11 and from an end-on perspective in Fig. 12. The former shows the preponderance of quasi-streamwise vortices in the high shear region adjacent to the boundary. Less organization of this kind is evident further from the wall. It is also interesting to observe, especially in the end-on perspec-

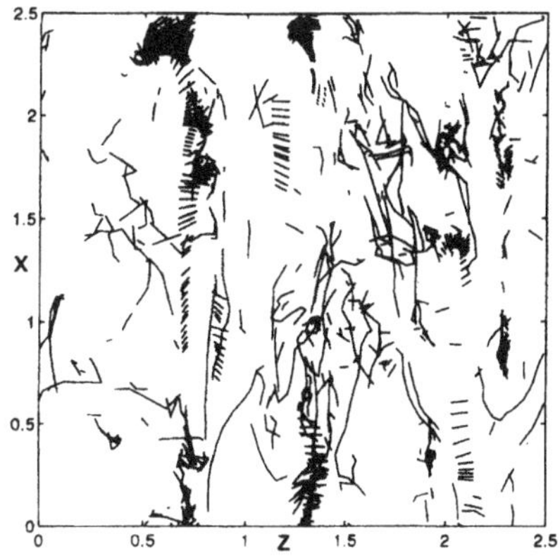

Figure 10. Contours of streamwise velocity fluctuation showing streaky structure.

tive, that the filaments often organize into larger structures. Some of these appear to have mushroom shapes reminiscent of those seen in smoke visualizations (Head & Bandyopadhyay, 1981). It also may be noted that the orientation of vortices in the central region tends to be more isotropic than near the wall, as is clear in Fig. 13 containing a plan view of vortices near the centerline.

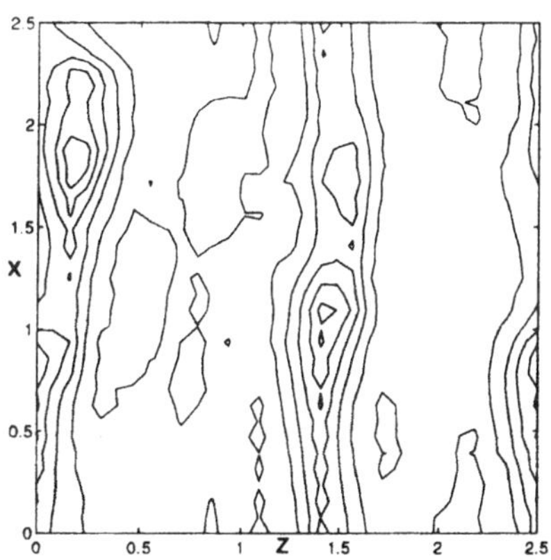

Figure 11. Side view of vortices in channel.

A summary of the important velocity statistics corresponding to the simulated channel flow is given in Fig. 14. Taken together, the computed mean velocity field and Reynolds shear stress plots in Figs. 14(a) and (b) show that the physical character of the computed field is distinctly turbulent in nature. In particular, the Reynolds shear stress takes on a significant correlation of the correct sign on each wall; a trend which is unlikely to be explained by anything other than a largely accurate rendition of the physics of vortex self-replication. The quantitative disagreement of $\overline{uv}$ with the DNS solutions is attributable to the coarseness of the sheet representation. In fact, for large sheets, the wall normal velocity calculated from (5) can have significant error unless neighboring sheets have nearly the same vorticity, an unlikely event in turbulent flow. The potential for error, however, drops off very rapidly with refinement of the sheet mesh.

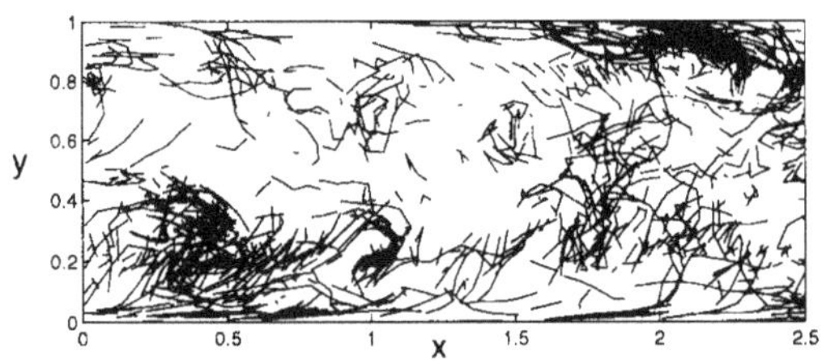

Figure 12. End-on view of vortices in channel.

Figures 14(a) and (b) may give the impression that two independent turbulent boundary layers have formed on the top and bottom channel walls, and that given additional time, these may join to form a legitimate channel flow. While this might partially be the case, a more likely explanation for the observed trends is an underrepresentation of the spanwise vorticity outside the wall regions. In fact, even though such vorticity is of low magnitude, it occurs over a relatively large region and can have precisely the effect on mean velocity seen in Fig. 14(a). Subsequent calculations with greater resolution may have a positive effect in alleviating this problem.

The three normal Reynolds stress distributions are given in Figs. 14(c) - (e). There is good qualitative similarity with DNS data in the case of $\overline{u^2}$, and away from the boundary for $\overline{w^2}$. In most aspects the quantitative agreement is quite reasonable with the exception that $\overline{v^2}$ and $\overline{w^2}$ are significantly overpredicted near the wall. The coarse sheet structure appears to be directly responsible for this behavior in $\overline{v^2}$, as discussed above. Moreover, the large surface panels used in forcing the non-penetration boundary

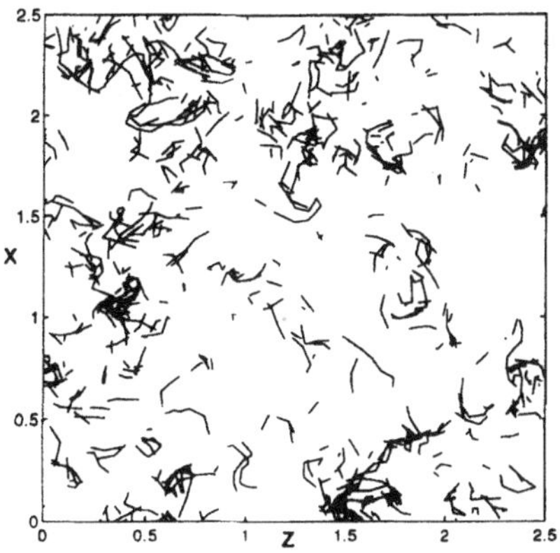

Figure 13. Plan view of vortices near channel center.

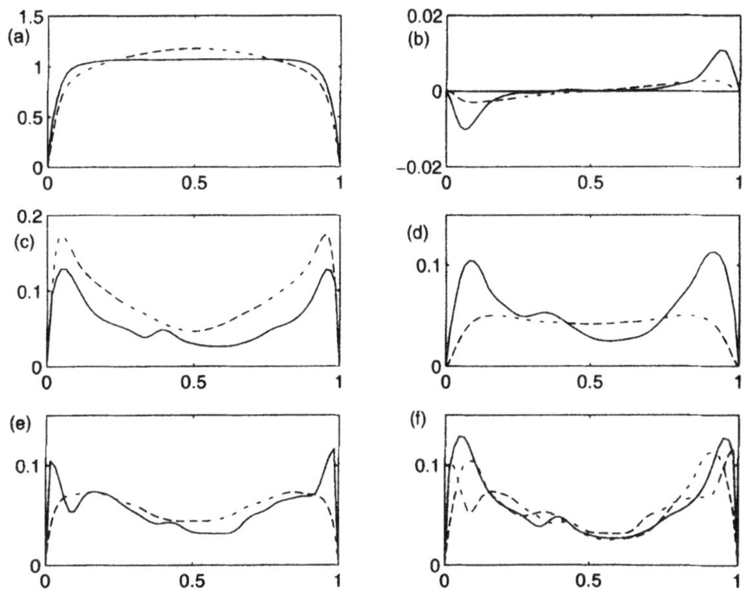

Figure 14. Velocity statistics (in (a) - (e), —, denotes the vortex method; – – –, denotes DNS): (a) $\overline{U}$; (b) $\overline{uv}$; (c), $\sqrt{\overline{u^2}}$; (d), $\sqrt{\overline{v^2}}$; (e), $\sqrt{\overline{w^2}}$; (f) —, $\sqrt{\overline{u^2}}$; – – –, $\sqrt{\overline{v^2}}$; – · –, $\sqrt{\overline{w^2}}$.

condition through a distribution of surface sources, will have a similar influence in exaggerating the prediction of $\overline{w^2}$. Finally, Fig. 14(f) establishes, in conformity with Fig. 13, that relative isotropy prevails near the channel centerline.

Improvements to the mean velocity and Reynolds stress predictions are likely to follow from several refinements to the approach. In particular, greater resolution should be a factor in eliminating the Reynolds stress overprediction near the wall and in improving the $\overline{U}$ field. A current round of computations employing an $O(N)$ velocity evaluation scheme (so N can be increased), together with several improvements to the basic algorithm, aim to show that all aspects of the turbulent physics can be successfully modeled, beyond those evident thus far.

8. Complex Flows

Application of this vortex method to complex external flows is straightforward: the most significant change being to generalize the steps taken in advancing the vorticity sheets according to (1). For example, a boundary-fitted coordinate system is used to impose a sheet mesh in a thin layer near the surface and each of the steps used in advancing the vortex sheets, namely, convection, interpolation and finite difference approximation of stretching and wall normal diffusion, are generalized to these coordinates. Furthermore, the velocity due to each individual sheet is determined from the previous formulas by appropriate coordinate rotation, and the non-penetration condition modeled using source boundary sheets.

As pointed out previously, the absence of periodicity in the flow direction suggests that external flows should depend less on accurately portraying vortex destruction than channel flow. To examine these issues, the present method has been applied to the flow past a prolate spheroid, with and without angle of attack. The considerable experimental data available in this case (Chesnakas & Simpson, 1994), including the effect of angle of attack on 3D flow separation, provides a stringent test of the algorithm in a practical setting.

While this work is just in its infancy, Fig. 15 gives a view of the flow after approximately 100 time steps after impulsive motion of a 8:1 prolate spheroid at 10° angle of attack. A layer of sheets (not visible) ten deep covers the surface and lies between the new vortex filaments appearing in the calculation and the body surface. The vortices in Fig. 15 cover a large part of the spheroid and are seen to be in the early stages of shedding off the rear into the wake. Animations of this flow may be viewed at the website http://www.krispintech.com.

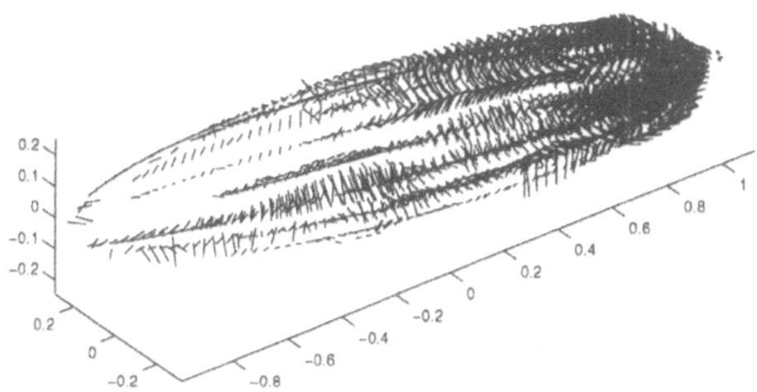

Figure 15. Start up of prolate spheroid flow.

9. Conclusions

A three-dimensional vortex sheet and filament method has been constructed for simulating turbulent flows. New filaments are created from sheets during ejection events initiated by parent vortices. The range of resolved scales is limited using Chorin's hairpin removal algorithm as a physically based subgrid model. Numerical studies of channel and prolate spheroid flows have been initiated to establish the capability of the scheme in predicting turbulent flow.

Long time calculations of channel flow show that the method successfully captures some of its most significant dynamical properties. Among these are a physical Reynolds shear stress and the quasi-streamwise vortex self-replication process. With provision for better resolution of vortices and some adjustments to the vortex destruction model, better quantitative predictions of mean velocity and Reynolds stress should follow. On the whole, the outlook is bright that a vortex method along the lines developed here will turn into an efficient and reliable means of predicting turbulent flows.

10. Acknowledgement

This research was supported in part by Krispin Technologies, Inc. DOE SBIR Phase I grant DE-FG02-97ER82413. Computations were performed in part on the Maui HPCC IBM-SP2 computer.

References

Bernard, P. S., 1995. "A deterministic vortex sheet method for boundary layer flow," *J. Comput. Phys.* **117**, pp. 132-145.

Bernard, P. S., Thomas, J. M., and Handler, R. A., 1993. "Vortex dynamics and the production of Reynolds stress," *J. Fluid Mech.* **253**, pp. 385-419.

Brooke, J. W. and Hanratty, T. J., 1993. "Origin of turbulence-producing eddies in a channel flow," *Phys. Fluids A* **5**, pp. 1011-1022.

Chesnakas, C. J. and Simpson, R. L., 1994. "Full three-dimensional measurements of the cross-flow separation region of a 6:1 prolate spheroid," *Exp. Fluids* **17**, pp. 68-74.

Chorin, A. J., 1973. "Numerical study of slightly viscous flow," *J. Fluid Mech.* **57**, pp. 785-796.

Chorin, A. J., 1982. "The evolution of a turbulent vortex," *Comm. Math. Phys.* **83**, pp. 517-535.

Chorin, A. J., 1993. "Hairpin removal in vortex interactions II," *J. Comput. Phys.* **107**, pp. 1-9.

Chorin, A. J., 1994. *Vorticity and Turbulence*, Springer-Verlag, New York.

Cottet, G. H. and Mas-Gallic, S., 1990. "A particle method to solve the Navier-Stokes system," *Numer. Math.* **57**, pp. 805-827.

Fishelov, D., 1990. "A new vortex scheme for viscous flows," *J. Comput. Phys.* **86**, pp. 211-224.

Greengard, L. and Rokhlin, V., 1987. "A fast algorithm for particle simulations," *J. Comput. Phys.* **73**, pp. 325-348.

Head, M. R. and Bandyopadhyay, P., 1981. "New aspects of turbulent boundary layer structure," *J. Fluid Mech.* **107**, pp. 297-338.

Hess, J. L. and Smith, A. M. O., 1967. "Calculation of potential flow about arbitrary bodies," *Prog. Aero. Sci.* **8**, pp. 1-138.

Hess, J. L., 1990. "Panel methods in computational fluid dynamics," *Ann. Rev. Fluid Mech.* **22**, pp. 255-274.

Kim, J., Moin, P., and Moser, R. D., 1987. "Turbulence statistics in fully-developed channel flow at low Reynolds number," *J. Fluid Mech.* **177**, pp. 133-166.

Leonard, A., 1975. "Numerical simulation of interacting, three-dimensional vortex filaments," *Lec. Notes in Phys.* **35**, pp. 245-250.

Miyake, Y., Ushiro, R., and Morikawa, T., 1997. "The regeneration of quasi-streamwise vortices in the near-wall region," *JSME Int'l J. Ser. B - Fluids Thermal Engrg.* **40**, pp. 257-264.

Ogami, Y. and Akamatsu, T., 1990. "Viscous flow simulation using the discrete vortex model – the diffusion velocity method," *Computers and Fluids* **19**, pp. 433-441.

Puckett, E. G., 1993. "Vortex methods: an introduction and survey of selected research topics," in *Incompressible computational fluid dynamics: Trends and advances*, edited by M. D. Gunzburger and R. A. Nicolaides, Cambridge University Press, Cambridge, pp. 335-407.

Rossi, L. F., 1995. "Resurrecting core spreading vortex methods: a new scheme that is both deterministic and convergent," *SIAM J. Sci. Comp.* **17**, pp. 370-388.

Russo, G., 1993. "A deterministic vortex method for the Navier-Stokes equations," *J. Comp. Phys.* **108**, pp. 84-94.

Shankar, S. and van Dommelen, L., 1996. "A new diffusion procedure for vortex methods," *J. Comput. Phys.* **127**, pp. 88-109.

Speziale, C. G., 1991. "Analytical methods for the development of Reynolds-stress closures in turbulence," *Annu. Rev. Fluid Mech.* **23**, pp. 107-157.

Strickland, J. H., 1994. "A prediction method for unsteady axisymmetric flow over parachutes," *J. Aircraft* **31**, pp. 637-643.

Strickland, J. H., 1997. Maven simulation videotape. Sandia National Laboratories Video Services.

Strickland, J. H. and Amos, D. E., 1992. "A fast solver for systems of axisymmetric ring

vortices," *AIAA J.* **30**, pp. 737-746.
Strickland, J. and Bray, R. S., 1994. "A three-dimensional fast solver for arbitrary vorton distributions," Report SAND93-1641, Sandia National Laboratory.
Winckelmans, G. S., Salmon, J. K., Leonard, A., and Warren, M. S., 1995. "Three-dimensional vortex particle and panel methods: fast tree-code solvers with active error control for arbitrary distributions/geometries," Proc. Forum on Vortex Methods for Engineering Applications, Albuquerque, NM, pp. 23-43.
Winckelmans, G. S., Salmon, J. K., Warren, M. S., Leonard, A., and Jodoin, B., 1996. "Application of fast parallel and sequential tree codes to computing three-dimensional flows with the vortex element and boundary element methods," Vortex Flows and Related Numerical Methods II, ESAIM: Proceedings, Vol. 1, pp. 225-240, http://www.emath.fr/proc/Vol.1/.

TWO-POINT CLOSURES AND STATISTICAL EQUILIBRIUM

Implications for Engineering Turbulence Models

TIMOTHY T. CLARK
Theoretical Division, Group T-3
Los Alamos National Laboratory
Los Alamos, New Mexico

Abstract. A two-point, or spectral, turbulence transport model describes the evolution of the two-point velocity covariance tensor, or its Fourier transform, the spectral tensor. Such a model describes the turbulent dynamics as functions of length-scale or wave-number. This permits a more general description of turbulence than is available with a one-point closure. This greater generality is useful in understanding the behavior of turbulent flows that are undergoing rapid transients and that are therefore not in "equilibrium". If the turbulent flow is in an "equilbrium" environment wherein the mean forces on the flow are relatively constant in time, the turbulent spectra tend toward self-similar forms. When applied to a specific spectral model (Besnard *et al.*, 1996) (Clark and Zemach, 1995), these self-similar forms may be exploited to reduce the model to the more familiar $R_{ij} - \epsilon$ and $K - \epsilon$ models. These one-point models have coefficients that are functions of the spectral distributions. We discuss the limits of validity of the two-point descriptions as well as the consequences of the equilibrium assumptions embedded in the one-point variants.

1. Introduction

For an incompressible flow of a fluid with constant viscosity ν and constant density ρ the pressure is considered to act through the flow domain in such a manner is to force the velocity to be divergence-free at all points. The pressure fluctuations at a point in space thus influence the entire flow-domain to some (perhaps infintesimal) extent. It is this "nonlocality" of the pressure effects that is a chief motivation for employing two-point closures. Consider the evolution equation for a velocity fluctuation in a statistically

M. D. Salas et al. (eds.), Modeling Complex Turbulent Flows, 183–202.

homogeneous turbulent flow. Using the usual Reynolds decomposition; $u_i = U_i + u'_i$ and $p = P + p'$, where the fluctuation is denoted by the prime, the equation is

$$\begin{aligned} &\frac{\partial u'_i(\mathbf{x},t)}{\partial t} + \frac{\partial}{\partial x_n}\left\{u'_i(\mathbf{x},t)U_n(t) + u'_n(\mathbf{x},t)U_i(t)\right\} \\ + &\frac{\partial}{\partial x_n}\left\{u'_i(\mathbf{x},t)u'_n(\mathbf{x},t) - R_{in}(t)\right\} - \nu\frac{\partial^2 u'_i(\mathbf{x},t)}{\partial x_n \partial x_n} \\ = &-\frac{\partial}{\partial x_i}\left\{\iiint_\Omega G(\mathbf{x},\mathbf{x}')\frac{\partial^2 \Gamma_{nm}(\mathbf{x}',t)}{\partial x'_n \partial x'_m}d\mathbf{x}'\right\} \end{aligned} \tag{1}$$

where Γ_{ij} is a symmetric tensor

$$\Gamma_{ij}(\mathbf{x},t) = \left[U_i(t)u'_j(\mathbf{x},t) + U_j(t)u'_i(\mathbf{x},t) + u'_i(\mathbf{x},t)u'_j(\mathbf{x},t) - R_{ij}(t)\right] \tag{2}$$

and $G(\mathbf{x},\mathbf{x}')$ is Green's function for an infinite domain;

$$G(\mathbf{x},\mathbf{x}') = \frac{1}{4\pi|\mathbf{x}-\mathbf{x}'|}. \tag{3}$$

The key feature to note is that the Navier-Stokes equation for the velocity fluctuation is in fact a two-point equation; evolution of the velocity at a point $\mathbf{x}$ is influenced instantaneously by the velocity at any point $\mathbf{x}'$. Two-point models attempt to describe the evolution of some features of the two-point correlations. Fourier techniques provide a computationally efficient representation of most two-point closures in a homogeneous enviroment since the solution of the pressure integral becomes an algebraic equation. Two-point closures in a Fourier representation are generally referred to as "spectral" models.

Since one-point models describe the evolution of only the single-point statistics, such models contain implicit assumptions regarding the nature of the two-point correlations arising from the pressure effects. Such assumptions can be thought of as implicit self-similarity, or "statistical equilibrium," assumptions and are embedded in a description of turbulence which describes a relatively small number of statistical "parameters", e.g., turbulent kinetic energy, and a length-scale, or a turbulent kinetic energy dissipation rate. When the actual statistics of a turbulent flow approach a statistical equilibrium (i.e., a statistical self-similarity) the two-point model can be reduced to a one-point model, thus illuminating the type and limitations of the self-similarity that is implicit in the one-point model.

2. Derivation of One-Point Models from the LWN Spectral Model

For the derivation of a one-point model from a two-point model, we shall use the "Local Wave Number" (LWN) model of Besnard *et al.* (1996) (see also (Clark and Zemach, 1995)) as the starting point. Due to the relative simplicity of the LWN representation of the turbulence cascade processes, this model is easy to study and analyze. In addition, it has been shown to produce results that are in good agreement with both theory and experiments (Clark and Zemach, 1995). However, it must be noted that the LWN model does not describe the evolution of the spectral tensor as a function of a full three-dimensional wave vector space. Instead, for ease of computation an implicit assumption of "angular" self-similarity was employed to reduce the description to wave number space. Thus the LWN model does not fully express the full dynamics of departures from self-similarity. However, as an illustrative tool, the LWN model can give insight into length-scale/wave-number departures from self-similarity and their implications for one-point models.

The particular form of the LWN model used for this present study differs slightly form the form original presented by Besnard *et al.* and by Clark and Zemach. Specifically, the time-scale used for the turbulent cascade and the "isotropization" terms have been modified. The modification leads to improved behavior in the dissipation range of the spectrum, but would have only minimal effects on the results presented here because the current circumstance is assumed to be at asymptotically high Reynolds number ($Re \to \infty$). For details regarding the model derivation, determination of coefficient values and model validity the reader is directed to Besnard *et al.* (1996), Clark and Zemach (1995) and Clark (1992). The form used herein (with the arguments (k, t) omitted for convenience) for homogeneous turbulence is

$$
\begin{aligned}
\frac{\partial E_{ij}}{\partial t} &= -2\nu k^2 E_{ij} - \left\{\left[S_{in}+W_{in}\right]E_{nj} + \left[S_{jn}+W_{jn}\right]E_{ni}\right\} \qquad (4)\\
&+ c_b\left\{\left[S_{in}+W_{in}\right]E_{nj} + \left[S_{jn}+W_{jn}\right]E_{ni} - \frac{2}{3}\delta_{ij}S_{nm}E_{mn}\right\}\\
&+ 8\left(c_b-\frac{3}{4}\right)\left\{\left[S_{in}-W_{in}\right]E_{nj} + \left[S_{jn}-W_{jn}\right]E_{ni} - \frac{2}{3}\delta_{ij}S_{nm}E_{mn}\right\}\\
&+ 6\left(\frac{11}{15}-c_b\right)S_{ij}E_{nn}\\
&+ 3\left(\frac{2}{3}-c_b\right)\left\{S_{in}\frac{\partial kE_{nj}}{\partial k} + S_{jn}\frac{\partial kE_{ni}}{\partial k} - \frac{2}{3}\delta_{ij}S_{nm}\frac{\partial kE_{nm}}{\partial k}\right\}\\
&+ 2\left(c_b-\frac{7}{10}\right)S_{ij}\frac{\partial kE_{nn}}{\partial k} + \frac{7}{2}\left(c_b-\frac{16}{21}\right)\delta_{ij}S_{nm}\frac{\partial kE_{nm}}{\partial k}\\
&+ \frac{\partial}{\partial k}\left\{\frac{k}{\tau(k)}\left[k\frac{\partial E_{ij}}{\partial k} - 2E_{ij}\right]\right\} + \frac{26}{9}\frac{1}{\tau(k)}\left\{\frac{1}{3}\delta_{ij}E_{nn} - E_{ij}\right\},
\end{aligned}
$$

where

$$S_{ij} = \frac{1}{2}\left\{\frac{\partial U_i}{\partial x_j} + \frac{\partial U_j}{\partial x_i}\right\}, \tag{5}$$

$$W_{ij} = \frac{1}{2}\left\{\frac{\partial U_i}{\partial x_j} - \frac{\partial U_j}{\partial x_i}\right\}, \tag{6}$$

and

$$\tau(k) = \frac{11}{2\sqrt{3}} C_K^{3/2} \left\{\int_0^k q^2 E_{nn}(q) dq\right\}^{-1/2}. \tag{7}$$

(Note that repeated Latin indices are summed and repeated Greek indices are not summed.) The coefficient C_K is the Kolmogorov constant and is taken to be equal to 1.5 for the present computations. Realizability considerations relate the coefficient c_b to the scaling of the spectral tensor in the limit $k \to 0$. If the spectral tensor scales as

$$\lim_{k \to 0} E_{ij}(k,t) = \mu_{ij}(t)\, k^n, \tag{8}$$

then

$$c_b \leq \frac{8n}{11n-1}. \tag{9}$$

Comparisons with experimental data indicate that the chosen value of c_b should be as large as possible, i.e., that the above relation should be treated as an equality (Clark and Zemach, 1995). The single-point Reynolds stress is related to the spectral tensor by

$$R_{ij}(t) = \left\langle u_i' u_j' \right\rangle = 2 \int_0^\infty E_{ij}(q,t) dq. \tag{10}$$

It has been speculated that in the limit of large time, the spectral tensor of a turbulence subjected to time-independent homogeneous mean-flow velocity gradients will asymptotically approach a self-similar form. For the LWN model this tendency is manifested in the computations (Clark and Zemach, 1995) (Clark, 1992), as will be shown later. The self-similar form produced by the LWN model in this regime is

$$E_{ij}(k,t) = K(t)\, L(t)\, f_{ij}(\xi), \tag{11}$$

where

$$\xi = kL(t), \tag{12}$$

subject to the scaling

$$\int_0^\infty f_{nn}(\xi)d\xi = 1. \tag{13}$$

Letting a tilde over a tensor denote the deviatoric part of the tensor, (e.g., $\tilde{f}_{ij} = f_{ij} - \delta_{ij} f_{nn}/3$), then the self-similar tensor f_{ij} is related to the anisotropy tensor b_{ij} by

$$\int_0^\infty \tilde{f}_{ij}(\xi)d\xi = b_{ij}, \tag{14}$$

where the anisotropy tensor is defined as

$$b_{ij} = \frac{R_{ij} - \frac{1}{3}\delta_{ij}R_{nn}}{R_{nn}} = \frac{\tilde{R}_{ij}}{R_{nn}}. \tag{15}$$

Substituting the self-similar form into the LWN model and integrating over all wavenumbers gives the K-equation;

$$\frac{\partial K}{\partial t} = -2\frac{\partial U_n}{\partial x_m} b_{nm} - \epsilon. \tag{16}$$

Note that at high turbulent Reynolds number the dissipation rate of turbulent kinetic energy, ϵ, is now identified as the flux of energy to infinity. For the LWN model the spectral energy flux is

$$F_f(k) = -\left\{\frac{k}{\tau(k)}\left[k\frac{\partial E_{nn}(k)}{\partial k} - 2E_{nn}(k)\right]\right\}, \tag{17}$$

and the turbulent energy dissipation rate is therefore

$$\epsilon = \lim_{k\to\infty} F_f(k) = \frac{\sqrt{3}}{2}\frac{2\sqrt{3}}{11}C_K^{-3/2}\frac{11}{3}E_\infty^{3/2} = \left(\frac{E_\infty}{C_K}\right)^{3/2}, \tag{18}$$

where, in the fully self-similar regime ($Re \to \infty$),

$$\lim_{k\to\infty} E_{nn}(k) = E_\infty k^{-5/3}. \tag{19}$$

From purely dimensional considerations the turbulent energy dissipation rate can be related to the turbulent kinetic energy and the turbulent length-scale by a constant of proportionality, α as

$$\epsilon(t) = \alpha\frac{K^{3/2}(t)}{L(t)}. \tag{20}$$

In the regime of asymptotically large Reynolds Number ($Re \rightarrow \infty$) the turbulent energy dissipation rate ϵ can be identified as the flux, or cascade of energy to infinite wavenumber. Recall that the length-scale L is the inverse of the wavenumber where the energy spectrum is maximum, and the turbulent kinetic energy is the integral over all wavenumbers of the energy spectrum. Thus since all the variables except α in the above equation can be specifically computed from the model without reference to the above equation, the value of α can be computed unambiguously. Note that Kolmogorov's scaling in the inertial range;

$$E_{nn}(k) = C_K \epsilon^{2/3} k^{-5/3} \tag{21}$$

implies that

$$f_{nn}(\xi) = C_K \left[\frac{L\epsilon}{K^{3/2}}\right]^{2/3} \xi^{-5/3} = C_K \alpha^{2/3} \xi^{-5/3} \tag{22}$$

in the inertial range. Letting

$$\lim_{\xi \to \infty} f_{nn}(\xi) = f_\infty \xi^{-5/3} \tag{23}$$

establishes the identity

$$f_\infty = C_K \alpha^{2/3}. \tag{24}$$

The scaled self-similar flux is

$$F_f(\xi) = -\left\{\frac{\xi}{\tau_{ss}(\xi)}\left[\xi \frac{\partial f_{nn}(\xi)}{\partial \xi} - 2 f_{nn}(\xi)\right]\right\}, \tag{25}$$

where

$$\tau_{ss}(\xi) = \frac{11}{2\sqrt{3}} C_K^{3/2} \left\{\int_0^\xi \varsigma^2 f_{nn}(\varsigma) d\varsigma\right\}^{-1/2}. \tag{26}$$

The scaled self-similar flux to infinity is equal to the parameter α via equation (24);

$$\lim_{\xi \to \infty} F_f(\xi) = \left(\frac{f_\infty}{C_K}\right)^{3/2} = \alpha \tag{27}$$

Hence α is the scaled self-similar energy flux, or "scaled dissipation rate."

To derive the ϵ-equation, we shall follow the procedure presented by Besnard *et al.* (1996) and begin by rewriting equation (20) in the following equivalent form;

$$\epsilon = \alpha K^{\frac{3m-2}{2m}} \left[\frac{K}{L^m}\right]^{\frac{1}{m}}. \tag{28}$$

Taking the logarithmic derivative of the above equation gives

$$\frac{\partial \epsilon}{\partial t} = \frac{\epsilon}{K}\left\{\left(\frac{3m-2}{2m}\right)\frac{\partial K}{\partial t} + \left(\frac{L^m}{m}\right)\frac{\partial}{\partial t}\left[\frac{K}{L^m}\right]\right\}. \tag{29}$$

Replacing $L(t)$ in this equation with $\alpha K(t)/\epsilon(t)$, the following form of the ϵ-equation can be obtained;

$$\frac{\partial \epsilon}{\partial t} = -\left[g_{\epsilon 0}(m) b_{nm} + g_{\epsilon 1}(m)\phi_{nm}(m)\right]\frac{\partial U_n}{\partial x_m}\epsilon - g_{\epsilon 2}(m)\frac{\epsilon^2}{K}, \tag{30}$$

where

$$g_{\epsilon 0}(m) = \frac{3m-2}{m}, \tag{31}$$

$$g_{\epsilon 1}(m) = \frac{21}{2}\left(c_b - \frac{16}{21}\right) + 3 - g_{\epsilon 0}(m), \tag{32}$$

$$g_{\epsilon 2}(m) = \frac{g_{\epsilon 0}(m)}{2} + \frac{1}{\alpha m}\left[\frac{J(m)}{I_{nn}(m)}\right], \tag{33}$$

$$I_{ij}(m) = \int_0^\infty \xi^m f_{ij}(\xi)\, d\xi, \tag{34}$$

$$\phi_{ij}(m) = \frac{\tilde{I}_{ij}(m)}{I_{nn}(m)}, \tag{35}$$

and

$$J(m) = \int_0^\infty \xi^m \frac{dF_f(\xi)}{d\xi} d\xi. \tag{36}$$

Again the tilde denotes a deviatoric tensor, e.g. $\tilde{I}_{ij} = I_{ij} - \delta_{ij} I_{nn}/3$. If the spectral tensor scales as k^n at $k \to 0$ and k^p, $(p \leq -5/3)$ in the limit $k \to \infty$, then the integrals for I_{ij} are bounded for $-(n+1) < m < 2/3$. From a computational aspect, at self-similarity the energy spectrum asymptotically approaches $k^{-5/3}$ from k^p,$(p > -5/3)$ as $k \to \infty$ and for practical numerical calculations of the $J(m)$-moment the value of m is restricted to $-(n+1) < m < 0$. Note that $m = 0$ leads to a singularity in the coefficients and is therefore excluded from consideration. Note that $\phi_{ij}(0) = \tilde{I}_{ij}(0) = b_{ij}$ since $I_{nn}(0) = 1$.

An algebraic equation for the anisotropy tensor can also be derived from the LWN model in a straightforward manner. The result is an algebraic relation (recall that in the self-similar limit the anisotropy tensor is time-independent);

$$\begin{aligned}\frac{\partial b_{ij}}{\partial t} &= (c_b-1)\left\{\frac{\partial U_i}{\partial x_n}b_{nj}+\frac{\partial U_j}{\partial x_n}b_{ni}-\frac{2}{3}\delta_{ij}\frac{\partial U_m}{\partial x_n}b_{mn}\right\} \\ &+ (8c_b-6)\left\{\frac{\partial U_n}{\partial x_i}b_{nj}+\frac{\partial U_n}{\partial x_j}b_{ni}-\frac{2}{3}\delta_{ij}\frac{\partial U_m}{\partial x_n}b_{mn}\right\} \\ &+ \left(\frac{11}{5}-3c_b\right)\left\{\frac{\partial U_i}{\partial x_j}+\frac{\partial U_j}{\partial x_i}\right\}+2\frac{\partial U_n}{\partial x_m}b_{nm}b_{ij} \\ &+ \frac{\epsilon}{K}\left\{b_{ij}-\frac{c_m}{\alpha}\beta_{ij}\right\}=0.\end{aligned} \tag{37}$$

where

$$c_m = \frac{26}{9}\frac{2\sqrt{3}}{11}C_K^{-3/2}, \tag{38}$$

and

$$\beta_{ij} = \int_0^\infty \left[\int_0^\xi \varsigma^2 f_{nn}(\varsigma)\,d\varsigma\right]^{1/2} \tilde{f}_{ij}(\xi)\,d\xi. \tag{39}$$

Equation (38) can not be directly inverted since it involves integral moments of the $\tilde{f}_{ij}$ tensor (the β_{ij} terms). To invert this equation, an approximation for β_{ij}, in terms of b_{ij} and (perhaps also in terms of the velocity gradients) must be assumed.

A second self-similar form associated with freely decaying anisotropic turbulence has also been identified. This form is a restriction of the previous form in that the individual deviatoric self-similar tensor components now have the same spectral form;

$$\tilde{f}_{ij}(\xi) = b_{ij}\tilde{f}_D(\xi), \tag{40}$$

subject to the constraint that

$$\int_0^\infty \tilde{f}_D(\xi)d\xi = 1. \tag{41}$$

This is similar to the form that Besnard *et al.* used to construct a $K-\epsilon$-model. However, for a more rigorous derivation using this restricted self-similar form, it must be noted that the mean-flow velocity gradients are zero in the limit in which this form applies. With this caveat, the derivation follows precisely the one outlined above, and the β-tensor and I-tensor definitions are now modified. For the β-tensor,

$$\beta_{ij} = \beta_D b_{ij}, \tag{42}$$

and in this case

$$\beta_D = \int_0^\infty \left[\int_0^\xi \varsigma^2 f_{nn}(\varsigma)\, d\varsigma\right]^{1/2} \tilde{f}_D(\xi)\, d\xi. \tag{43}$$

Likewise for the I-tensor;

$$\tilde{I}_{ij}(m) = \tilde{I}_D(m) b_{ij}, \tag{44}$$

where

$$\tilde{I}_D(m) = \int_0^\infty \xi^m \tilde{f}_D(\xi)\, d\xi. \tag{45}$$

Note that in this regime the b_{ij}-equation reduces to

$$c_m \frac{\beta_D}{\alpha} = 1. \tag{46}$$

The term on the left side of this equation can be identified with the so-called Rotta constant for the return to isotropy (Clark, 1992), and that a value of unity corresponds to no return to isotropy. This is consistent with the discussion of Clark and Zemach (1995). The asymptotic approach of $c_m\beta_D/\alpha$ to a value of unity for the decay of an anisotropic turbulence as computed from the LWN model will be demonstrated in Section 3.2.

Note that in the self-similar limit a specific mean-flow velocity gradient tensor will produce a specific anisotropy. This assertion is difficult to prove since a direct inversion of the general b_{ij}-equation is not possible due to the presence of the β_{ij} term. It is also apparent that a specific value of b_{ij} (i.e., $\tilde{I}_{ij}(0)$) does not uniquely determine β_{ij} or other moments of b_{ij} (i.e., $\tilde{I}_{ij}(m), m \neq 0$). Hence the total dynamical description of the spectral distributions is not embodied in a finite set of single-point statistics except in the self-similar limit. In this limit the number of degrees of freedom required to characterize the Reynolds stress tensor $R_{ij}(t) = 2K(t)\,(b_{ij} + \delta_{ij}/3)$ can be reduced to the single-point descriptive level using only a few moments; β_{ij}, $\tilde{I}_{ij}(m)$ and $J_{nn}(m)$. The absolute validity of the model is then restricted to the "equilibrium" regime, but an approximate validity might be approached for the non-equilibrium regime if the value of the parameter m in the moments is chosen judiciously and suitable parameterizations for the moments are chosen which are, in some sense, well behaved. In the subsequent section we demonstrate the behavior of the various moments in both equilibrium and non-equilibrium regimes.

3. Illustrative Calculations

3.1. EVOLUTION OF THE SELF-SIMILAR FORMS

The LWN model will be used to demonstrate the emergence of the self-similar forms for both the regime of turbulence subjected to homogeneous mean-flow velocity gradients as well as freely decaying anisotropic turbulence. Using the LWN model, an initially isotropic turbulence will be subjected to a mean-flow velocity gradient until approximate self-similarity is achieved. After this point in time, the mean-flow gradient will be "removed" and the anisotropic turbulence will be permitted to freely decay until a self-similar state is achieved. This test problem is not intended to match any experiment. Indeed, the total strains applied will vastly exceed any experiment. The choice of problem parameters is based on didactic considerations alone.

The specific values of the strain rates are based on an experiment by LePenven, Gence and Comte-Bellot (Le Penven *et al.*, 1985). The intent of the experiment was to study the dependence of the rate of "return-to-isotropy" on the third invariant of the anisotropy tensor b_{ij}. However, the purpose of these LWN calculations is to illustrate the emergence of self-similar forms within the LWN model, and thus the straining and "free-decay" phases will be followed for much longer times than is possible within an experiment. The nonzero components of the mean-flow velocity gradient tensor are

$$\frac{\partial U_1}{\partial x_1} = 0.78825; \quad \frac{\partial U_2}{\partial x_2} = -0.20974; \quad \frac{\partial U_3}{\partial x_3} = -0.57851.$$

This strain is applied during the first stage of the calculation ($0 < t \leq 64$) to an initially isotropic turbulence. During the second stage of the calculation ($64 < t \leq 1088$) the turbulence is allowed to freely decay. Figure 1 shows the evolution of the turbulent kinetic energy. Note that during the first stage the energy is tending towards an exponential growth in time and during the second stage the energy is decaying as a power-law in time. Figure 2 shows the evolution of the principal components of the anisotropy tensor. This figure clearly shows that during both stages of the computation the principal components of the anisotropy tensor are tending towards time-independent nonzero values. In this figure b_{22} is nearly zero at the end of the straining stage of the calculation, but grows in magnitude after the strain is released. This phenomenon can be better understood by observing the evolution of the corresponding spectra.

Figures 3 and 4 show the evolution of the spectral shape functions $\tilde{f}_{nn}$ and $\tilde{f}_{22}$, respectively. The observed evolution of $\tilde{f}_{11}$ and $\tilde{f}_{33}$ are qualitatively like that of $\tilde{f}_{22}$, but somewhat simpler since they do not change sign as does $\tilde{f}_{22}$. These functions are computed by assuming that the self-similar form given by equations (11)-(14) is obtained. Then the associated scalings are applied to the predicted spectra. These figures show the evolution toward the self-similar spectral distributions. Figure 4 shows that the $\tilde{f}_{22}$-component changes sign between low-ξ and high-ξ. Upon integration over ξ the high-ξ part of $\tilde{f}_{22}$ partially cancels the low-ξ part leading to the resulting relatively small value of b_{22} observed in figure 2. Figure 5 shows all three principal components of $\tilde{f}_{ij}$ at the end of the first stage of the computation, when the self-similarity is well-approximated. This figure demonstrates that each of these three components has a distinctive shape.

Figures 6 and 7 show the evolution of the spectral shape functions $\tilde{f}_{nn}$ and $\tilde{f}_{22}$ during the second stage which is a free-decay of anisotropic turbulence. (Again, the evolution of $\tilde{f}_{11}$ and $\tilde{f}_{33}$ converge relatively simply to the self-similar form.) These functions are constructed by assuming that the self-similar form given by equations (11)-(13) and equations (40)-(41) is obtained and applying the associated scalings to each principal component of the computed spectral tensor. Figure 7 shows that the high-ξ part of the spectrum is changing sign. The effects of cancellation of high-ξ and low-ξ parts of the spectrum observed in the first stage of the calculation is diminishing in time with the result that the absolute magnitude of the b_{22} component appears to grow during the free-decay stage, as shown in figure 2. Figure 8 shows that at the end of the second stage the principal components of $\tilde{f}_{ij} = b_{ij}\tilde{f}_D$ all have the same shape. At this point the self-similar form for freely-decaying anisotropic turbulence is well approximated.

3.2. EVOLUTION OF SPECTRAL MOMENTS

As shown in the previous subsection, during the homogeneous straining stage and during the free-decay stage the predicted turbulence spectra tended towards two different self-similar states. In either of these states, a rigorous reduction of the spectral model two a single-point engineering model can be made. However, many circumstances of practical significance are not in equilibrium and thus no rigorous reduction from the two-point closure to an engineering closure may be possible. Thus it is of interest how the spectral moments which comprise the one-point closure will vary from there equilibrium values during a nonequilibrium transient. One of the simplest nonequilibrium phenomenon is the free-decay of an anisotropic turbulence from a state of spectral equilibrium.

Figures 9 and 10 show the evolution of the principal components of

$c_m\beta_{ij}/\alpha$ and $g_{\epsilon 0}(m)b_{ij} + g_{\epsilon 1}(m)\phi_{ij}(m)$ (for $m = -3$) respectively. Figure 11 shows the evolution of $g_{\epsilon 0}(m)b_{22} + g_{\epsilon 1}(m)\phi_{22}(m)$ for $m = -1, -2, -3$, and illustrates the sensitivity to the choice of m. Since both β_{ij} and $\phi_{ij}(m)$ are traceless and integral functions of $\tilde{f}_{ij}(\xi)$ and since b_{ij} is simply the integral of $\tilde{f}_{ij}(\xi)$ it seems reasonable to seek a parameterization of these terms as linear function of b_{ij}. Figures 12 and 13 show $[c_m\beta_{\nu\nu}/\alpha]/b_{\nu\nu}$ and $[g_{\epsilon 0}(m)b_{\nu\nu} + g_{\epsilon 1}(m)\phi_{\nu\nu}(m)]/b_{\nu\nu}$ (for $m = -3$). For a simple linear parameterization of these tensors;

$$c_m \frac{\beta_{ij}}{\alpha} = C_R b_{ij}, \tag{47}$$

and

$$g_{\epsilon 0}(m)b_{\nu\nu} + g_{\epsilon 1}(m)\phi_{\alpha\alpha}(m) = C_\phi(m)b_{ij}. \tag{48}$$

the variations in coefficients C_R and $C_\phi(m)$ are shown in figures 12 and 13 respectively. Note that since the lines are not "straight," a simple constant fit to determine the value of the coefficients necessarily introduces error. However, since β_{ij} and $\phi_{ij}(m)$ are manifestly linear in $\tilde{f}_{ij}(\xi)$, the introduction of models that are nonlinear in b_{ij} (which is also linear in $\tilde{f}_{ij}(\xi)$) seems unmotivated by either physical or mathematical considerations. The regimes in the calculation where self-similarity is well approximated would be represented by straight (constant) lines on these graphs. The departures form constant values are associated with "nonequilibrium" effects where the reduction to a one-point closure can only be approximate. For more general circumstances where an exact self-similarity might not be achieved, graphs analogous to figures 11 and 12 may not coincide with those shown here. In the non-equilibrium regime precise agreement between a one-point closure and experiment (or two-point closures, or simulations) should not be expected. Note that all three lines in Figure 12 tend toward a value of unity, in agreement with equation (46), and suggest a lack of a long-term return toward isotropy (Clark and Zemach, 1995).

Next consider the behavior of the coefficient $g_{\epsilon 2}(m)$. Figure 14 shows the behavior of $g_{\epsilon 2}(m)$ versus time. During the straining period, the values of $g_{\epsilon 2}(m)$ associated with various values of m diverge from each other. This divergence preserves consistency properties with the other terms in the ϵ-equation so that the evolution of ϵ is independent of the choice of m. During the free-decay, the various values of $g_{\epsilon 2}(m)$ tend towards a single value. Elementary symmetry considerations indicate that for an isotropic turbulence at infinite Reynolds number the turbulent kinetic energy asymptotically decays as

$$K(t) = K(t_0)\left[1 + \frac{t}{t_0}\right]^{-\gamma}, \tag{49}$$

where

$$\gamma = \frac{2(n+1)}{n+3}, \tag{50}$$

where n is given by equation (8). The $K-\epsilon$ equations for decaying isotropic turbulence give

$$\gamma = \frac{1}{g_{\epsilon 2}(m) - 1}. \tag{51}$$

Hence, at late stages of the decay we should obtain

$$g_{\epsilon 2}(m) = \frac{3n+5}{2(n+1)}, \tag{52}$$

independent of the choice of m. In the present calculations $n = 3$ and hence $\gamma = 7/4$ and indeed figure 14 indicates that in the late stages of the free-decay shown in $g_{\epsilon 2}(m) \rightarrow 7/4$ for $m = -1, -2, -3$.

4. Conclusions

In a state of statistical equilbrium a two-point closure can be reduced to a one-point closure by exploiting the self-similarities that emerge within the two-point model. These self-similarities and the consequent moments used to construct the one-point closures may have some relevance to the more general non-equilibrium circumstance. However, since a finite set of integral moments of the spectrum may not fully specifiy the spectrum in a non-equilibrium flow, the appropriate choice of moments to incorporate into a one-point model for such circumstances necessarily remains an "art".

Acknowledgements

This work was supported by the Los Alamos National Laboratory's Nuclear Weapons Technology Office.

References

Besnard, D., Harlow, F., Rauenzahn, R. and Zemach, C., 1996. "Spectral transport model of turbulence," *Theoret. Comput. Fluid Dyn.* **8**, pp. 1-35.

Clark, T. and Zemach, C., 1995. "A spectral model applied to homogeneous turbulence," *Phys. Fluids* **7**, pp. 1674-1690.

Clark, T., 1992. "Spectral Self-Similarity of Homogeneous Anisotropic Turbulence," Ph.D. Dissertation, University of New Mexico, 1991. Also *Los Alamos National Laboratory Report LA-12284-T.*

Le Penven, L., Gence, J. N., and Comte-Bellot, G., 1985. "On the Approach to Isotropy of Homogeneous Turbulence: Effect of the Partition of Kinetic Energy Among the Velocity Components," in *Frontiers in Fluid Mechanics*, Ed. S.H. Davis and J.L. Lumley, pub. Springer-Verlag, Berlin. pp. 1-21.

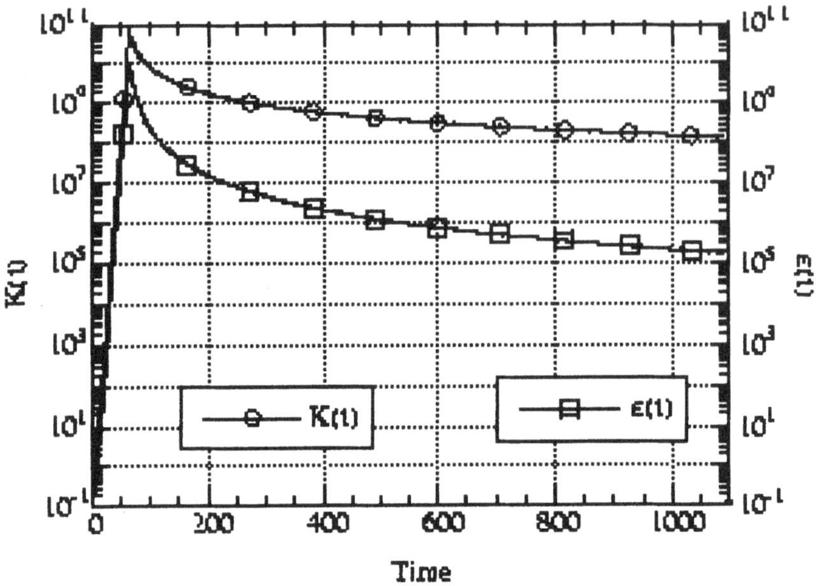

Figure 1. Evolution of the turbulent kinetic energy, $K(t)$ and its dissipation rate, $\epsilon(t)$.

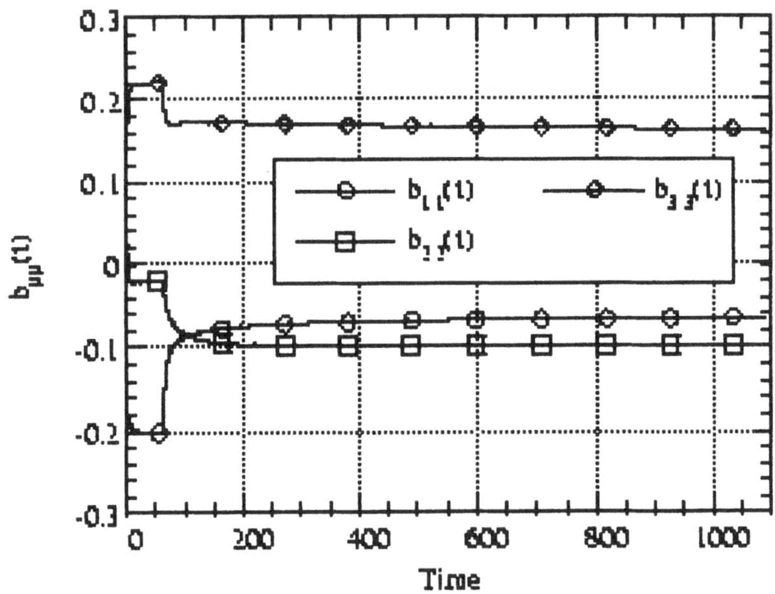

Figure 2. Evolution of the principal components of the anisotropy tensor b_{ij}.

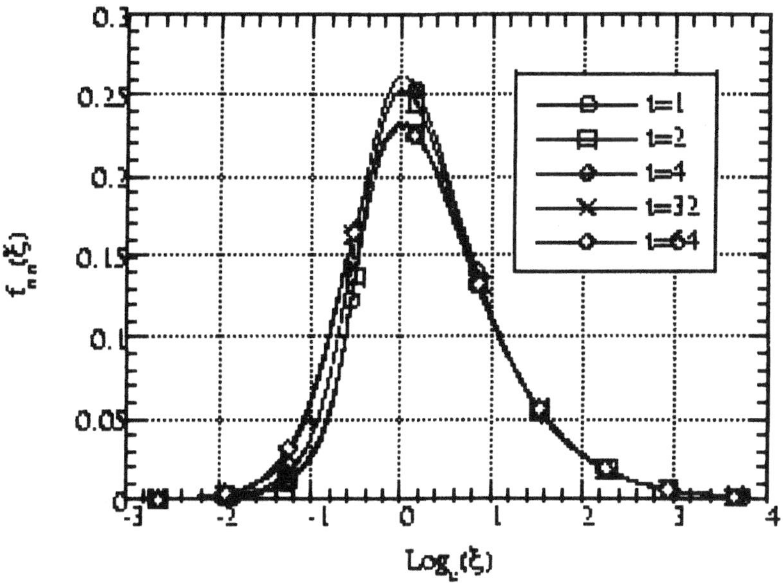

Figure 3. Evolution of the energy spectral shape function $f_{nn}(\xi)$ during the mean-flow straining.

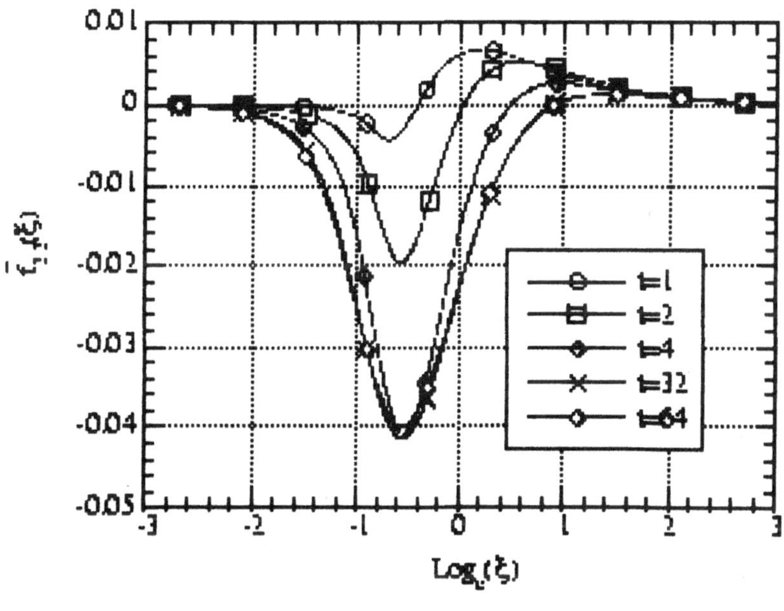

Figure 4. Evolution of the energy spectral shape function $\tilde{f}_{22}(\xi)$ during the mean-flow straining.

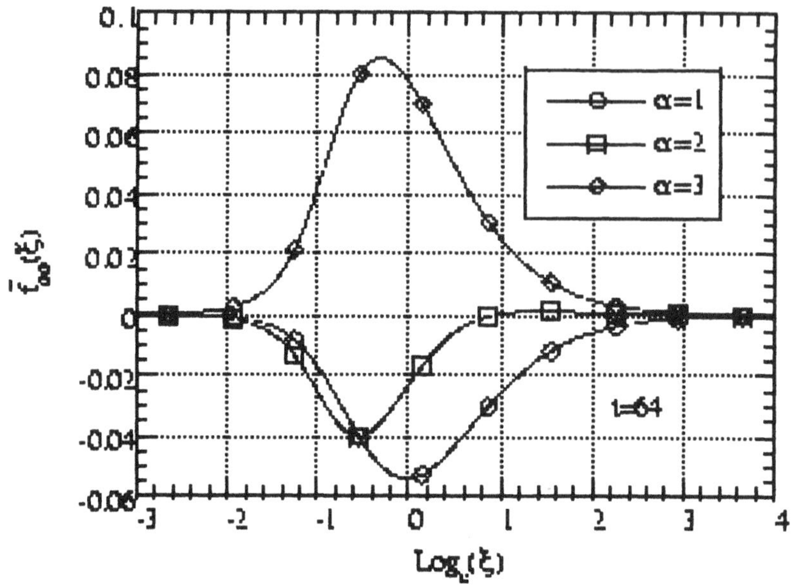

Figure 5. Self-similar deviatoric spectral shape functions $\tilde{f}_{\sigma\sigma}(\xi)$ at equilibrium at the end the mean-flow straining ($t = 50$).

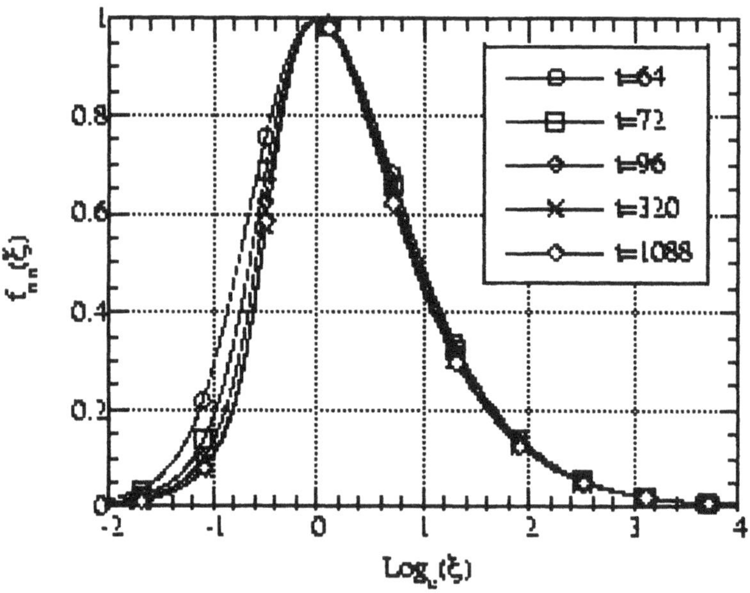

Figure 6. Evolution of the energy spectral shape function $f_{nn}(\xi)$ during free-decay.

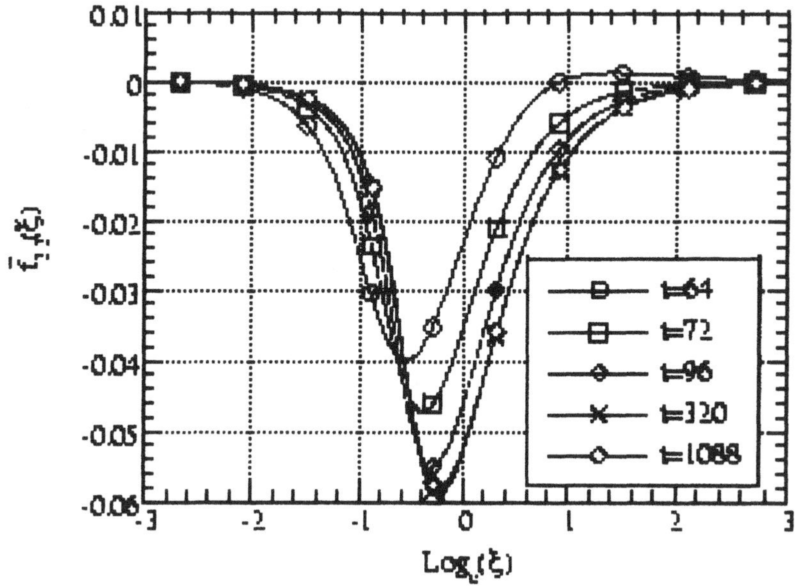

Figure 7. Evolution of the energy spectral shape function $\tilde{f}_{22}(\xi)$ during free-decay.

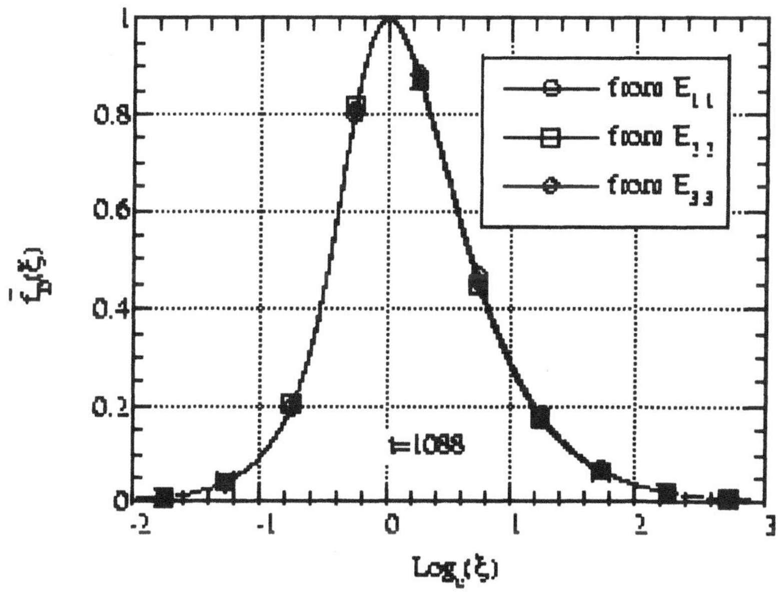

Figure 8. Self-similar deviatoric spectral shape functions $\tilde{f}_{\sigma\sigma}(\xi) \propto b_{\sigma\sigma}\tilde{f}_D(\xi)$ at equilibrium at the end the free-decay ($t = 1088$).

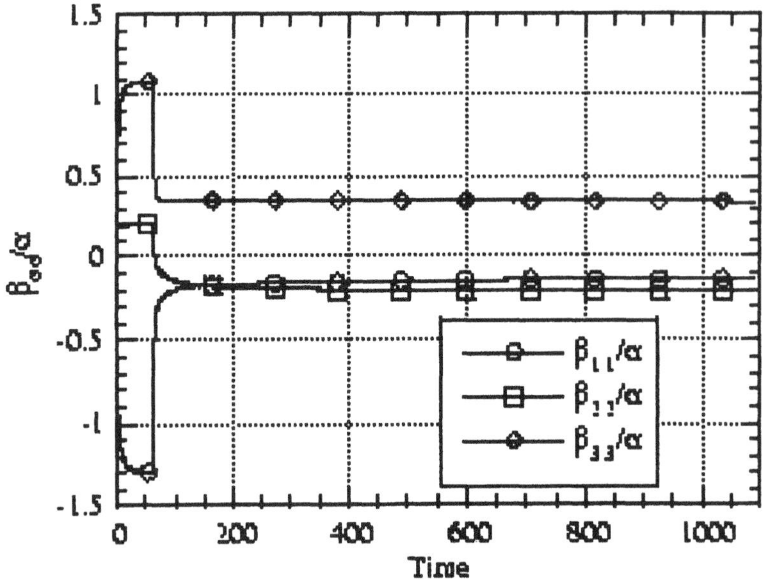

Figure 9. Evolution of the principal components of $\beta_{\sigma\sigma}/\alpha$.

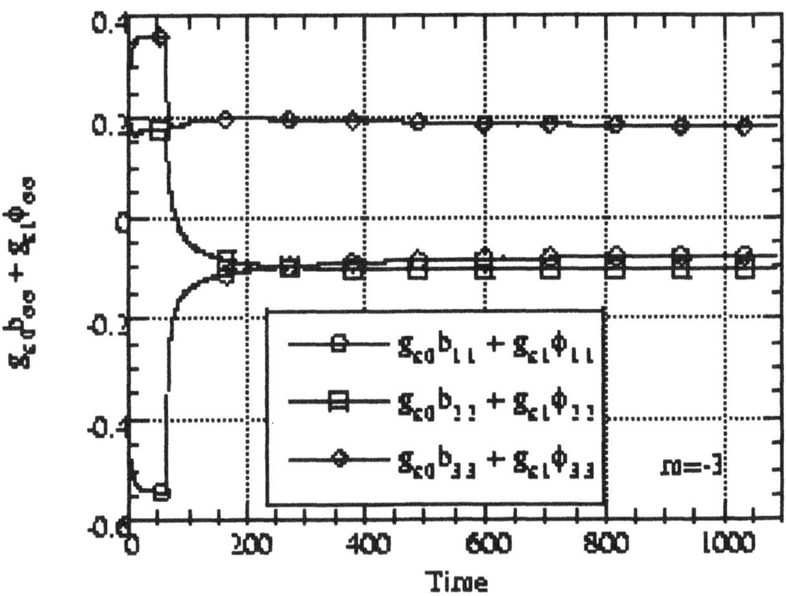

Figure 10. Evolution of $g_{\epsilon 0}(m)b_{\sigma\sigma} + g_{\epsilon 1}(m)\phi_{\sigma\sigma}(m)$ for $m = -3$.

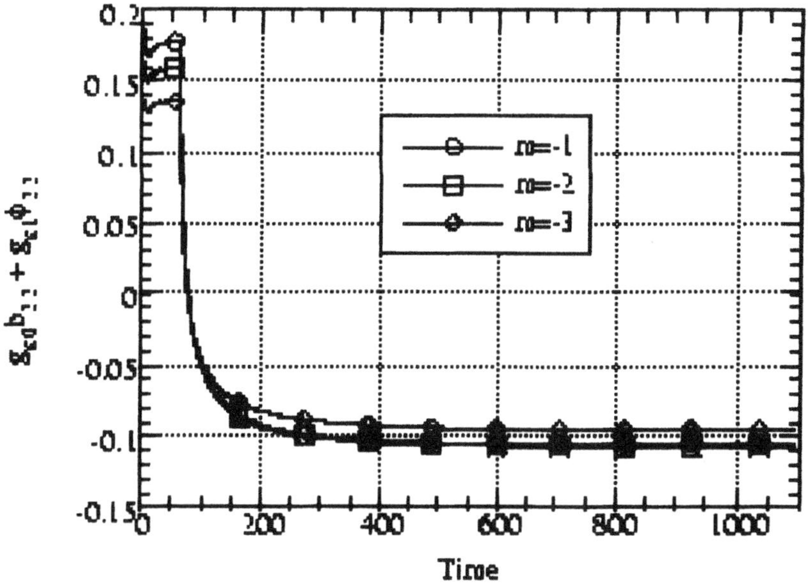

Figure 11. Evolution of $g_{\epsilon 0}(m)b_{22} + g_{\epsilon 1}(m)\phi_{22}(m)$ for $m = -1, -2, -3$.

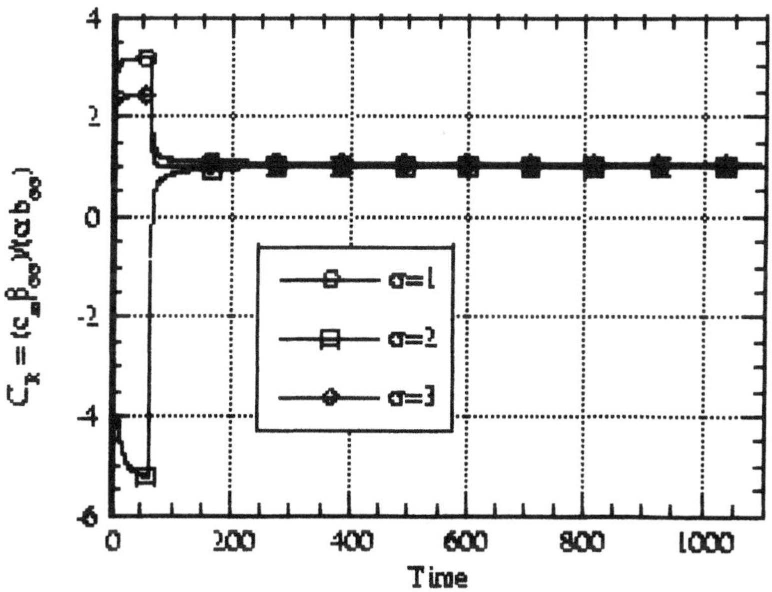

Figure 12. Evolution of $[c_m\beta_{\sigma\sigma}/\alpha]/b_{\sigma\sigma}$ demonstrating predicted variations in the coefficient of a "standard" Rotta formulation for the tendency towards isotropy.

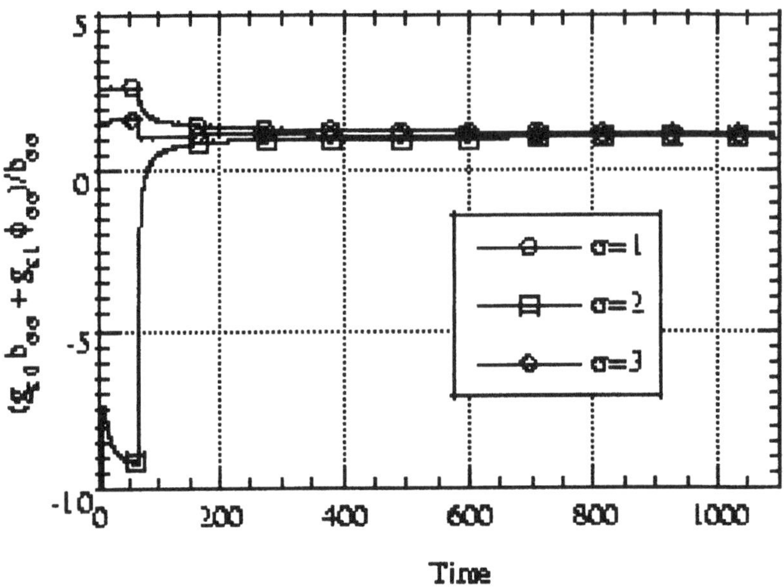

Figure 13. Evolution $[g_{\epsilon 0}(m)b_{\sigma\sigma} + g_{\epsilon 1}\phi_{\sigma\sigma}(m)]/b_{\sigma\sigma}$ for $m = -3$ demonstrating predicted variations in the coefficient of a "standard" $K - \epsilon$ formulation.

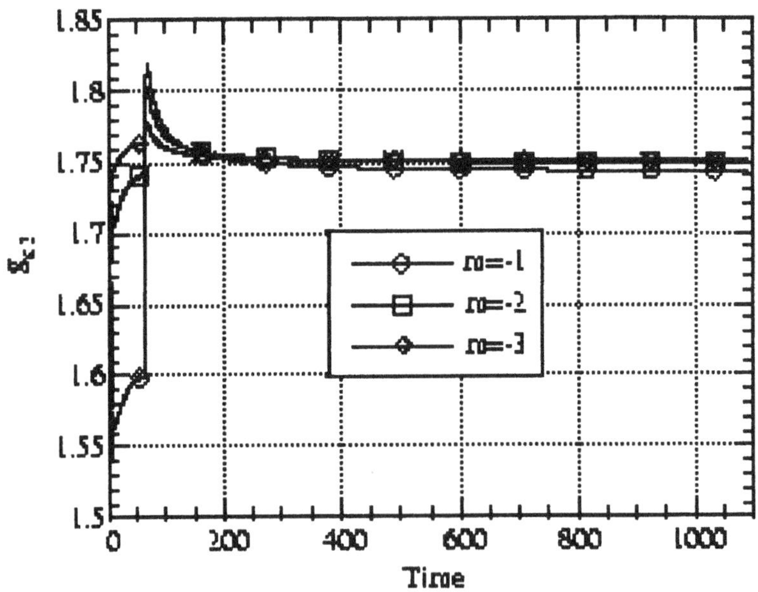

Figure 14. Evolution of $g_{\epsilon 2}(m)$ for $m = -1, -2, -3$. At the end of the free-decay the values converge to a value consistent with more general theoretical considerations.

MODELING THE TURBULENT WALL FLOWS SUBJECTED TO STRONG PRESSURE VARIATIONS

K. HANJALIĆ AND I. HADŽIĆ
Faculty of Applied Physics, Delft University of Technology, Lorenzweg 1, 2628 CJ Delft, The Netherlands

S. JAKIRLIĆ
Lehrstuhl für Strömunsgmechanik, University of Erlangen, Cauerstrasse 4, 91058 Erlangen, Germany

AND

B. BASARA
Advanced Simulation Technology Dept., AVL LIST GmbH, Kleiststrasse 48, 8020 Graz, Austria

Abstract. Mean pressure gradients affect turbulence mainly through the modulation of the mean rate of strain. Modification of the turbulence structure feeds, in turn, back into the mean flow. Particularly affected is the near wall region (including the viscous sublayer) where the pressure gradient invalidates the conventional boundary-layer 'equilibrium' assumptions and inner-wall scaling. Accurate predictions of such flows require application of advanced turbulence closures, preferably at the differential second-moment level with integration up to the wall. This paper aims at substantiating the above arguments by revisiting some of the recent experimental and DNS results and by presenting a series of computations relevant to low-speed external aerodynamics. Several attached and separated flows, subjected to strong adverse and favorable pressure gradient, as well as to periodic alternation of the pressure gradient sign, all computed with a low-Re-number second-moment closure, display good agreement with experimental and DNS data. It is argued that models of this kind (in full or a truncated form) may serve both for steady or transient Reynolds-Averaged Navier-Stokes (RANS, TRANS) computations of a variety of industrial and aeronautical flows, particularly if transition phenomena, wall friction and heat transfer are in focus.

M. D. Salas et al. (eds.), Modeling Complex Turbulent Flows, 203–222.

1. Pressure gradient and implications on turbulence modeling

It is known that in incompressible flow the pressure gradient affects the turbulence properties and their budget only *indirectly*, through the modulation of the mean strain. Experiments and direct numerical simulations (DNS) in attached boundary layers at moderate-to-strong pressure gradients (e.g. (Samuel & Joubert, 1974), (Simpson *et al.*, 1981), (Nagano *et al.*, 1993), (Spalart & Watmuff, 1993), (Spalart & Coleman, 1997)) reveal that the strongest effects are noticed in the wall region, permeating even through the viscous sublayer up to the wall and invalidating the 'equilibrium' inner wall scaling (U_τ, ν/U_τ) for turbulence properties. The free stream fluctuations remain largely unaffected, except, possibly much further downstream (depending on the imposed pressure gradient) where the turbulent diffusion and lateral convection due to thickening of the boundary layer may modify the rest of the flow away from the wall. Moreover, because of indirect ('chain') effects on the stress production, redistribution and turbulent transport, different stress components respond at different rates, thus modifying the anisotropy of the stress field and, through a feed-back, the mean flow field. Favorable pressure gradient thickens the viscous sublayer, but also suppresses the wall-normal velocity fluctuations, increasing thus the near-wall stress anisotropy, expanding the extent of the region where the turbulence is closer to the two-component limit, (Jakirlić, 1997). Adverse pressure gradient acts opposite and shifts the anisotropy maximum away from the wall. The extent of the affected region increases with an increase in the adverse pressure gradient. This is illustrated in Fig. 1, which shows the near-wall distribution of Lumley's two-componentality ('flatness') parameter $A = 1 - 9/8(A_2 - A_3)$ for boundary layers at zero, favorable and adverse pressure gradients. ($A_2 = a_{ij}a_{ji}$, $A_3 = a_{ij}a_{jk}a_{ki}$ are the second and third invariants of the turbulent stress anisotropy $a_{ij} = \overline{u_i u_j}/k - 2/3\delta_{ij}$). Of course, the modulation in the turbulence anisotropy is not confined only to the stress-bearing large-scale motion, but extends to the dissipative scales, modulating the anisotropy of the dissipation rate tensor irrespective of the bulk Reynolds number.

These are only some of the facets of the effects of the mean pressure gradient on the turbulence fluctuations in the near-wall region, indicating a necessity for a turbulence model to account for separate contribution of *each* stress components to the momentum balance, as well as to the dynamics of the turbulence scale and of the stress tensor itself. Best prospects for accurate predictions of turbulent flows with strong pressure gradients have the second-moment models which can mimic better the dynamics of the turbulent stress field and evolution of each stress component. Furthermore, the departure from local equilibrium and conventional boundary layer scaling, calls for resolving in full the near-wall layer with all necessary implications on model modifications for the wall-proximity and viscous effects.

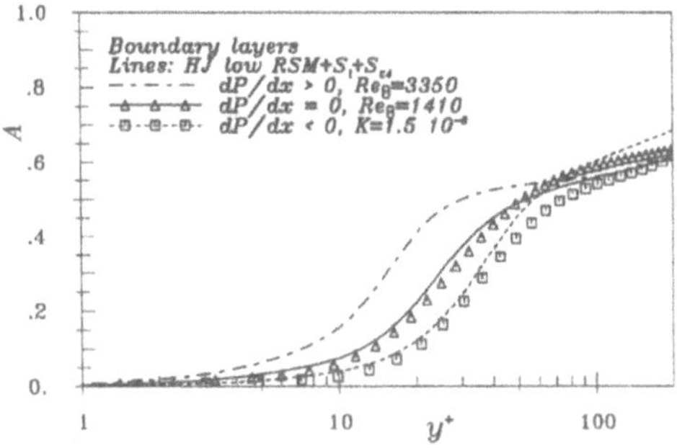

Figure 1. Lumley's two-componentality ('flatness') parameter $A = 1 - 9/8(A_2 - A_3)$ for boundary layers at zero, favorable and adverse pressure gradients.

This paper presents some results of a systematic testing of a version of the second-moment closure with low-*Re*-number and wall-vicinity modifications. Some modifications of ε-equations are considered, which are applicable also to the high-*Re*-number models. Arguments in favor of these modifications, as well as the complete low-*Re*-number model will be substantiated by presenting some results for a series of attached and separating wall flows. For illustration, some cases obtained with the high-*Re*-number second-moment and with the standard $k - \varepsilon$ closures will also be shown. Considered cases include flows in strong favorable pressure gradient (including laminarization), by-pass transition on a finite-thickness plate, flows in adverse pressure gradient (non-separating and separating) and flows subjected to periodic alternation (in time or space) of positive and negative pressure gradients. Also, some results of computational study of mean flow and turbulence field in flows separating on sharp edges (backstep) and on curved surfaces (airfoil and a vehicle), will be presented.

The presented model serves more as an illustration of a potential, rather than a proposal of a 'superior' model to be used for all purposes. The rationale of the model has been published elsewhere (e.g. (Hanjalić *et al.*, 1997)) and it is summarized in the Appendix. The model demands a fine grid near walls, which may hinder its wider application to very complex three-dimensional flows at high *Re*-numbers. It is argued, however, that this approach may be unavoidable if transition phenomena and accurate wall friction and heat transfer are in focus, such as in the problem of gas turbine blade cooling. Successful reproduction of the near wall second-moment statistics (in addition to mean flow), even in separation bubbles, around reattachment and in subsequent recovery, qualifies this approach (in full or a truncated form) both for steady or transient Reynolds-Averaged-Navier-Stokes (RANS, TRANS) computations of a variety of industrial and aeronautical flows.

2. Illustrations

2.1. TRANSITIONAL FLOWS

While the problem of development of laminar flow instability and natural transition to turbulence remains still intractable to the Reynolds averaging approach, the reverse transition at severe flow acceleration served from early days as a major test for validating the model of low-Re-number effects in wall bounded flows. Most current models failed to reproduce laminarization at appropriate conditions in accord with experiments or DNS, be it in the case of a self-preserving sink flow or in more complex cases with variable acceleration or other laminarizing mechanisms. Major reason for failure lies in the practise to model jointly the viscous and non-viscous (wall-blockage) effects in terms of damping functions employing viscosity and local wall-distance. In more complex situations, where a strong variation of pressure gradient or other flow conditions impose a different degree of anisotropy and affect the process of stress redistribution, the turbulence field could hardly be simulated by an isotropic eddy diffusivity model. The present model proved capable of reproducing the effect of favorable pressure gradient very well, including the prediction of laminarization of a sink flow at the appropriate acceleration parameter (Hanjalić *et al.*, 1997).

More recent studies revealed that the RANS approach can be used also to predict some forms of forward transition. The first category are the laminar-like flows with some background turbulence ('turbulescence'), which is too weak to influence the laminar-like character of the mean flow, but sufficient to be 'revived' when a flow deformation is imposed, or reaches a sufficient strength to interact with the background turbulence. The second category are the initially laminar flows with a continuous source of turbulence somewhere in the flow or at its edge, from where the turbulence will diffuse (be entrained) into the rest of the non-turbulent flow.

The first example is the oscillating boundary layer around a zero mean, where in a wide range of Reynolds numbers ($Re_{\delta_S} = U_\infty \delta_S/\nu$ from ≈ 700 to 3000, where δ_S is the Stokes thickness) both the forward and reverse transitions occur within a single cycle. Once the fully developed conditions are achieved, the flow becomes phase-self-similar and independent of initial conditions, and is very convenient for testing the dynamics of response of turbulence models. Of course, the forward transition in this case has no resemblance with the actual process of development of Tollmien-Schlichting type of instabilities in an originally purely laminar flow. It is merely a switch-over from a 'laminar-like' regime with slowly decaying turbulence remnants. This transition to fully developed turbulence occurs at the onset of deceleration phase and is characterized by a sudden rise of turbulent stresses, reflected also in a steep increase in the wall shear stress to a value pertaining to fully turbulent wall-boundary layers. Prediction of this phenomenon at an appropriate phase angle, which varies with Reynolds number, is a major challenge to turbulence models, because it reflects the ability of a model to reproduce the full dynamics of a strongly anisotropic turbulence field subjected to periodic perturbations. Because different interactions have their own dynamics (preferential production into the streamwise stress component, subsequent redistribution by pressure-strain, nonhomogeneous decay over the part of the cycle around the flow reversal when the mean flow becomes almost stagnant) each stress component will exhibit a different degree of hysteresis in the response to the mean flow perturbation. The reproduction of the decay process is a major prerequisite for the prediction of sudden transition at the appropriate phase angle. For that reason this flow is a sensitive indicator of the model performances in dealing with low-Re-number phenomena. Fig. 2 shows a very successful reproduction of DNS, (Justesen & Spalart, 1990) of the selected components of the turbulent stress anisotropy in an oscillating boundary layer with a "steep" variation of free stream velocity with a resting period between the acceleration and deceleration. For comparison, the modulation of the imposed nondimensional strain rate Sk/ε (where $S = \sqrt{S_{ij}S_{ji}}$) at corresponding wall distances is also presented, indicating no direct phase relationship with the modulation of the stress anisotropy.

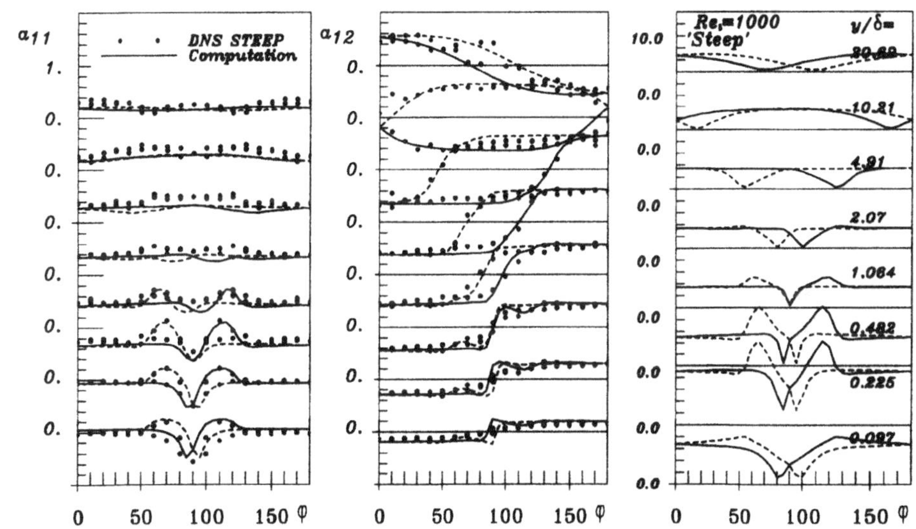

Figure 2. Hysteresis of a_{11}, a_{12} and Sk/ε across an oscillating boundary layer with "steep" variation of free stream velocity, Re_{δ_S}=1000; Symb.: DNS Justesen & Spalart; Lines: computations Hanjalić *et al.*; —— 0 – 180°, – – – 180 – 360°.

The same model gave equally satisfactory results for other Reynolds numbers over a broader range, as well as in an oscillating flow in a pipe of finite length where the pipe-flow *Re*-number is additional parameter governing the transition (Hanjalić *et al.*, 1995).

Predicting the 'by-pass transition' poses different problems. Here the laminar-to-turbulent transition is promoted by turbulence penetration into the laminar boundary layer from the outer stream with a uniform turbulence field. Unlike in preceding cases, the major prerequisite for a successful reproduction of the transition for different levels of free-stream turbulence depends on the model ability to mimic the turbulent diffusion. The review by (Savill, 1996) of the performances of various models in predicting the by-pass transition on a flat plate with different levels of free stream turbulence revealed that models which do not use the local wall distance in damping functions, perform generally better and that the second-moment closures are generally more powerful than the two-equation models. None of the models was able to reproduce the cases with the free-stream turbulence below 3%, what is probably the lower limit tractable by the RANS models.

Transition on bodies with finite thickness and in non-uniform pressure field involves additional difficulties. Inability to reproduce the proper turbulence level and anisotropy in the stagnation region leads usually to very erroneous results. An illustrative example is the transition in a laminar boundary layer developing over a finite thickness plate with round leading edge. Experiments indicate that a thin laminar separation bubble appears shortly behind the leading edge. The transition to turbulence occurs at the rear end of the separation bubble, very close to the wall, followed by a gradual diffusion of turbulence into the outer flow region. Predicting the correct shape and size of the separation region, which is crucial for predicting correctly the transition, requires the application of both an advanced turbulence model and an accurate numerical method (higher order convection schemes) combined with a very fine numerical grid. Fig. 3 compares computations with the low-*Re*-number $k-\varepsilon$ model (Launder and Sharma) and with the present second-moment closure. The $k-\varepsilon$ model produces the transition and an excessive turbulence level already in the stagnation region, causing a strong mixing, which prevents the separation. The low-*Re*-number second-moment closure (using locally refined block-structured grid) reproduced the flow pattern with laminar separation bubble, the location of the transition and the subsequent development of the turbulence field in good agreement with experiments.

2.2. BOUNDARY LAYERS IN ADVERSE PRESSURE GRADIENT.

The computation of an oscillating boundary layer demonstrated the ability of a model to respond to the imposed alternating favorable and adverse

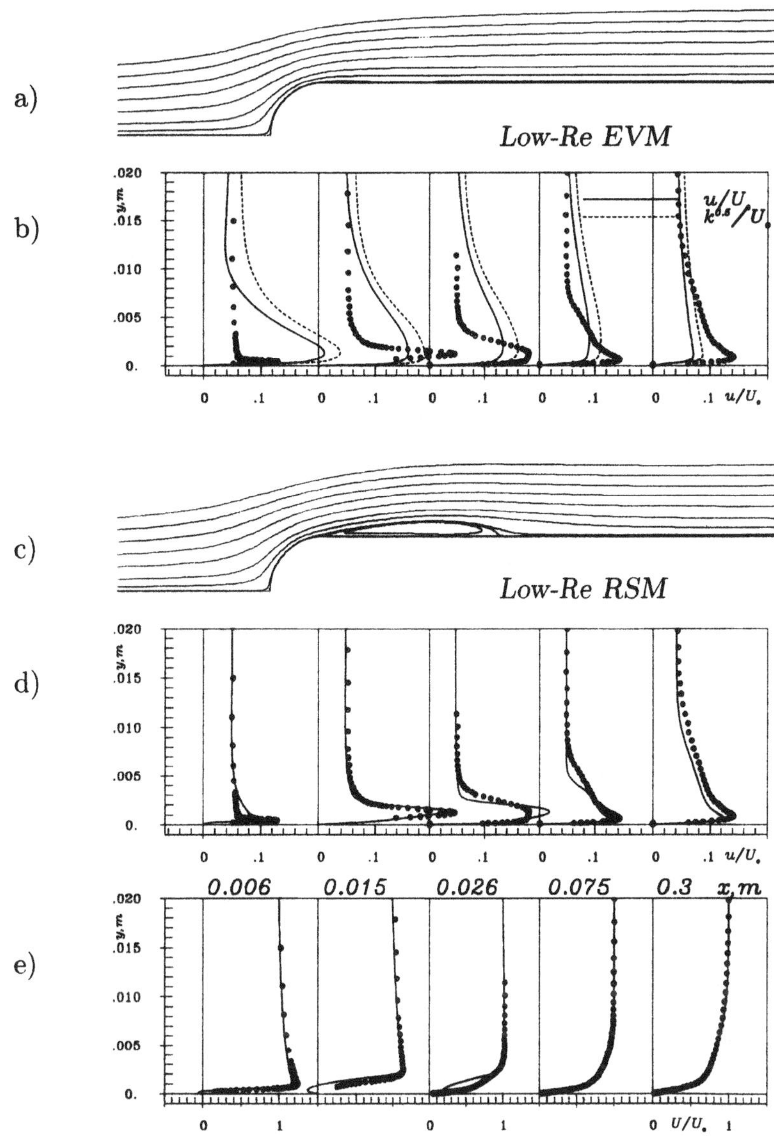

Figure 3. **By-pass transition on a flat plate with circular leading edge (radius $R = 5\,mm$, $U_o = 5\,m/s$, and $\sqrt{\overline{u^2}}/U_o = 0.05$). Stream-lines and streamwise velocity fluctuations: a) & b) - EVM, and c) & d) - RSM; e) - streamwise mean velocity obtained with RSM.** Symbols: Experiments Rolls-Royce Applied Science Laboratory (1995).

pressure gradient. However, a prolonged action of a strong or increasing adverse pressure gradient, which may lead to flow separation, poses additional challenge since the time lag in response of turbulence field to the pressure variation shows a marked influence only at a later stage downstream. In the 1980/81 AFOSR/HTTM-Stanford Conference on Complex Turbulent Flows one of the test cases was such a flow investigated by (Samuel & Joubert, 1974). Most participants reproduced well the initial variation of the skin-friction coefficient, but many failed to do so for the last portion of the flow. It should be recalled that over much of the flow length both dP/dx and d^2P/dx^2 are positive (increasingly adverse). Here the turbulence intensities in the outer 80% of the flow at all stations retained essentially the same values and similar profile shape irrespective of the imposed pressure gradient. However, at the last measuring station, where the pressure gradient was decreasingly adverse, the turbulence intensity in the outer region showed a sudden increase. (Samuel & Joubert, 1974) argued that the reason for a sudden change in the turbulence level was caused by the rapid thickening of the layer after imposing a decreasingly adverse pressure gradient. More recent analysis of other data seems to suggest that a prolonged evolution of stress anisotropy due to effects of strong pressure gradient (irrespective whether increasingly or decreasingly adverse) can be a major cause for the downstream modulation of the turbulence field. This would explain the failure of two-equation models to predict the friction factor further downstream. Surprisingly, the standard second-moment closure gave only marginal improvements, indicating at possible inadequacy of the simple form of the standard ε equation for reproducing the scale dynamics in strongly evolving flows. Similar deficiencies were discovered later also in other nonequilibrium flows, primarily in separated regions and around reattachment. An indication of the role of scale variable was that somewhat better results were claimed with an ω instead of an ε equation. Because the ε-equation is more frequently used, various modifications have been proposed in the literature to accommodate non-equilibrium and separating flows. An additional term emphasizing the role of irrotational straining in the production of ε, proposed in 1980 (Hanjalić & Launder, 1980), produced the desired improvements of flows with a strong pressure variation.

For illustration we consider in parallel the cases investigated by (Samuel & Joubert, 1974) and (Nagano *et al.*, 1993). Unlike the Samuel & Joubert case, the flow of Nagano *et al.* was subjected to a more sudden, though moderate (constant) pressure gradient. Experimental data of Nagano *et al.* seem more consistent and have been obtained closer to the wall, revealing also some other features of the pressure gradient effect. For this reason we present some results of computation of Nagano *et al.* flow with the low-*Re*-number second-moment closure. The main features of the two flows are displayed in Fig. 4a, showing the evolution of the non-dimensional pressure gradient (Clauser parameter $\beta = (\delta_1/\tau_w)dP/dx$ and $P^+ = \nu(dP/dx)/\rho U_\tau^3$). Predicted friction factors for the two cases, Fig. 4b, show good agreement

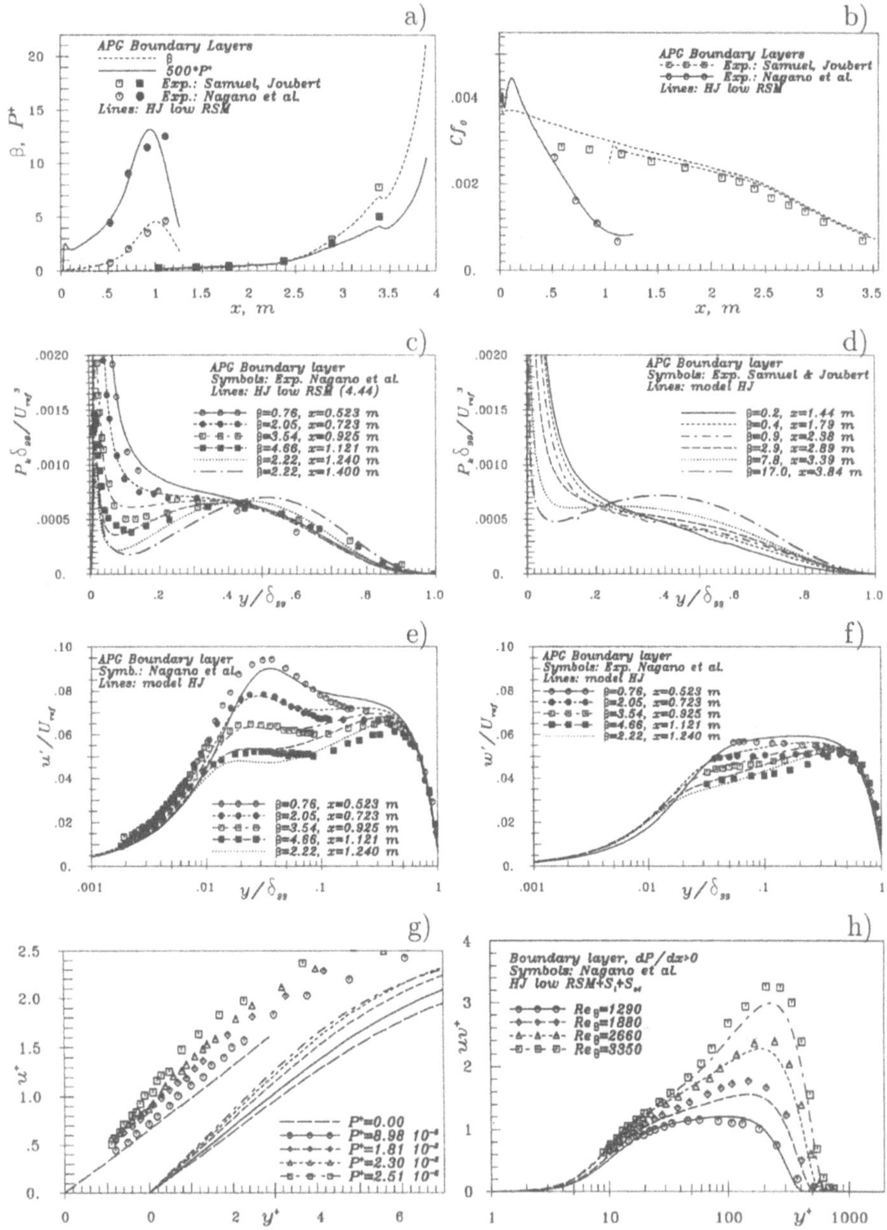

Figure 4. **Boundary layers in increasing adverse pressure gradient; a) - nondimensional pressure gradient parameters and b) - friction factor for Nagano** *et al.* **and Samuel & Joubert; e) to h) - turbulence intensities and shear stresses at various locations for Nagano** *et al.* flow. Symbols: experiments; Lines: computations.

with experiments, illustrating the importance of initial conditions at the start of computations (dotted lines). Fig. 4c and 4d show the production of kinetic energy for the two cases. It should be noted that the chain lines in both figures show the computed results at the last stations, for which data are not available. For the case of Nagano *et al.* the results were computed by retaining the constant $\beta = 2.2$ from the previous station (not correspondent to the experiment) in order to see the effects of prolonged action of the adverse pressure gradient. The profiles at other stations in the outer region collapsed all on one curve indicating no effect of pressure gradient on the absolute level of the kinetic energy (see normalization with constant reference velocity). However, the last profile shows an increase in the production, fully in accord with the findings of Samuel and Joubert. It should be noted that the pressure gradient in the Nagano *et al.* case was decreasing already from $x \approx 0.6$ and the increase in the kinetic energy production was visible only further downstream due to a cumulative transport.

The same effect is noticeable in the Samuel & Joubert case, where the chain curve (obtained by extrapolating β further downstream), Fig. 4a, departs visibly from the rest of the curves. Figures 4e to 4h show the results of computations with the low-*Re*-number second-moment model, with additional term in the ε equation, in good agreement with the measurements along the whole flow length covered by the experiment. An interesting feature of this flow is that the inner wall scaling proved to be inadequate both for the mean and turbulence properties, causing not only a departure from the logarithmic law of the wall, but also the variation of the slope of the normalized turbulent stresses in the limit when the wall is approached, Fig. 4g. It is interesting to note that the model described here can reproduce reasonably well this effect.

2.3. SEPARATION BUBBLE ON A PLANE WALL

The next illustration is a 'rapid' separation and reattachment of a turbulent flow on a flat wall, for which Direct Numerical Simulations were performed recently by (Spalart & Coleman, 1997). The pressure gradient was created by imposed suction and blowing along the opposing flow boundary. We have reproduced the flow conditions by adopting the same solution domain and the prescribed transpiration velocity through the upper boundary, as in DNS. Due to Reynolds number limitation (the incoming boundary layer at $Re_{\theta,in} \approx 300$ was not well developed) there was an uncertainty in reproducing the exact inflow conditions (particularly the dissipation profile, which was not provided by DNS), to which the downstream flow pattern is very sensitive (Spalart & Coleman, 1997) and (Dengel & Fernholz, 1990).

The present computations were performed with the low-*Re*-number second-moment closure with and without the term S_l (see Appendix), yielding similar flow patterns in reasonable agreement with the DNS. However, without S_l the dividing streamline shows an anomalous forward bending in the separation point and backward bending at reattachment. This deficiency of the standard second-moment closure was detected earlier in studies of backward facing step flow, particularly at low *Re*numbers (for discussion see (Hanjalić & Jakirlić, 1997)). While the use of wall functions and placing the first grid point at a relatively large distance from the wall concealed the anomaly, it becomes visible when using models which allow the integration up to the wall and the application of finer numerical grids. The introduction of an S_l term into the dissipation equation, which compensates for excessive growth of the length scale in the stagnation zones, eliminates the anomaly both when used in conjunction with the standard high-*Re*-number second-moment model with wall functions, or with low-*Re*-number models and integration up to the wall.

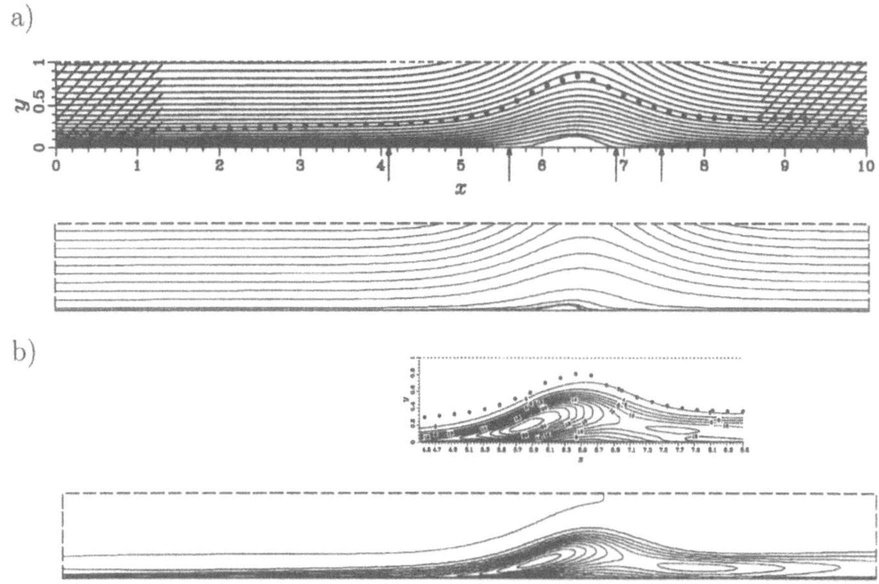

Figure 5. Streamlines (a) and kinetic energy contours (b) in a separation bubble. Top figures: DNS (Spalart & Coleman). Bottom figures: computations low-*Re*-number RSM.

The computed separation bubbles, Fig. 5, are somewhat thinner than in DNS, which may be a consequence of inadequate inflow conditions. The streamline anomaly without S_l makes the distance between separation and reattachment shorter. The inclusion of S_l rectifies the anomaly (Fig. 5) and extends the bubble length at the wall, though both the separation and reattachment points are predicted more upstream than in DNS. This is seen in the plot of the friction factor, Fig. 7, which shows also some discrepancy in the recovery zone. The mean velocity profiles, however, agree very well with the DNS, Fig. 6. The computed components of turbulence intensity

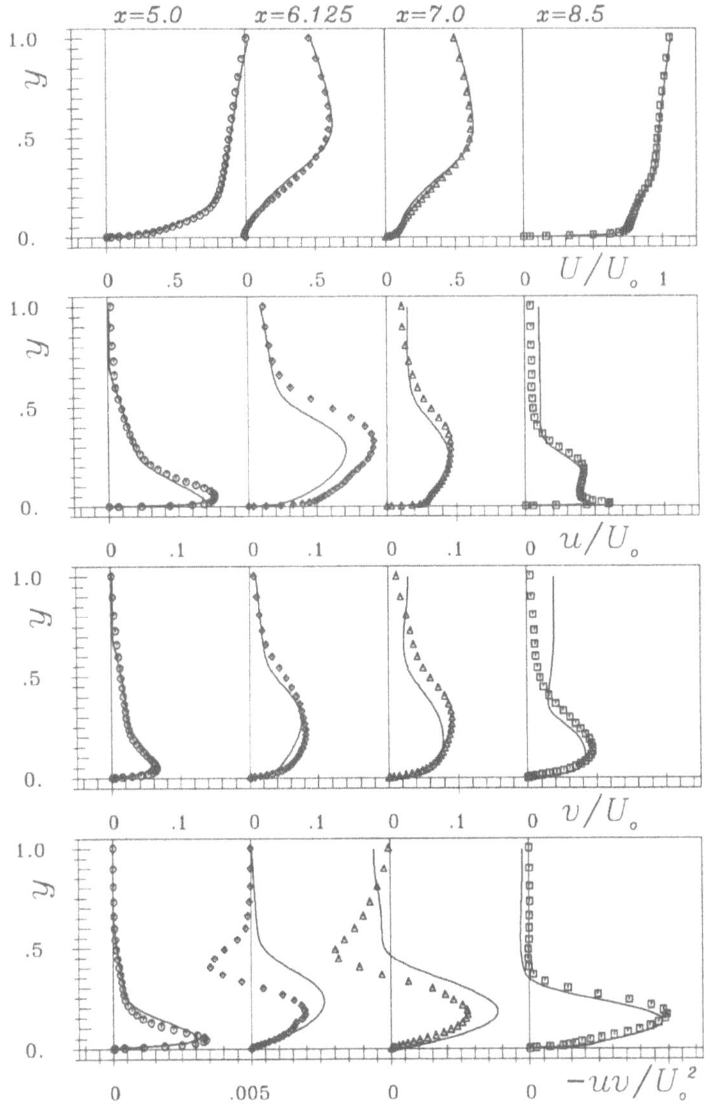

Figure 6. Mean-velocity and Reynolds-stress tensor components at characteristic positions ($x = 5$ before separation, $x = 6.125$ in the middle of separation, $x = 7$ just beyond separation and $x = 8.5$ recovery of boundary layer). Symbols: DNS by Spalart and Coleman (1997); Lines: low-*Re*-number RSM.

are similar to DNS (see also the contours of kinetic energy, Fig. 5), though with some discrepancy, which is most noticeable in the profiles of the shear stress.

It should be noted that DNS was performed with $600 \times 200 \times 256$ nodes and the statistical sample contained 429 fields, requiring 800 CRAY 90

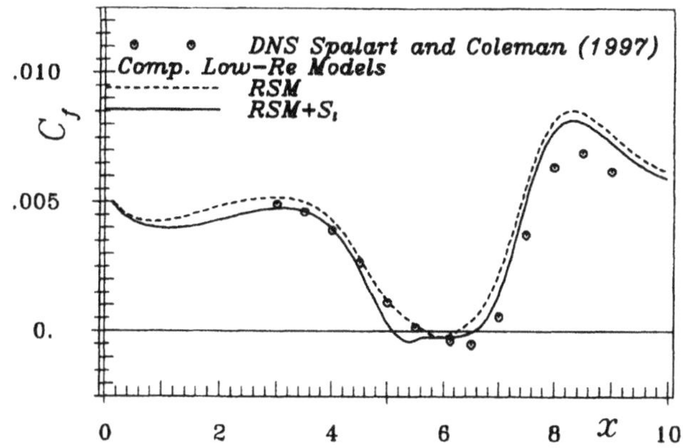

Figure 7. Friction factor along the wall in the flow with separation bubble.

hours. Despite such effort, (Spalart & Coleman, 1997) noted several defects in the results which could not be eliminated because of high costs. In contrast, the present RANS computations were performed with a $100 \times 100 \times 1$ grid in a steady mode, requiring 2 CRAY hours.

2.4. BACKWARD-FACING STEP AND SUDDEN EXPANSION

A turbulent flow behind a backward facing step contains several features pertinent to real complex flows: abrupt change of boundary conditions at the step which leads to boundary layer separation, a curved free shear layer and its bifurcation at the reattachment, primary and secondary recirculation, reattachment and subsequent recovery of the wall boundary layer. Flows with a smaller expansion ratio (downstream/upstream flow-width) with a milder effect of pressure gradient seem to be more difficult to reproduce by statistical turbulence models due to a stronger effect of stress field anisotropy (as compared with the pressure gradient) on the flow evolution and recovery.

The standard linear two-equation $k - \varepsilon$ model gives a too short recirculation length and a poor reproduction not only of mean velocity and turbulence profiles, but even more of the friction coefficient. Unsatisfactory predictions are not located only around reattachment, but also elsewhere in the recirculating and recovery region. Second moment closures account better for streamline curvature and differentiate the sign of extra strain rate effects through the exact treatment of the stress production term. Indeed, the standard high-*Re*-number second moment closure produces the desired elongation of the separation bubble. Some improvements of the agreement with the experimentally obtained velocity profiles in the recirculation region upstream from the reattachment were reported, but opposite effects were also found in some cases (for more details see (Hanjalić

& Jakirlić, 1997)). The present model yielded very good agreement with the DNS and experiments for a range of Reynolds numbers and expansion ratios. Improvements of the streamline pattern and of velocity profiles around reattachment with S_l term produced other benefits, particularly in the recovery zone further downstream. This is illustrated best by the plot of mean velocity profiles at selected station in semi-logarithmic coordinates, Fig. 8 indicating a strong departure from the universal logarithmic low and inadequacy of the wall functions. Good agreement is also achieved in the recirculation region, Fig. 8. An overall effect of the integration up to the wall with the low-Re-number model is demonstrated by the plot of the friction coefficient along the wall, compared with the computations with the standard high-Re-number model, Fig. 9, showing a substantial improvement.

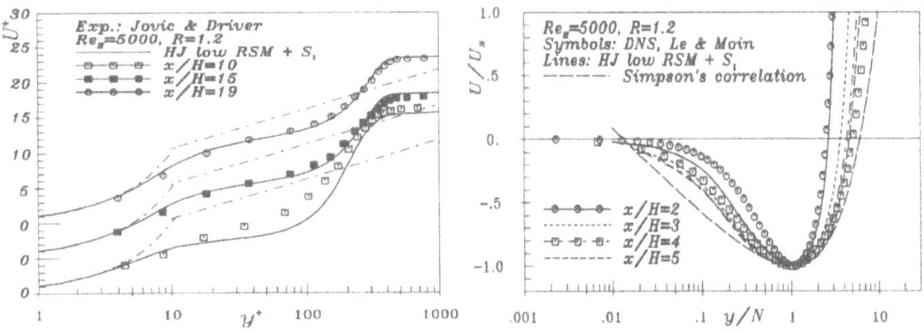

Figure 8. a) - Semi-logarithmic plot of streamwise mean velocity in the recovery region. b) - Mean velocity in the recirculation region. Symbols: experiments and DNS; Lines: computations.

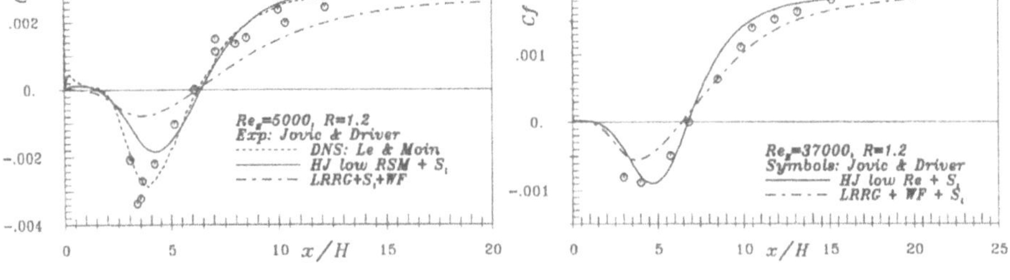

Figure 9. Back-step flow: wall friction factor for low- and high-Re-numbers.

2.5. AIRFOIL

The next example is the flow over the NACA 4412 airfoil at maximum lift, with the incidence angle of 13.87^o (Coles & Wadcock, 1979). Prediction of the leading edge transition (imposed experimentally at $0.023 \leq x/c \leq 0.1$) and of the trailing edge (oscillatory) separation has posed difficulties to standard one-and two-equation turbulence models used in aeronautics, (e.g. (Guilmineau *et al.*, 1997)). In most computations reported in the literature, the transition is imposed artificially at the prescribed location adopted from experiment by switching on the turbulence model (or, alternatively, the production of the kinetic energy). In this case the transition length is very short and the treatment of the transition proved to have little influence on the overall results. However, predicting the transition location in unknown flows without an empirical input is the major criterion in judging the RANS turbulence model for transitional flows.

We have performed computations with a four-block O-grid with both the high- and low-Re-number second-moment closures (398×88 and 398×128 nodes, respectively), without imposing any artificial transition. The nearest grid point to the wall was at y^+ between 0.1 and 1.5 for the fine mesh and between 10 and 50 for the coarse mesh. The high-Re-number model gave almost identical results both for 1.5% and for 5% free stream turbulence. Because of the high bulk flow Reynolds number (the cord-base $Re = 1.52 \times 10^6$), the application of the low-Re-number model due to a need for a fine and highly non-uniform numerical grid near walls becomes more demanding (slower convergence), particularly with low free stream turbulence. It should be noted that all computations were performed in a steady mode, which may not be fully suitable, particularly for a low free-stream turbulence.

We present here results obtained with 5% free stream turbulence. The pressure coefficient obtained by both models agree well with the experiments (Fig. 10a). A difference between the two sets of results is more visible in the mean velocity and shear stress profiles, and, particularly, in the size of the separation bubble, Fig. 10b to 10f. Unlike in the backstep flow, where both models result in a similar streamline pattern in the recirculation zone, here the low-Re-number models yields a substantially thinner bubble, though of almost the same length.

Because of insufficient data, it is difficult to judge which streamline patter is closer to reality. The mean velocity and shear stress profiles obtained by both models show also close agreement with experiments at most locations along the airfoil and also in the near wake, but the low-Re-number model seems to be slightly superior.

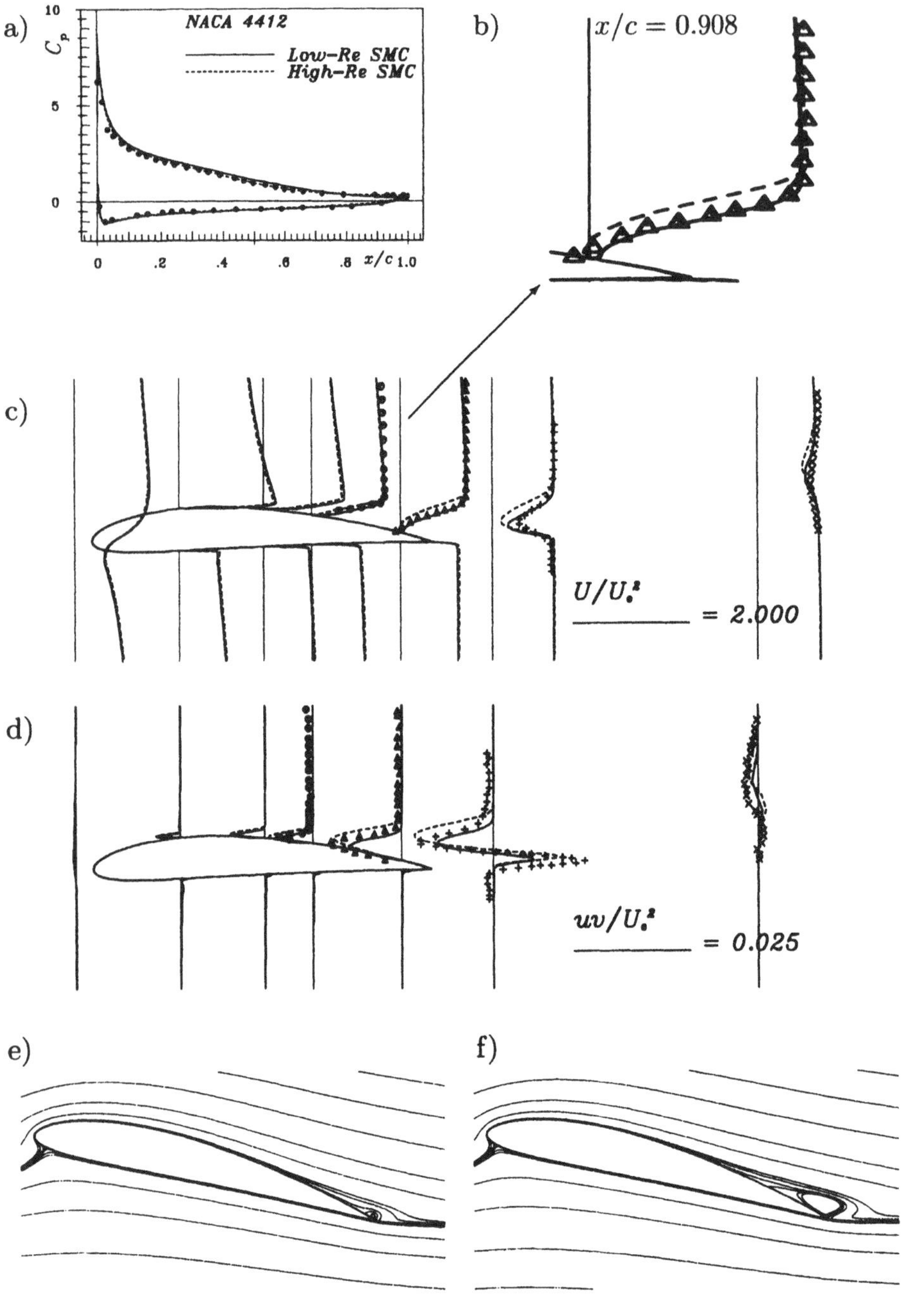

Figure 10. NACA 4412 flow: a) - pressure coefficient C_p; b) - profiles of chordwise mean velocity U/U_o at $x/c = 0.908$; c) and d) - profiles of chordwise mean velocity U/U_o and turbulent shear-stress $\overline{uv}/U_o^2$ at *x/c = -0.05, 0.25, 0.5, 0.642, 0.908, 1.1747, 1.95146* (full lines low-Re RSM, dashed lines high-Re RSM); e) streamlines obtained with low-Re RSM and f) with high-Re RSM + WF.

In view of the fact that some simpler models such as the zonal $k - \omega$ model of Menter, see (Guilmineau *et al.*, 1997), can give satisfactory reproduction of available experimental mean flow parameters for a two-dimensional high-load airfoils, this example may not be adequate for illustrating arguments in favor of second-moment closures. Nonetheless we have included this flow in the present discussion mainly to demonstrate that the second-moment closures (including their low-*Re*-number variants with integration up to the wall) may be successfully used to compute complex flows over curved surfaces and with strong pressure variations.

2.6. THREE-DIMENSIONAL SEPARATION ON A BLUFF BODY

The last case considered provides a comparison between the standard high-*Re*number $k-\varepsilon$ and second-moment model in the computation of a separating flow over a complex 3-dimensional bluff body at high Reynolds number. This illustration is aimed first at demonstrating that the second-moment closure can be successfully applied to a general 3-dimensional flow with only moderate and affordable increase in computing time, compared with $k - \varepsilon$ model. Second, it clearly demonstrates improvement in predictions. Fig. 11 shows the velocity distribution around an automobile with focus on rear window separation obtained using the FIRE commercial code (Basara *et al.*, 1996). Profiles of all three component of the mean velocity at position (4) (rear-window bottom end), obtained with the second-moment $\overline{u_i u_j} - \varepsilon$ model using a quasi-linear SSG pressure-strain model (Speziale *et al.*, 1991) are visibly better as compared with the $k-\varepsilon$ model. It is interesting to note that the new term S_l in the dissipation equation showed no effect, probably due to cross-flow (3-dimensionality effects) around separation and reattachment, but this aspect needs still to be clarified.

3. Concluding Remarks

There is a view in the Computational Fluid Dynamics (CFD) community that the conventional Reynolds-Averaged Navier-Stokes (RANS) turbulence models will soon be replaced by Large-Eddy Simulation (LES) for a wide industrial applications. While such prospects may sound feasible for specific branches of industry, a wider application of LES for real high-*Re*-number flows in aeronautics seem still very distant, as argued recently by (Spalart *et al.*, 1997). The conventional RANS models have not indeed fulfilled the early expectations, but they remain still the only viable means for complex industrial computations. Of course, improvements are needed and possible, and this article outlines some prospects in that direction.

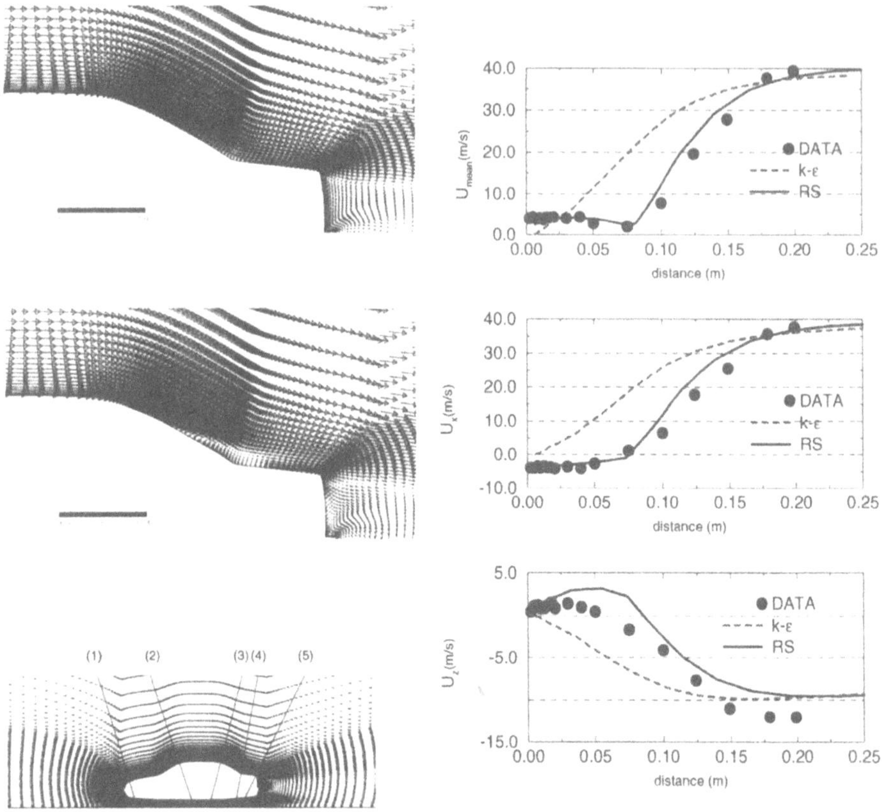

Figure 11. **Velocity vectors around the rear of a car body computed with $k-\varepsilon$ and SSG models, and comparison of velocity profiles at position (4) (Basara *et al.*).**

Some recent advancements in refinements of second-moment closure models have been presented, which brought improvements in reproducing several types of turbulent flows relevant to external aerodynamics. The illustrations focussed on effect of pressure gradient, but also on some other phenomena (transition, separation, reattachment, recovery), which for long could not be satisfactory resolved even with the higher order models, casting a shadow over the RANS approach to model turbulence in complex flows. The considered model refinements, as well as others based on different theoretical approaches, but still within the framework of RANS (non-linear realizable models of pressure-strain based on tensorial expansion, elliptic relaxation method, and others) open new prospects and give new impetus for further improvement and a wider application of higher-order turbulence models in the aerospace industry.

Acknowledgement: We thank P. R. Spalart for providing DNS data for several flows.

4. Appendix: The low-*Re*-number Second-Moment Closure

$$\frac{D\overline{u_iu_j}}{Dt} = \mathcal{D}_{ij} - \left(\overline{u_iu_k}\frac{\partial U_j}{\partial x_k} + \overline{u_ju_k}\frac{\partial U_i}{\partial x_k}\right) + \Phi_{ij} - \varepsilon_{ij}$$

$$\frac{D\varepsilon}{Dt} = \mathcal{D}_\varepsilon - C_{\varepsilon 1}\frac{\varepsilon}{k}\overline{u_iu_j}\frac{\partial U_i}{\partial x_j} - C_{\varepsilon 2}f_\varepsilon\frac{\varepsilon\tilde{\varepsilon}}{k} + C_{\varepsilon 3}\nu\frac{k}{\varepsilon}\overline{u_ju_k}\frac{\partial^2 U_i}{\partial x_j\partial x_l}\frac{\partial^2 U_i}{\partial x_k\partial x_l} + S_\Omega + S_l$$

The pressure-strain model:

$$\Phi_{ij,1} = -C_1\varepsilon a_{ij} \qquad \Phi_{ij,2} = -C_2\left(P_{ij} - \frac{2}{3}P_k\delta_{ij}\right)$$

$$\Phi^w_{ij,1} = C^w_1 f_w\frac{\varepsilon}{k}\left(\overline{u_ku_m}n_kn_m\delta_{ij} - \frac{3}{2}\overline{u_iu_k}n_kn_j - \frac{3}{2}\overline{u_ku_j}n_kn_i\right)$$

$$\Phi^w_{ij,2} = C^w_2 f_w\left(\Phi_{km,2}n_kn_m\delta_{ij} - \frac{3}{2}\Phi_{ik,2}n_kn_j - \frac{3}{2}\Phi_{kj,2}n_kn_i\right)$$

$$C_1 = C + \sqrt{A}E^2 \qquad C = 2.5AF^{1/4}f \qquad F = \min\{0.6; A_2\}$$

$$f = \min\left\{\left(\frac{Re_t}{150}\right)^{3/2}; 1\right\} \qquad f_w = \min\left[\frac{k^{3/2}}{2.5\varepsilon x_n}; 1.4\right]$$

$$C_2 = 0.8A^{1/2} \qquad C^w_1 = \max(1-0.7C; 0.3) \qquad C^w_2 = \min(A; 0.3)$$

$$A = 1 - \frac{9}{8}(A_2 - A_3) \qquad A_2 = a_{ij}a_{ji} \qquad A_3 = a_{ij}a_{jk}a_{ki} \qquad a_{ij} = \frac{\overline{u_iu_j}}{k} - \frac{2}{3}\delta_{ij}$$

$$E = 1 - \frac{9}{8}(E_2 - E_3) \qquad E_2 = e_{ij}e_{ji} \qquad E_3 = e_{ij}e_{jk}e_{ki} \qquad e_{ij} = \frac{\varepsilon_{ij}}{\varepsilon} - \frac{2}{3}\delta_{ij}$$

The model of the stress dissipation rate:

$$\varepsilon_{ij} = f_s\varepsilon^*_{ij} + (1 - f_s)\frac{2}{3}\delta_{ij}\varepsilon$$

$$\varepsilon^*_{ij} = \frac{\varepsilon}{k}\frac{[\overline{u_iu_j} + (\overline{u_iu_k}n_jn_k + \overline{u_ju_k}n_in_k + \overline{u_ku_l}n_kn_ln_in_j)f_d]}{1 + \frac{3}{2}\frac{\overline{u_pu_q}}{k}n_pn_qf_d}$$

$$f_s = 1 - \sqrt{A}E^2 \qquad f_d = (1+0.1Re_t)^{-1} \qquad f_\varepsilon = 1 - \frac{C_{\varepsilon 2} - 1.4}{C_{\varepsilon 2}}\exp\left[-\left(\frac{Re_t}{6}\right)^2\right]$$

Additional terms in ε-equation:

$$S_\Omega = -C_{\varepsilon 4}f_4k\Omega_k\Omega_k \qquad S_l = \max\left\{\left[\left(\frac{1}{C_l}\frac{\partial l}{\partial x_n}\right)^2 - 1\right]\left(\frac{1}{C_l}\frac{\partial l}{\partial x_n}\right)^2; 0\right\}\frac{\tilde{\varepsilon}\varepsilon}{k}A$$

The basic coefficients take the following values:

$$C_s{=}0.22 \quad C_\varepsilon{=}0.18 \quad C_{\varepsilon 1}{=}1.44 \quad C_{\varepsilon 2}{=}1.92 \quad C_{\varepsilon 3}{=}0.25 \quad C_l{=}2.5$$

References

Basara, B., Plimon, A., Bachler, G. and Brandstätter, W., 1996. Calculation of flow around a car body with various turbulence models, *Proc. First MIRA Int. Vehicle Aerodynamic Conference*, Birmingham, UK.

Coles, D. & Wadcock, A.J., 1979. Flying-hot-wire study of flow past a NACA 4412 airfoil at maximum lift, *AIAA J.* **17**(4), pp. 321-329.

Dengel, P. & Fernholz, H.H., 1990. An experimental investigation of an incompressible turbulent boundary layer in the vicinity of separation, *J. Fluid Mech.* **212**, pp. 615-636.

Gibson, M.M. & Launder, B.E., 1978. Ground effects on pressure fluctuations in the atmospheric boundary layer, *J. Fluid Mech.* **86**(3), pp. 491-511.

Guilmineau, E., Piquet, J. & Queutey, P., 1997. Two-dimensional turbulent viscous flow simulation past airfoils at fixed incidence, *Comp. & Fluids* **26**(2), pp. 135-162.

Hanjalić K., 1994. Advanced turbulence closure models: a view of current status and future prospects, *Int. J. Heat & Fluid Flow* **15**(3), pp. 178-203.

Hanjalić K. & Launder, B.E., 1980. Sensitizing the dissipation equation to irrotational strains, *ASME J. Fluids Eng.* **102**, pp. 34-40.

Hanjalić K., Jakirlić & Durst, F., 1994. A computational study of joint effects of transverse shear and streamwise acceleration on three-dimensional boundary layers, *Int. J. Heat & Fluid Flow* **15**(4), pp. 269-282.

Hanjalić, K., Jakirlić, S. & Hadžić, I., 1995. Computation of oscillating turbulent flows at transitional Re-numbers. *Turb. Shear Flows* **9**, Durst F. *et al.*, eds., Springer Berlin, pp. 323-342.

Hanjalić, K. & Hadžić, I., 1996. Modelling the transitional phenomena with statistical turbulence closure models, *Transitional Boundary Layers in Aeronautics*, Henkes, R.A.W.M. & Van Ingen, J.L., eds., North Holland Amsterdam, pp. 283-294.

Hanjalić, K., Jakirlić, S. & Hadžić, I., 1997. Expanding the limits of 'equilibrium' second-moment turbulence closures, *Fluid Dynamics Research* **20**, pp. 25-41.

Hanjalić, K. & Jakirlić, S., 1997. Contribution towards the second-moment closure modelling of separating turbulent flows, *Comp. & Fluids*, (in press).

Jakirlić, S., 1997. *Reynolds-Spannungs-Modelierung komplexer turbulenter Strömungen.* Ph.D. thesis, Univ. Erlangen Germany, UTZ Verlag, Munich.

Justesen, P. & Spalart, P.R., 1990. Two-Equation Turbulence Modeling of Oscillatory Boundary Layers, *AIAA-90-0496.*

Nagano, Y., Tagawa, M. & Tsuji, T., 1993. Effects of adverse pressure gradients on mean flows and turbulence statistics in a boundary layer, *Turb. Shear Flows* **8**, Durst F. *et al.*, eds., Springer Berlin, pp. 7-21.

Samuel, A.E. & Joubert, P.N., 1974. A boundary layer developing in an increasingly adverse pressure gradient, *J. Fluid Mech.* **66**(3), pp. 481-505.

Savill, A.M., 1996. Transition predictions with turbulence models. *Transitional Boundary Layers in Aeronautics*, Eds. Henkes, R.A.W.M. & Van Ingen, J.L., North Holland Amsterdam, pp. 311-319.

Simpson, R.L., Chew, Y.-T. & Shivaprasad, B.G., 1981. The structure of a separating turbulent boundary layer. Part 1. Mean flow and Reynolds stresses, *J. Fluid Mech.* **113**, pp. 23-51.

Spalart, P.R., Jou, W-H., Strelets, M. & Allmaras, S.R., 1997. Comments on the feasibility of LES for wings, and on hybrid RANS/LES approach, 1st AFOSR Int. Conf. on DNS and LES, Louisiana Tech. Univ. Ruston, LA.

Spalart, P.R. & Watmuff, J.H., 1993. Experimental and numerical study of a turbulent boundary layer with pressure gradient, *J. Fluid Mech.* **249**, pp. 337-371.

Spalart, P.R. & Coleman, G.N., 1997. Numerical study of a separation bubble with heat transfer, *Eur. J. Mech., B./Fluids* **16**, pp. 169-189.

Speziale C.G., Sarkar S. & Gatski T.B., 1991. Modeling the pressure-strain correlation of turbulence: An invariant dynamical systems approach. *J. Fluid Mech.* **227**, pp. 245-272.

SOME STRUCTURAL FEATURES OF PRESSURE-DRIVEN THREE-DIMENSIONAL TURBULENT BOUNDARY LAYERS FROM EXPERIMENTS

ROGER L. SIMPSON, M. SEMIH ÖLCMEN, J.L. FLEMING AND D.S. CIOCHETTO
Department of Aerospace and Ocean Engineering
Virginia Polytechnic Institute and State University
Blacksburg, Virginia

Abstract. A variety of experiments have revealed some generalizations about the behavior of three-dimensional turbulent boundary layers. These include the anisotropic nature of the eddy viscosity and the lags between the mean flow gradients and the turbulence. The v' rms normal-to-wall velocity fluctuation appears to be an important velocity scale. The parameters $S = v'^2/|\tau/\rho|$, $B_2 = \overline{v^3}/[(\overline{uv^2})^2 + (\overline{v^2w})^2]^{1/2}$ and $\overline{v^3}/(\overline{u^2v} + \overline{v^3} + \overline{w^2v})$ are independent of rotation about the normal to wall axis and have nearly universal relationships in the outer region of many non-equilibrium 3D boundary layers. Comparisons of some second-order turbulence models with results deduced from detailed measurements indicate that better modeling of pressure and turbulence diffusion and pressure/rate-of-strain terms is needed, especially near the wall. Some work underway is providing v spatial correlation data that is needed for estimating the pressure fluctuation related terms.

1. Introduction

Three-dimensional pressure-driven turbulent boundary layers (3DTBL) occur in many practical cases, which in most cases are non-equilibrium flows. The imposed spanwise pressure gradient causes a change in direction of the flow as a function of distance from the wall. The purpose of this article is to give some features of the structure of such flows as revealed from many recent experiments, which can increase the understanding and improve the modeling of these flows.

M. D. Salas et al. (eds.), Modeling Complex Turbulent Flows, 223–243.

Simpson (1996) presented a perspective on 3DTBL behavior and reviewed many related referenced works. Here we will summarize first the key insights and conclusions from that work and augment them with some new observations. While results from many other flow cases will be included, the principal example geometry will be the flat-wind-tunnel-wall 3D turbulent boundary layer (approach momentum thickness Reynolds number $Re_\theta = 5940$) and the horseshoe-vortex flow around the junction with a 3:2 elliptic nose, NACA 0020 tail wing (Figure 1), which was experimentally examined by the authors. These wind tunnel results are compared with some recent water tunnel results for this geometry at much lower Re_θ values.

Using these and other experimental data sets, the strong relationship between the normal to the wall rms fluctuation v' and the shearing stress magnitude will be discussed, along with other observed relationships among important velocity fluctuation triple products in some stress transport equations. A summary will be given of comparisons between some second-order modeling equations with experimental results. Finally, some comments will be made on needed and future work.

2. Summary of Some Previous Insights

Simpson (1996) and Johnston and Flack (1996) implicitly show that many recently revealed features of 3DTBLs are due to the use of three-velocity-component laser-Doppler anemometry. In our work a specially-designed fine-spatial-resolution ($30\mu m$ measurement volume diameter) fiber-optic laser-Doppler velocimeter (LDV) (5-velocity-component) is used for pointwise and two-point spatial correlation measurements as close to the wall as $y^+ = 3$ (Ölçmen & Simpson, 1995b). Although one should be able to infer the wall shearing stress τ_w from the nearest-wall LDV velocity profile, Johnston and Flack suggest that there is a continuing need for improved direct measurement of τ_w.

Examination of a number of data sets shows that no universal law-of-the-wall mean-velocity profile exists. Local near-wall equilibrium along some flows cause the mean velocity profiles to be self-similar in wall coordinates, but without a universal mean-velocity profile law. A law-of-the-wall semi-logarithmic profile with universal constants only seems to exist for the beginning stages of 3D flow. Without a relationship with universal constants, the law-of-the-wall concept loses much of its usefulness.

The eddy viscosity is highly anisotropic in 3D flows. Algebraic turbulence models that relate shear stresses to mean velocity gradients through eddy viscosities and mixing lengths continue to be widely used in aerodynamic calculations, largely because of their relative simplicity compared to other models and the collective experience of users of these models. These models are crude approximations to the flow physics and are not robust for strongly 3D cases, such as separated flows. Accelerating portions of 3D flows with weak crossflows may be adequately calculated with this approach, but adverse pressure gradient flows are not well calculated. The observed reduction in eddy viscosity magnitude, as compared to algebraic eddy viscosity models, is due to non-equilibrium lags in the turbulence structure, which usually accompany the strong adverse pressure gradients, even in 2D cases.

The mean turbulent shearing stresses generally lag the mean velocity gradient direction, so that an isotropic eddy viscosity turbulence model cannot reflect the correct physics of the stress-producing structures. There appears to be a reduction in the correlation between the shearing stress and the turbulent kinetic energy due to the different histories of the outer layer and inner layer flow structures — they come from different upstream directions (Figure 2). In the wall shear-stress co-ordinates, there appears to be a nearest-wall region in which the mean flow direction does not change much with the distance from the wall. However, the turbulent shear-stress direction is generally different from this direction and the mean-velocity gradient direction.

The eddy viscosity for the spanwise direction ν_z is not a constant factor N_e times the streamwise eddy viscosity ν_x and depends upon the co-ordinate system. N_e appears to be about constant in the outer layer but varies much near the wall. In more detail than presented here, Simpson (1996) discusses the data and suggests that $N_e = 0.6$ in a wall-shearing-stress co-ordinate system or local streamline co-ordinate system is a prescription for crudely accounting for anisotropy in eddy viscosity calculation methods in many flows. This prescription of co-ordinate system forces the spanwise shearing stress near the wall to approach zero either with a constant or varying N_e since the crossflow velocity gradient $\partial W/\partial y$ is zero at the wall. While a number of experiments have about this value of N_e, other flows examined by Ciochetto and Simpson (1997) have N_e values away from the wall that are closer to unity, e.g., Pompeo *et al.* (1992; 1993) and Schwarz and Bradshaw (1992; 1993).

This prescription can also be justified from a physical viewpoint; the relatively large wall-region turbulent shear stresses control the flow more than the outer layer stresses and need to be more accurately modeled. The nearest-wall flow seems to approximate a mean 2D flow. The wall-stress co-ordinates are more closely aligned than local free-stream co-ordinates with the mean shear stress direction (SSA) in this region. Therefore, near-wall estimates of ν_x in WC will be similar to mean 2D cases and errors in ν_z will have less impact on the calculated total Reynolds shear stress.

3. Some Features of an Example Reynolds-averaged 3D Flowfield

The well-documented wing/body junction low-speed flow of Ölçmen and Simpson (1995a; 1996a) can be used as a computational test case of turbulence models. The mean 2D upstream flow develops into a non-equilibrium 3D flow and then relaxes back toward a mean 2D flow far downstream. In addition to measurements along the path shown in Figure 3(a), which is outside the region near the wing where large mean flow streamwise vortices are dominant, the mean flowfield and Reynolds stresses have been obtained over large regions of this geometry: (1) the nose separated flow and juncture vortex around the side of the wing, which is briefly discussed below (Devenport & Simpson, 1990a; Devenport & Simpson, 1990b; Devenport & Simpson, 1990c); (Ölçmen & Simpson, 1997b); (2) the mean 2D flow entrance conditions (Ölçmen & Simpson, 1990; Ölçmen & Simpson, 1995a); and (3) the downstream wake region (Fleming *et al.*, 1991). Skin friction magnitude and direction were directly measured using oil-film interferometry (Ailinger & Simpson, 1990). In addition, convective heat-transfer and the temperature field have been measured for this flowfield (Lewis *et al.*, 1993; Lewis & Simpson, 1996; Lewis & Simpson, 1998).

Figure 3(b) shows mean velocity profiles at these stations in local wall shear stress coordinates. In upstream flow direction or wind tunnel coordinates (Simpson, 1996), the streamwise wind tunnel mean flow velocity U first slightly decelerates due to the adverse pressure gradient and then accelerates around the wing, while the crossflow W magnitude increases along the flow. The maximum W magnitude in tunnel coordinates is located progressively further from the wall. In the outer region of a pressure-driven flow the Reynolds shearing stresses persist with the upstream 2DTBL magnitude and direction until the 3D effects propagate to that location away from the wall. It appears that transport equations are needed to account for this nature of the lag within the flow.

Figure 4 shows a comparison of wind tunnel data ($Re_\theta = 9520$) at Station 5 with the low Reynolds number ($Re_\theta = 500, 760$, and 890) water tunnel data of Fleming and Simpson (1997) at the same station for flow around the same shaped wing. Both flowfields are subjected to the same non-dimensional pressure distributions around the wing. The U/U_e and W/U_e profiles in local freestream direction coordinates scale fairly well in terms of y/δ in the outer layer, apparently due to the turbulence-dominated transport and skewing of the vorticity in the outer layer. The largest Reynolds number effect is the increase in mean crossflow velocity with Reynolds number. The relative increase in viscous forces in the near-wall region at low Reynolds numbers tends to keep the boundary layer mean flow more closely aligned with the free-stream direction.

Using multiple probes in the wind tunnel flow, Ha and Simpson (1993) found that the coherency of larger-scaled structures appear to be reduced by the effects of three-dimensionality. This length scale reduction occurs with the growth of 3D turbulence effects near the wall that are unrelated to the outer region flow that originated from another direction in the upstream flow (Figure 2). Length scales in the outer region showed no change along the flow, indicating strong persistence of the coherency of outer region structures along the local free-stream direction. In the low Reynolds number cases, length scale results from autocorrelations revealed 50% shorter integral length scales for the 3D flow as compared to a 2D flow, which is consistent with the higher Reynolds number wind tunnel results.

The skewing angle of the coherent structures closely follows the local mean velocity angle especially in the inner region, indicating that the coherent structures are convected in the mean flow direction. In the outer layer at this location, equally coherent motions occur over a wider angular range between the far upstream 2D flow direction and the local flow direction. This indicates that the direction of the far upstream outer region coherent motions lags for a long distance, even after being subjected to the skewing effects of local turbulence transport equation mechanisms.

The turbulent kinetic energy (TKE), which is frequently used in turbulence modeling, also remains about frozen for Stations 1–4. This suggests that the TKE is close to equilibrium in which the production equals the sum of dissipation, net diffusion, and convection at each of these stations. As the flow accelerates at Stations 5–7, the near-wall production of TKE causes a great increase of TKE for $10 < y^+ < 60$, while the TKE decreases slightly in the outer region.

The fluctuation terms are rather sensitive to Reynolds number. Figures 5–8 show the u', v' and w' rms fluctuations and shearing Reynolds stresses for both the high and low Reynolds number cases at Station 5. While the u', v' and w' distributions have similar shapes over these Reynolds numbers, the profiles for a given stress cannot be scaled on either inner or outer variables to produce a collapse.

Like the magnitude of the shearing stress $|\tau/\rho|$, the TKE and the ratio of the two parameters ($|\tau/\rho|/TKE = 2a_1$), another parameter which is independent of rotation about the y-axis is $|\tau/\rho|/v'^2 = 1/S$ and its reciprocal. For 2D flows, S is near unity in the semi-log mean velocity profile region. For this 3D flow, S is nearly a constant over an order of magnitude of y for a given profile. Ölçmen and Simpson (1995a) examined nine other pressure-driven and shear-driven datasets and concluded that S is also about constant for a given profile for $y^+ > 50$ and $y/\delta < 0.6$, with values between 1 and 2. At stations where the 3D effects are largest, S is higher than obtained for 2D flows, which is due to less correlation between the u and v velocity fluctuation components. Figure 9 shows almost no effect of Reynolds number on S. (The departure of the nearest wall data from a common correlation is believed to be due to high measured values of v' as the wall is approached.) Figures 5, 7, and 8 and Fleming and Simpson (1997) show that the profiles of v'^2, $-\overline{uv}$, and $-\overline{vw}$, over the large Reynolds number range do not correlate in inner variables or outer variables, while $1/S$ collapses in inner variables. This is in contrast to a_1, which varies much with 3D effects and Reynolds number.

On physical grounds, v' should be an important turbulence velocity scale, especially near the wall since the shear-stress producing ejections and sweeps strongly determine v'. The a_1 parameter contains the u'^2 in the TKE, which reflects some of the inactive low frequency motions that do not contribute to the shear-stress producing motions. The parameter S directly relates this scale v' to the desired shearing stress $|\tau/\rho|$ to help provide closure on a turbulence model.

Another phenomenon occurs for the flowfield shown in Figure 1. For a sufficiently blunt nose, such as this wing, the horseshoe vortex structure between the vortex and the flat wall shows a periodic low-frequency chaotic switching between velocity states that produce double-peaked (bimodal) velocity histograms of the velocity component spanwise to the vortical core direction. This self-induced large-scale unsteadiness of wing/body junction nose separations is responsible for high surface pressure fluctuations and high heat transfer rates around the nose. These chaotic vortices are produced by the 3D separation in front of the nose; multiple unsteady vortices are present in this region that can merge and are stretched around the wing. These chaotic intermittent vortical structures also produce very large apparent Reynolds-averaged stress values because their unsteady flow effects are averaged with smaller-scaled turbulent contributions. In one mode the flow is nearly tangent to the wing while in the other mode higher velocity flow is swirled down near the wing from near the freestream by the vortical motion. Near the flat wall, the latter mode flow is at a large angle to the wing surface. Simpson (1996) in his review and Ölçmen and Simpson (1997b) give many more details and examples of the features of this flow behavior, which appears to be present in all turbulent wing/body junction cases with sufficient nose bluntness.

Clearly, it is inappropriate to combine the effects of these large-scale separation-induced chaotic vortices with the turbulence structure. A seemingly proper way to model the horseshoe vortex flow around the wing is to use a large-eddy simulation for the chaotic scales, which are more coherent, and a subgrid model for the less coherent turbulence that accompanies the boundary layer that approaches the wing.

4. Some Modeling Equations

It appears clear that the isotropic eddy viscosity approach to 3DTBL modeling does not capture the flow physics. As a result of the limitations of algebraic models to mimic the observed lags within these flows, a 3-Reynolds-stress transport equation model is suggested with the normal-to-wall rms velocity fluctuation v' as the velocity scale, since v'^2 correlates well with the shearing stress magnitude $|\tau/\rho|$. Clearly, this model cannot mimic the nature of the low-frequency chaotic horseshoe vortices mentioned above.

Observations presented in this paper suggest that transport equations are required to account for: (1) the variable anisotropy of the eddy viscosities, (2) the lags between the mean flow and the turbulence field, and (3) the strong relation between the important shearing stresses and v':

x-direction momentum equation:

$$\frac{DU}{Dt} = U\frac{\partial U}{\partial x} + V\frac{\partial U}{\partial y} + W\frac{\partial U}{\partial z} = \frac{-1}{\rho}\frac{\partial P}{\partial x} + \nu\frac{\partial^2 U}{\partial y^2} - \frac{\partial \overline{uv}}{\partial y} \tag{1}$$

z-direction momentum equation:

$$\frac{DW}{Dt} = U\frac{\partial W}{\partial x} + V\frac{\partial W}{\partial y} + W\frac{\partial W}{\partial z} = \frac{-1}{\rho}\frac{\partial P}{\partial z} + \nu\frac{\partial^2 W}{\partial y^2} - \frac{\partial \overline{vw}}{\partial y} \tag{2}$$

Continuity equation:

$$\frac{\partial U}{\partial x} + \frac{\partial V}{\partial y} + \frac{\partial W}{\partial z} = 0 \tag{3}$$

In addition to the momentum and continuity equations, transport equations for uv, $-vw$, and v^2 are required. Transport of any $-uv$ stress:

$$\frac{D(-\overline{uv})}{Dt} = \overline{v^2}\frac{\partial U}{\partial y} - \overline{\frac{p'}{\rho}\left(\frac{\partial u}{\partial y} + \frac{\partial v}{\partial x}\right)} + \frac{\partial}{\partial y}\left(\frac{\overline{p'u}}{\rho} + \overline{uv^2}\right) - \nu\left(\overline{v\,\nabla^2 u + u\,\nabla^2 v}\right) \tag{4}$$

Transport of $-vw$ stress:

$$\frac{D(-\overline{vw})}{Dt} = \overline{v^2}\frac{\partial W}{\partial y} - \overline{\frac{p'}{\rho}\left(\frac{\partial w}{\partial y} + \frac{\partial v}{\partial z}\right)} + \frac{\partial}{\partial y}\left(\frac{\overline{p'w}}{\rho} + \overline{v^2 w}\right) - \nu\left(\overline{w\,\nabla^2 v + v\,\nabla^2 w}\right) \tag{5}$$

Transport of $-v^2$ stress:

$$\frac{1}{2}\frac{D(\overline{v^2})}{Dt} = -\overline{v^2}\frac{\partial V}{\partial y} - \frac{\partial}{\partial y}\left(\frac{\overline{v^3}}{2}\right) - \overline{\frac{v}{\rho}\frac{\partial p'}{\partial y}} + \nu\left(\overline{v\,\nabla^2 v}\right) \tag{6}$$

Algebraic relations among some of the parameters in these equations have been revealed from the database of 3D experiments examined by Ciochetto and Simpson (1995; 1997). Eleven datasets encompassing several different test geometries were examined: Pompeo *et al.* (1992; 1993) plane of symmetry flows with spanwise rates-of-strain; Schwarz and Bradshaw (1992; 1993) 30° bend flow; Ölçmen and Simpson (1995a) wing/body junction flow; Baskaran *et al.* (1990) pressure-driven flows with wall curvature; Chesnakas and Simpson (1994; 1992; 1997) Kreplin and Stäger (1993), and Barberis and Molton (1993) leeside flows on axisymmetric bodies at angles of attack; Devenport and Simpson (1990b) and McMahon *et al.* (1982) wing/body vortex flows; and Driver and Johnston (1990) and Littell and Eaton (1991) shear-driven flows.

In these datasets, the $S = v'^2/|\tau/\rho|$ parameter correlates fairly well from station to station within a flow and between experiments, *as long as embedded mean flow vortices are not present.* The ratio $1/S$ approximates a constant for $0.3 - 0.4 < y/\delta < 0.7 - 0.8$ ranging from values of $0.5 - 0.8$ with an overall average of approximately 0.7. At $y/\delta = 1.0, 1/S$ appears to have a mean value, for the data presented, or approximately 0.3. The parameter appears to be mildly affected by different 3D effects, however the effects were of the order of the uncertainty in the experiments. Such effects include a noticeable decrease in the crossflow decay region of the Schwarz and Bradshaw experiment (Figure 10), a slight rise at the aft end of the prolate spheroid flows of Kreplin and Stäger and Chesnakas and Simpson due to separation vortices, and a lower value in the decay region of Driver. It did, however, maintain consistent values for the Baskaran *et al.* flows (Figure 11) and the experiments of Littell and Eaton. All of the other parameters investigated for the experiment of Littell and Eaton had completely different behavior from the other 3D experiments, only $1/S$ exhibited behavior consistent with all of the other 3DTBLs. The experiments in the horseshoe vortex region of the wing-body junction failed to produce good behavior of the $1/S$ parameter. This is surmised to result from the significant wall-normal component of velocity and the chaotic vortical structure.

Some algebraic parameters describing the turbulent triple products were less sensitive to three-dimensional effects than $1/S$. They also maintained a constant value for the outer part of the 3DTBL. The $B_2 = \overline{v^3}/\left[\left(\overline{uv^2}\right)^2 + \left(\overline{v^2w}\right)^2\right]^{1/2}$ parameter is invariant to rotation about the y-axis and relates the turbulent transport of the instantaneous stresses $\overline{v^2}$, $-\overline{uv}$, and $-\overline{vw}$ in the y direction. This parameter tends to approximate a constant value of 0.85 ± 0.15 for $0.3 \le y/\delta \le 0.7 - -1.0$. The values for each station are a very good approximation to a constant in this region for the experiments examined by Ciochetto and Simpson. This strongly indicates that the same outer region intermittent coherent structure flow phenomena produce these triple products, even for the convex curved wall and line of symmetry flows (Figure 12).

The parameter $\overline{v^3}/\left(\overline{u^2v}+\overline{v^3}+\overline{w^2v}\right)$, which is also invariant to y-axis rotation, maintains better correlation across the TBL and from experiment to experiment than the B_2 parameter. It seems to hold even for the leeside vortex flow of Kreplin and Stäger. It maintains a value of $0.3-0.4$ for $Y/\delta > 0.4$ and tends to extend farther towards the wall for the upstream stations in a given experiment. These results indicate that the turbulent transport of TKE is closely related to the v transport of $\overline{v^2}$ and could simplify outer region modeling. Figure 13 shows that surface convex curvature does not have much effect on this conclusion.

5. Examination of Other Model Equations

The datasets used above provide some insights into some algebraic relations, mainly in the outer logarithmic velocity profile regions, between v'^2 and $|\tau/\rho|$ and among several triple-product turbulent diffusion terms which are strongly correlated in a variety of pressure-driven 3D non-equilibrium turbulent flows. However, additional relationships are needed to complete the closure of a calculation method. The viscous dissipation term in Equation 6 needs to be modeled, while the viscous terms in each of Equations 4 and 5 are almost negligible except near the wall. The pressure diffusion and pressure/rate-of-strain terms in Equations 4–6 need to be modeled.

Ölçmen and Simpson (1996b; 1996c; 1996d; 1997a) examined several second-order closure models in light of their detailed experimental data (Ölçmen & Simpson, 1997b) obtained for a 2DTBL and several locations shown in Figure 1 for the wing/body junction wind tunnel flow. Enough spatial data were obtained to determine the convective terms from data and perform a term-by-term examination of the Reynolds-stress transport equation budgets (Ölçmen & Simpson, 1996b). All terms except the dissipation (ϵ), pressure diffusion, and pressure/rate-of-strain (PRS) terms were evaluated directly using data.

Using the TKE transport equation and Lumley's (1978) pressure diffusion model, the dissipation rate ϵ was obtained from the difference among other terms. The Hallbäck *et al.* (1990) anisotropic dissipation distribution was used and produced stress transport equation budgets that were in better agreement with low Reynolds number direct numerical simulation (DNS) budgets than the isotropic dissipation case.

The resulting pressure/rate-of-strain (PRS) terms for each transport equation and flow location were then compared by Ölçmen and Simpson (1996b; 1996c; 1996d) with seven models. The Fu *et al.* (1987) pressure/rate-of-strain model (labeled FLT1) as it was modified by Shih and Lumley (1993) (labeled FLT2) worked best among the tested models for the 2D flow normal stresses. The Shih-Lumley/Choi-Lumley (SLCL) (1984) PRS model worked best for the $-\overline{uv}$ shearing stress transport equation for the 2D flow.

The FLT2 PRS model for the v'^2 transport equation worked best for the 3D flow stations that were tested which were away from the chaotic bimodal vortices that were mentioned above. For the same stations, the PRS terms in the shear stress transport equations were best described by the FLT1 and SLCL models. There is still a need for improved PRS models for 3D flows, especially near the wall in the $-\overline{vw}$ shearing stress transport equation.

Ölçmen and Simpson (1997a) used several turbulent diffusion models to compare with the experimentally measured profiles. All models that were tested failed to capture the magnitude of the triple products, apparently due to the scaling factor q^2/ϵ. Note that since the experimentally measured turbulent diffusion terms were used in the transport equation budgets, Ölçmen and Simpson's conclusions about the PRS models do not depend upon a turbulent diffusion model.

6. Concluding Comments and Future Work

Because of the non-equilibrium nature of 3D turbulent boundary layers, it is necessary to use Reynolds-averaged transport equations which can mimic the lags between the mean flow and the shearing stress structure. Work discussed and referenced here show that v' is closely related to the shear stress magnitude by S in a variety of non-equilibrium 3D experiments over a range of Reynolds numbers. The parameters $B_2 = \overline{v^3} / \left[\left(\overline{uv^2} \right)^2 + \left(\overline{v^2 w} \right)^2 \right]^{1/2}$ and $\overline{v^3} / \left(\overline{u^2 v} + \overline{v^3} + \overline{w^2 v} \right)$ are nearly constant across the outer boundary layer in these flows.

These parameters alone are not enough for closure and relationships for the pressure diffusion, pressure/rate-of-strain (PRS), turbulent diffusion, and dissipation are needed. The work quoted here shows that several uncertainties exist in the modeling of these terms, especially in the near-wall region. Better models for the turbulent diffusion are needed. The relationships among pressure and velocity fluctuations remain an important modeling issue. Data are needed for the pressure diffusion and PRS in order to verify existing models and to develop better models.

Since the direct measurement of the pressure fluctuation within these flows is presently unfeasible with a fine-resolution sensor, one must use the Poisson volumetric integral of the turbulent velocity fluctuation correlation contributions to the pressure fluctuation (Chou, 1945). To this end, recently considerable $v_i - v_j$ and $v_i - v_j^2$ LDV spatial correlation wind tunnel data have been obtained for a 2DTBL and at Station 5 of the wing/body junction wind tunnel flow. These data are being used in the Poisson pressure/fluctuation volume integral in order to better understand the effects of 3D skewing on the pressure fluctuation structure.

7. Acknowledgements

Portions of the referenced work at the Virginia Polytechnic Institute and State University have been supported by: Office of Naval Research (current Grant N00014-94-1-0092; Dr. L.P. Purtell, Program Manager), Defense Advanced Research Projects Agency, Air Force Office of Scientific Research, and the National Aeronautics and Space Administration. The authors gratefully acknowledge the support of these agencies.

References

Ailinger, K.G. and Simpson, R.L. (1990). "Measurements of Surface Shear Stresses Under a Three-dimensional Turbulent Boundary Layer Using Oil-film Laser Interferometry," Virginia Polytechnic Institute and State University Report VPI-AOE-173; DTIC Report ADA2294940XSP.

Barberis, D. and Molton, P. (1993). "Experimental Study of 3-D Separation on a Large Scale Model," AIAA 24th Fluid Dynamics Conference, Orlando, FL, AIAA-93-3007.

Baskaran, V., Pontikis, Y.G., and Bradshaw, P. (1990). "An Experimental Investigation of a Three-dimensional Turbulent Boundary Layer on 'Infinite' Swept Curved Wings," *Journal of Fluid Mechanics* **211**, pp. 95–122.

Chesnakas, C.J. and Simpson, R.L. (1992). "An Investigation of the Three-dimensional Turbulent Flow in the Cross-flow Separation Region of a 6:1 Prolate Spheroid," Sixth International Symposium on Applications of Laser Techniques to Fluid Mechanics, Lisbon, Portugal; *Exp. in Fluids* **17**, (1994), pp. 68–74.

Chesnakas, C.J. and Simpson, R.L. (1997). "Detailed Investigation of the Three-dimensional Separation About a 6:1 Prolate Spheroid," *AIAA Journal* **35**, No. 6, pp. 990–999.

Chesnakas, C.J., Simpson, R.L., and Maddem M.M. (1994). "Three-dimensional Velocity Measurements on a 6:1 Prolate Spheroid at 10deg Angle of Attack," Virginia Polytechnic Institute and State University Report VPI-AOE-202REV; DTIC Report ADA2764850XSP.

Choi, K.S. and Lumley, J.L. (1984). "Turbulence and Chaotic Phenomena in Fluids," *Proceedings IUTAM Symposium*, Kyoto, Japan, T. Tatsumi, ed., North-Holland, Amsterdam, p. 267.

Chou, P.Y. (1945). "On Velocity Correlations and the Solutions to the Equations of Turbulent Motions," *Quarterly Applied Math.* **3**, pp. 38–54.

Ciochetto, D.S. and Simpson, R.L. (1995). "An Investigation of 3-D Turbulent Shear Flow Experiments and Modeling Parameters," *Turbulent Shear Flows 10*, Penn. State, pp. 7-25–7-30.

Ciochetto, D.S. and Simpson, R.L. (1997). "Analysis of Three-dimensional Turbulent

Shear Flow Experiments with Respect to Algebraic Modeling Parameters," Virginia Polytechnic Institute and State University Report VPI-AOE-248; submitted to DTIC.

Devenport, W.J. and Simpson, R.L. (1990a). "A Time-dependent and Time-averaged Turbulence Structure Near the Nose of a Wing-body Junction," *Journal of Fluid Mechanics* **210**, pp. 23–55.

Devenport, W.J. and Simpson, R.L. (1990b). "The Flow Past a Wing-body Junction — An Experimental Evaluation of Turbulence Models," 18th Symposium on Naval Hydrodynamics, Ann Arbor, MI; *AIAA Journal* **30**, (1992), pp. 873–881.

Devenport, W.J. and Simpson, R.L. (1990c). "An Experimental Investigation of the Flow Past an Idealized Wing-body Junction," Virginia Polytechnic Institute and State University Report VPI-AOE-172; DTIC Report ADA2296028XSP.

Driver, D.M. and Johnston, J.P. (1990). "Experimental Study of a Three-dimensional Shear-driven Turbulent Boundary Layer with Streamwise Adverse Pressure Gradient," NASA TM-102211; also Report MD-57, Thermosciences Division, Department of Mechanical Engineering, Stanford University.

Fleming, J.L. and Simpson, R.L. (1997). "Experimental Investigation of the Near Wall Flow Structure of a Low Reynolds Number 3-D Turbulent Boundary Layer," Virginia Polytechnic Institute and State University Report VPI-AOE-247; submitted to DTIC.

Fleming, J., Simpson, R.L., and Devenport, W.J. (1991). "An Experimental Study of a Turbulent Wing-body Junction and Wake Flow," Virginia Polytechnic Institute and State University Report VPI-AOE-179; DTIC Report ADA2433886XSP; AIAA-92-0434; *Exp. Fluids* **14**, (1993), pp. 366–378.

Fu, S., Launder, B.E., and Tselepidakis, D.P. (1987). "Accomodating the Effects of High Strain Rates in Modeling the Pressure-strain Correlation," UMIST Mechanical Engineering Department Report TFD/87/5.

Ha, S. and Simpson, R.L. (1993). "An Experimental Investigation of a Three-dimensional Turbulent Boundary Layer Using Multiple-sensor Probes," Ninth Symposium on Turbulent Shear Flows, Kyoto, Japan, pp. 2-3–(1-6).

Hallbäck, M., Groth, J., and Johansson, A.V. (1990). "An Algebraic Model for Non-isotropic Turbulent Dissipation Rate in Reynolds Stress Closure," *Physics of Fluids A* **2**, No. 10, pp. 1859–1866.

Johnston, J.P. and Flack, K.A. (1996). "Review — Advances in Three-dimensional Turbulent Boundary Layers with Emphasis on the Wall-layer Regions," *Journal of Fluids Engineering* **118**, No. 2, pp. 219–232.

Kreplin, H.P. and Stäger, R. (1993). "Measurements of the Reynolds Stress Tensor in the Three Dimensional Boundary Layer of an Inclined Body of Revolution," Ninth Symposium on Turbulent Shear Flows, Kyoto, Japan, pp. 2-4–(1-6).

Lewis, D.J. and Simpson, R.L. (1996). "An Experimental Investigation of Heat Transfer in Three-dimensional and Separating Turbulent Boundary Layers," Virginia Polytechnic Institute and State University Report VPI-AOE-229; submitted to DTIC.

Lewis, D.J. and Simpson, R.L. (1998). "Turbulence Structure of Heat Transfer Through a Pressure-driven Three-dimensional Turbulent Boundary Layer," *AIAA Journal of Thermophysics and Heat Transfer* **12**, No. 2, pp. 248–255.

Lewis, D.J., Simpson, R.L., and Diller, T. (1993). "Time-resolved Surface Heat Flux Measurements in the Wing/Body Junction Vortex," AIAA-93-0291, AIAA 31st Aerospace Sciences Meeting, Reno, NV; *AIAA Journal of Thermo. and Heat Trans.* **8**, (1994), pp. 656–663.

Littell, H.S. and Eaton, J.K. (1991). "An Experimental Investigation on the Three-dimensional Boundary Layer on a Rotating Disk," Report No. MD-60, Thermosciences Division, Department of Mechanical Engineering, Stanford University.

Lumley, J.L. (1978). "Computation and Modeling of Turbulent Flows," *Advances in Applied Mech.* **18**, pp. 124–176.

McMahon, H., Hubbartt, J., and Kubendran, L. (1982). "Mean Velocities and Reynolds Stresses in a Juncture Flow," NASA CR-3605.

Ölçmen, S.M. and Simpson, R.L. (1990). "An Experimental Investigation of Pressure-driven Three-dimensional Turbulent Boundary Layer," Virginia Polytechnic Institute and State University Report VPI-AOE-178; DTIC Report ADA2294957XSP.

Ölçmen, S.M. and Simpson, R.L. (1995a). "An Experimental Study of Three-dimensional Pressure-driven Turbulent Boundary Layer, *J. of Fluid Mechanics* **290**, pp. 225–262.

Ölçmen, S.M. and Simpson, R.L. (1995b). "A 5-Velocity-Component Laser-Doppler Velocimeter for Measurements of a Three-dimensional Turbulent Boundary Layer, paper 4.2," Seventh International Symposium on Applications of Laser Techniques to Fluid Mechanics, Lisbon, Portugal; revised version invited paper, *Measurement Science and Technology* **6**, (1995), pp. 702–716. Highlighted in *The Year of Review*, Fluid Dynamics, pp. 20–21, *Aerospace America* **33**, No. 12, (1995).

Ölçmen, S.M. and Simpson, R.L. (1996a). "An Experimental Study of a Three-dimensional Pressure-driven Turbulent Boundary Layer: Data Bank Contribution," *Journal of Fluids Engineering* **118**, pp. 416–418.

Ölçmen, S.M. and Simpson, R.L. (1996b). "Experimental Transport-rate Budgets in Complex Three-dimensional Turbulent Flows at a Wing/Body Junction," 27th AIAA Fluid Dynamics Conference, AIAA-96-2035, New Orleans, LA.

Ölçmen, S.M. and Simpson, R.L. (1996c). "Theoretical and Experimental Pressure-strain Comparison in a Pressure-driven Three-dimensional Turbulent Boundary Layer," 1st AIAA Theoretical Fluid Mechanics Meeting, AIAA-96-2141, New Orleans, LA.

Ölçmen, S.M. and Simpson, R.L. (1996d). "Experimental Evaluation of Pressure-strain Models in Complex 3-D Turbulent Flow Near a Wing/Body Junction," Virginia Polytechnic Institute and State University Report VPI-AOE-228; DTIC Report ADA3071164XSP.

Ölçmen, S.M. and Simpson, R.L. (1996e). "Higher Order Turbulence Results for a Three-dimensional Pressure-driven Turbulent Boundary Layer," Virginia Polytechnic Institute and State University Report VPI-AOE-237; submitted to DTIC.

Ölçmen, S.M. and Simpson, R.L. (1997a). "Experimental Evaluation of Turbulent Diffusion Models in Complex 3-D Flow Near a Wing/Body Junction," AIAA-97-650, 35th AIAA Aerospace Sciences Meeting.

Ölçmen, S.M. and Simpson, R.L. (1997b). "Some Features of a Turbulent Wing-body Junction Vortical Flow," AIAA-97-0651, 35th AIAA Aerospace Sciences Meeting; Virginia Polytechnic Institute and State University Report VPI-AOE-238; submitted to DTIC.

Ölçmen, S.M. and Simpson, R.L. (1997c). "Higher Order Turbulence Results for a Three-dimensional Pressure-driven Turbulent Boundary Layer," under revision for *Journal of Fluid Mechanics.*

Pompeo, L.P. (1992). "An Experimental Study of Three-dimensional Turbulent Boundary Layers," Ph.D. dissertation, ETH No. 9780, Swiss Federal Institute of Technology, Zurich.

Pompeo, L.P., Bettelini, M.S.G., and Thomann, H. (1993). "Laterally Strained Turbulent Boundary Layers Near a Plane of Symmetry," *Journal of Fluid Mechanics* **257**, pp. 507–532.

Schwarz, W.R. and Bradshaw, P. (1992). "Three-dimensional Turbulent Boundary Layer in a 30 Degree Bend: Experiment and Modeling," Report No. MD-61, Thermosciences Division, Stanford University.

Schwarz, W.R. and Bradshaw, P. (1993). "Measurements in a Pressure Driven Three-dimensional Turbulent Boundary Layer During Development and Decay," *AIAA Journal* **31**, No. 7, pp. 1207–1214.

Shih, T.H. and Lumley, J.L. (1993). "Critical Comparison of Second-order Closures with Direct Numerical Simulations of Homogeneous Turbulence," *AIAA Journal* **31**, No. 4, pp. 663–670.

Simpson, R.L. (1996). "Aspects of Turbulent Boundary Layer Separation," *Prog. Aerospace Sci.* **32**, pp. 457–521.

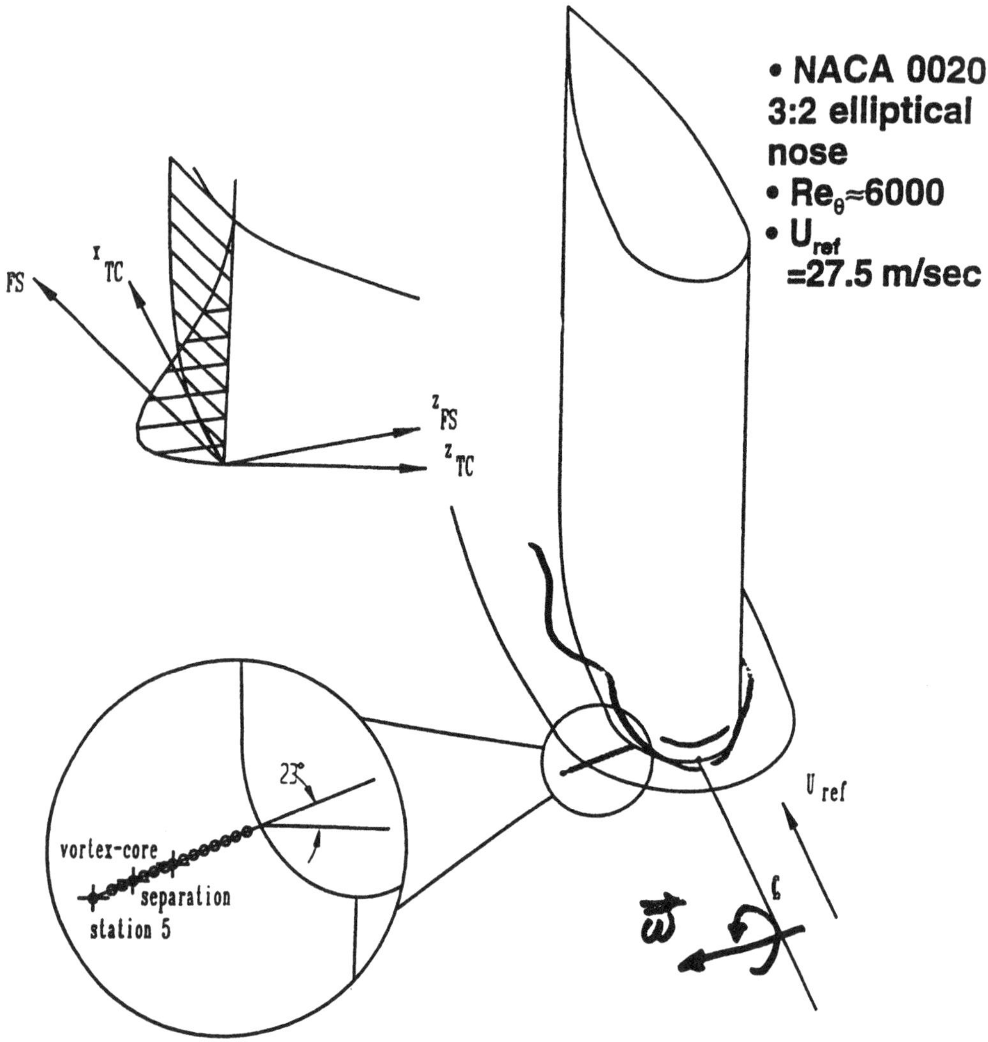

Figure 1. Schematic view of the wing and the measurement locations, and description of the coordinate systems. $()_{FS}$ free-stream coordinates, $()_{TC}$ tunnel coordinates.

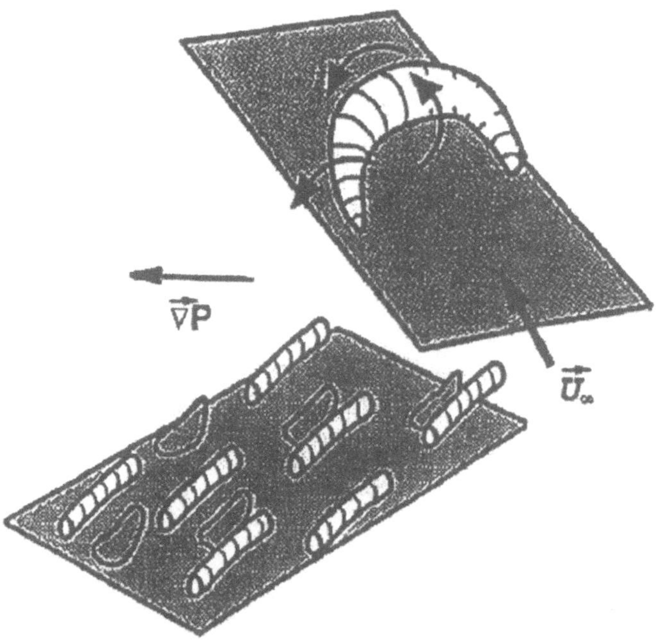

Figure 2. Coherent structures in a skewed 3D turbulent boundary layer. Outer layer "croissant-shaped" structures that entrain freestream fluid move in different directions than the near wall streamwise vortices.

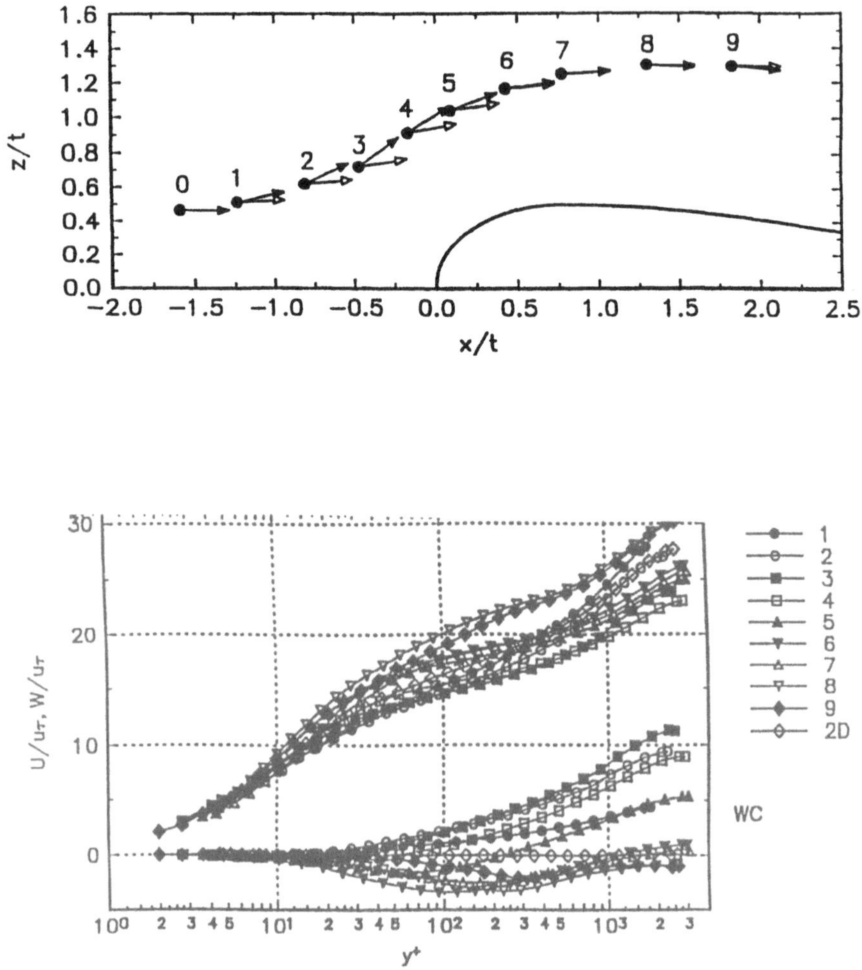

Figure 3. (a) Wing shape and measurement locations. Full arrows are in wall-stress direction. Empty arrows are in free-stream direction. (b) $U/U, W/U$, and V/U, mean velocity components in wall-stress coordinates. The uncertainties are on the order of the symbol size.

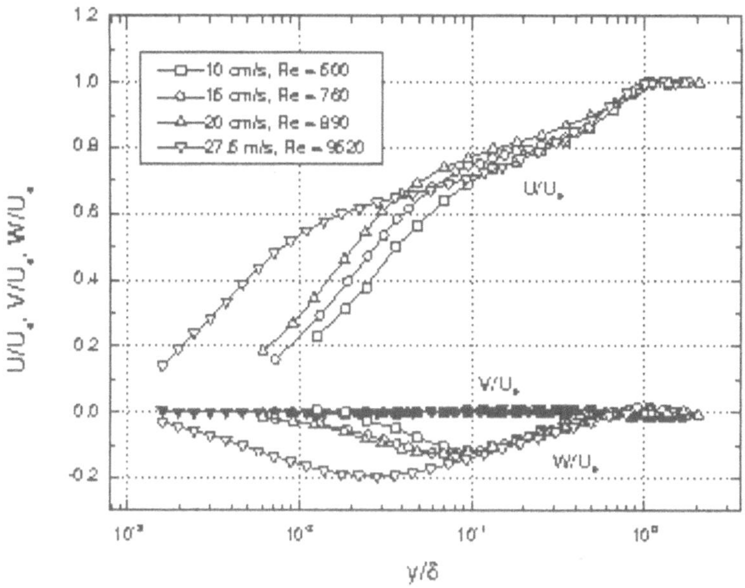

Figure 4. Three-dimensional mean velocity profiles $U/U_e, V/U_e$, and W/U_e using outer scaling in local freestream coordinates, *et al.*, the x-axis is aligned in the U_e direction. Data from the three low speed water tunnel experiments of Fleming and Simpson (1997) and Station 5 of the Ölçmen and Simpson (1995a) wind tunnel experiments.

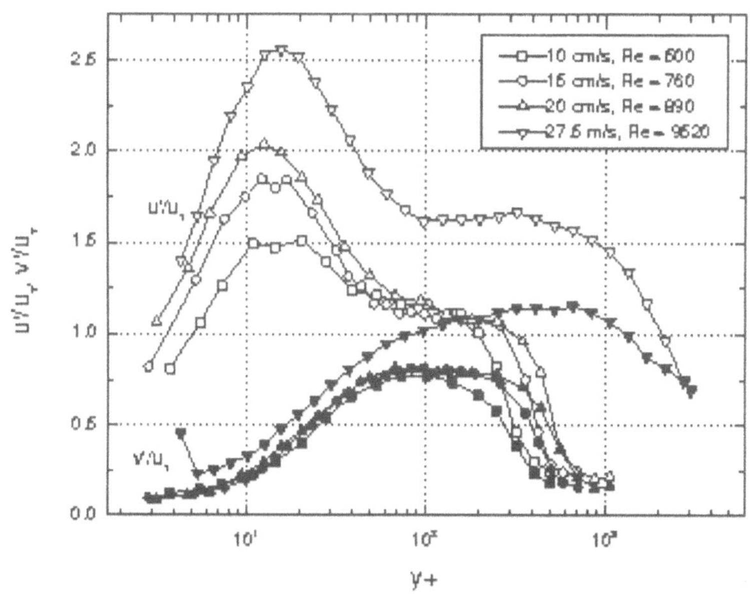

Figure 5. Three-dimensional turbulence intensity results for the conditions in Figure 4, normalized on wall variables in local freestream coordinates; u' results, open symbols; v' results, solid symbols.

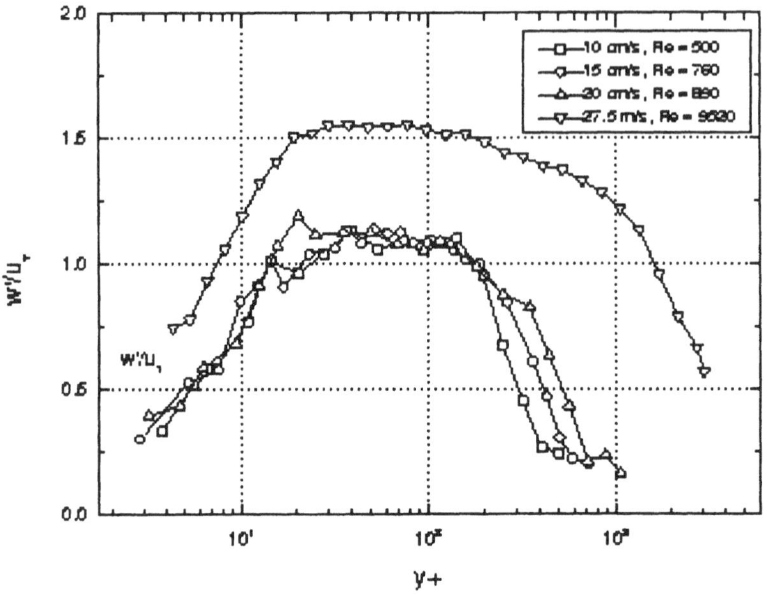

Figure 6. Spanwise intensity w' results for the flow conditions described in Figures 4 and 5, normalized on wall variables in local freestream coordinates.

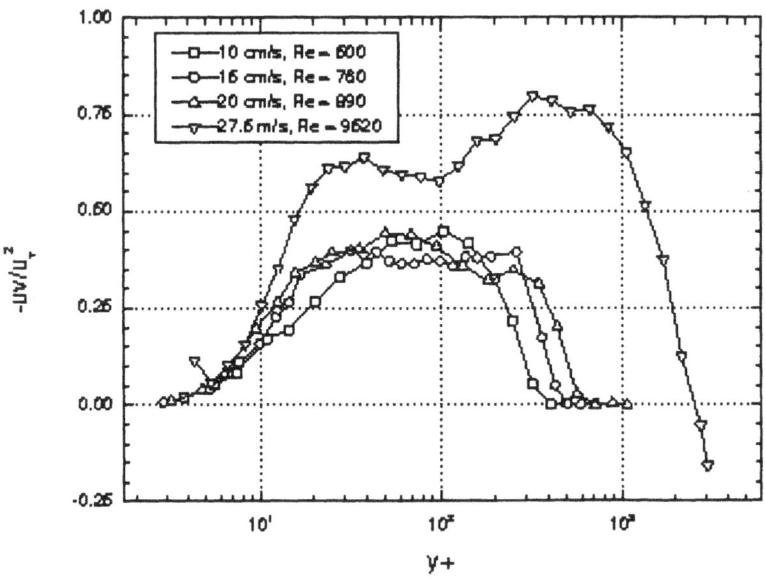

Figure 7. Three-dimensional flow streamwise Reynolds shearing stress profiles in local freestream coordinates for the conditions described in Figure 4.

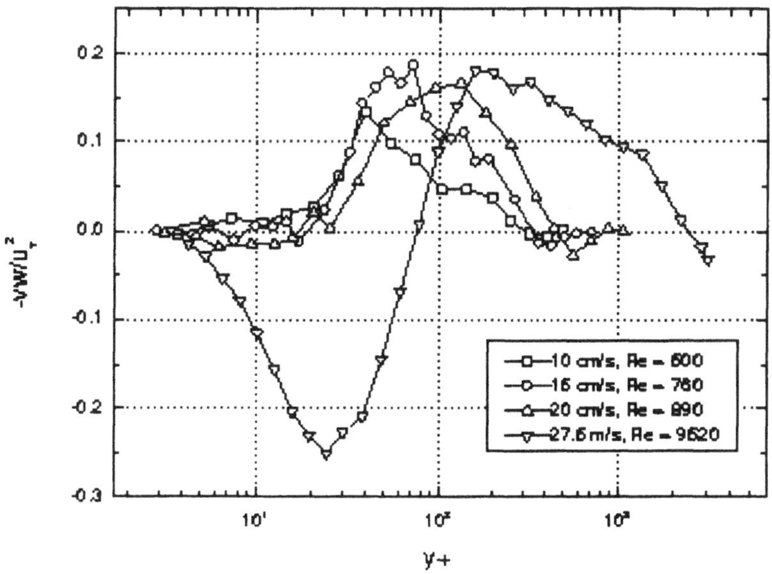

Figure 8. Three-dimensional flow spanwise Reynolds shearing stress profiles in local freestream coordinates for the conditions described in Figure 4.

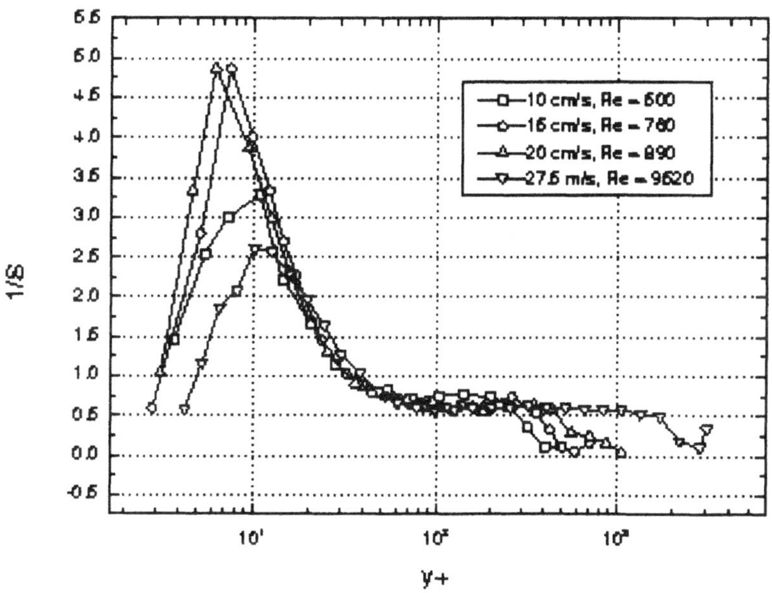

Figure 9. $S = v'^2/|\tau/\rho|$ profiles in inner or wall variables shown for the three-dimensional flow conditions and data shown in Figures 4–8. Note that S is invariant to coordinate system rotation about the y-axis.

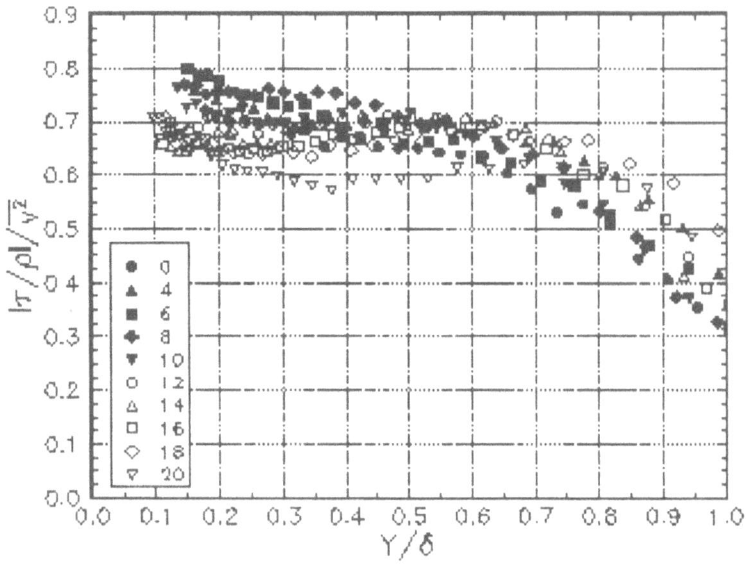

Figure 10. $1/S$ for the Schwarz and Bradshaw (1992, 1993), 30° bend flow. Stations in crossflow development region and decay region are the closed and open symbols, respectively.

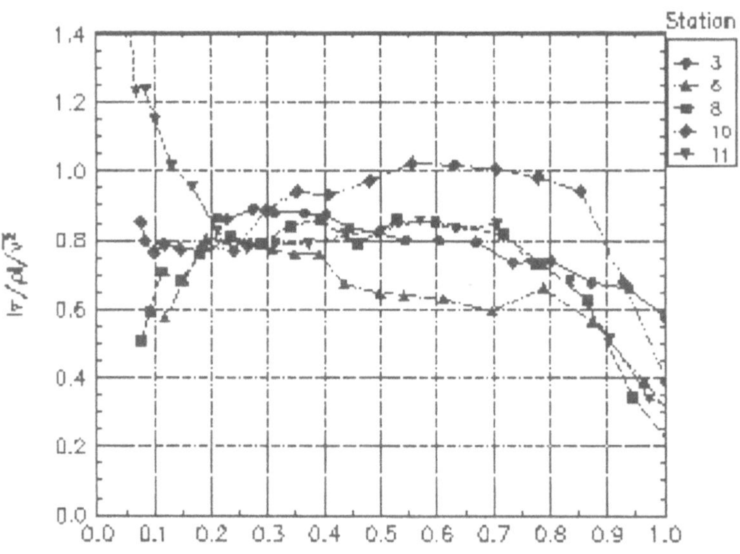

Figure 11. $1/S$ for the convex wall curvature case of Baskaran *et al.* (1990).

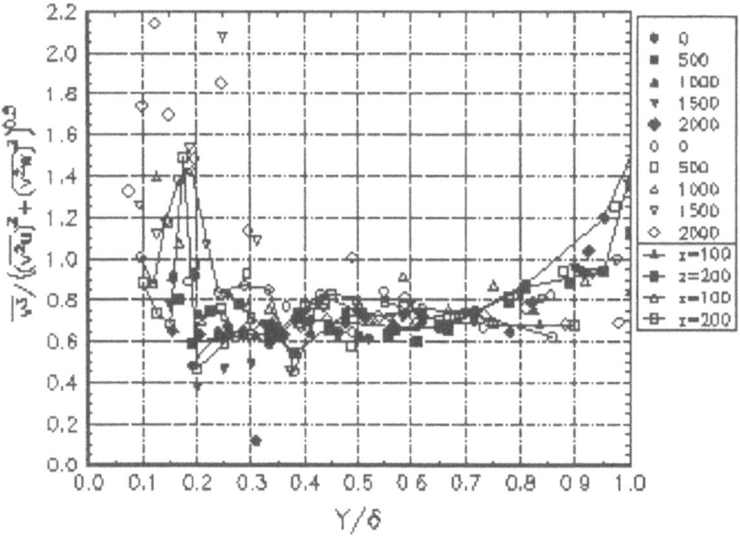

Figure 12. B_2 for the flows of Pompeo *et al.* (1992, 1993). Centerline stations (symbols only) have the x-location shown in mm. Off centerline stations (lines) were at $x = 1000mm$, the z-location is shown in mm. Closed and open symbols are for the diverging and converging duct, respectively.

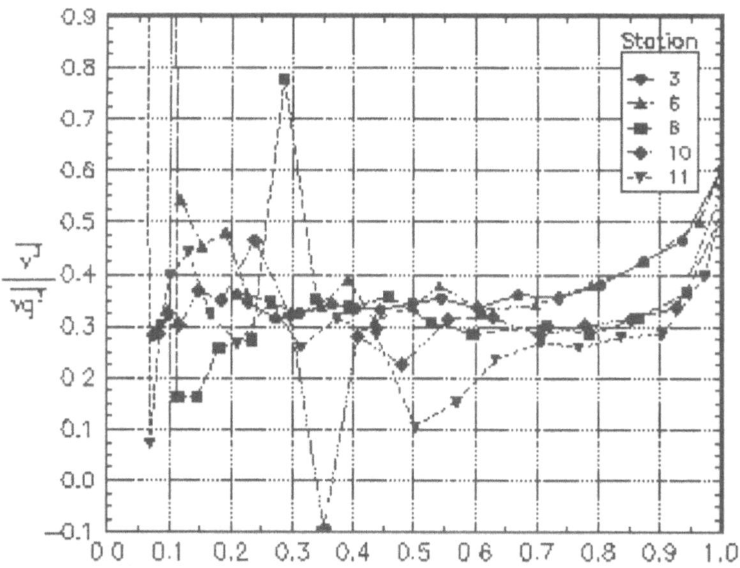

Figure 13. The normalized $\overline{v^3}/\overline{vq^2}$ parameter for the convex wall curvature case of Baskaran *et al.* (1990).

PHYSICS AND COMPUTATIONS OF FLOWS WITH ADVERSE PRESSURE GRADIENTS

P. G. HUANG
Department of Mechanical Engineering
University of Kentucky
Lexington, Kentucky

Abstract. For flows with small to mild pressure gradients, turbulence models that satisfy the law of the wall generally result in better agreement with experiments. When the pressure gradients are large, most simple two-equation models fail to predict the onset and the size of the flow separation. The cause was found to be associated with the failure of the models to respond to non-equilibrium flows when large ratios of the turbulent kinetic-energy generation to the energy dissipation rate were encountered. Recent improved models, such as SST and algebraic stress models, have corrected this deficiency by introducing an adjustment to the values of the turbulence model coefficient such that the resultant eddy viscosity decreases as the ratios of the generation of turbulent kinetic energy to the energy dissipation rates increase. Although this type of model adjustment has been found to be very successful for flows under strong adverse pressure gradients, it was not able to predict the overshoot of turbulent shear stress in the flow recovery region.

1. Introduction

Almost all turbulence models are calibrated to match the law of the wall for the inner layer of zero-pressure-gradient boundary layer flows. Although the inner layer occupies only a small fraction of the boundary layer (approximately 10%), it is responsible for most of the growth of the velocity to the freestream value. Therefore, the adjustment of the turbulence models to satisfy the law of the wall for flows with a zero pressure gradient is one of the most important anchors used for near-wall turbulence modeling.

M. D. Salas et al. (eds.), Modeling Complex Turbulent Flows, 245–258.

The turbulence models are then *extrapolated* to predict other flow features, such as pressure gradient, curvature and flow separation. The general consensus is that there is no universal model available to predict all flow features. As a result, corrections are frequently needed for the models to provide a better prediction of some certain identifiable flow physics best representing the underlying flow feature. These model corrections can generally be provided by (or, sometimes, by a combination of) two different approaches - one is by building more physics into the models through rigorous mathematical and physical constraints and the other one, done purely empirically, is by adjusting the model in an ad-hoc manner such that it matches the experimental trend. Irrespective of whichever approach is used, the key to the success of the modifications is to identify, understand and then capture the most important physics of the flows.

In this article, we shall first address the key physics observed in flows under pressure gradients. In section 2, the law of the wall for flows with a zero pressure gradient is introduced. In section 3, the law of the wall for flows with adverse pressure gradients is discussed. Not all models show a good match to the law of the wall for adverse pressure gradients. Models satisfying the law of the wall for adverse pressure gradients yield a better prediction for flows with weak adverse pressure gradients.

In section 4, we present an interesting "slingshot" effect of the turbulent shear stress for flows under strong pressure gradients. When the pressure gradient is large enough to cause flow separation, satisfying the law of the wall alone is not good enough to predict the flow. In the adverse pressure gradient flow region, the predicted turbulent shear stresses were found to be too large when they were compared with experimental data and thus the calculation led to a delay of the onset of flow separation. Recent improvements of the models can be viewed as a modification designed to limit the magnitude of the eddy viscosity. This modification appears very successful in predicting flows with strong pressure gradients. In contrast, in the recovery region of the flow, the observation from experiments shows that the turbulent shear stress overshoots rapidly and then recovers back to the flat-plate level. None of the models we tested can predict the overshoot of the turbulent shear stress.

2. The law of the wall under a zero pressure gradient

The dimensionless analysis of the near-wall region yields,

$$\frac{u}{u_\tau} = f\left(\frac{u_\tau y}{\nu}\right) \tag{1}$$

where $u_\tau = \sqrt{\tau_w/\rho}$. Alternatively, (1) can be written as,

$$\frac{\partial u}{\partial y} = \frac{u_\tau^2}{\nu} f'\left(\frac{u_\tau y}{\nu}\right) = \frac{u_\tau}{y} g\left(\frac{u_\tau y}{\nu}\right). \tag{2}$$

For large y^+ ($\equiv u_\tau y/\nu$) we expect ν not to affect flow locally, so the function g must tend to a constant, such as $1/\kappa$. Hence, (2) becomes,

$$\frac{\partial u}{\partial y} = \frac{u_\tau}{\kappa y} \tag{3}$$

Integration of (3) yields the logarithmic law of the wall,

$$\frac{u}{u_\tau} = \frac{1}{\kappa} \ln \frac{u_\tau y}{\nu} + C. \tag{4}$$

Equation (4) was found to be valid for regions which are close to the wall (within 10% of the boundary layer) but nevertheless sufficiently far away from the viscous sublayer ($y^+ > 30$) for boundary layer flows with not-too-strong pressure gradients or in pipe or channel flows to appear. From experimental observation. $\kappa \approx 0.41$ and $C \approx 5.2$.

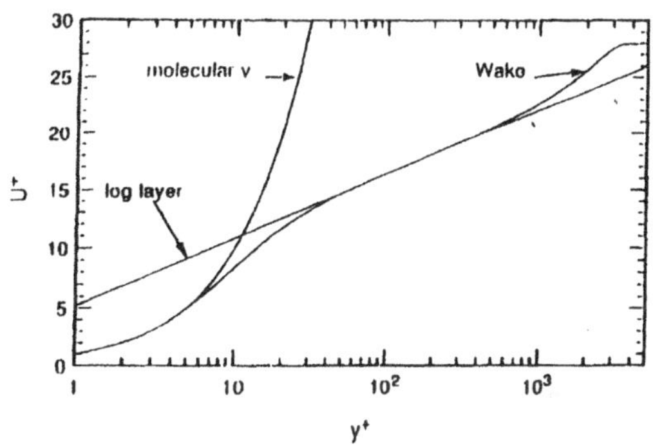

Figure 1. Prediction of the law of the wall for a boundary layer flow with a zero pressure gradient.

The high-Reynolds-number versions of turbulence models are first adjusted to satisfy (3) in region where y^+ is sufficiently large ($y^+ > 30$), while far away from the influence of the freestream ($y/\delta < 0.1$, where δ is the boundary layer thickness). The low Reynolds number corrections are then incorporated to ensure that the value of the intercept C is approximately 5.2 when the velocity approaches the law of the wall line, as shown in Figure 1. The wake region ($y/\delta > 0.1$), occupying most of the boundary layer, is represented by the rise of the velocity profile away from the law of the wall line. This rise of the profile is largely attributed to the convective contribution of the momentum equations.

For the eddy viscosity models, the turbulent shear stress is related to the velocity gradient according to:

$$-\overline{u'v'} = \nu_t \frac{\partial u}{\partial y}. \tag{5}$$

The inner layer analysis assumes that the shear stress is constant and thus,

$$\nu_t \frac{\partial u}{\partial y} = \frac{\tau_w}{\rho} = u_\tau^2. \tag{6}$$

By substituting (3) into (6), it can be shown that

$$\nu_t = u_\tau \kappa y \tag{7}$$

in the logarithmic region.

For example, the mixing length model assumed,

$$\nu_t = l^2 |\frac{\partial u}{\partial y}|. \tag{8}$$

By substituting (3) into (8) and comparing with (7), a formula for the mixing length emerged,

$$l = \kappa y. \tag{9}$$

The high Reynolds number form of the one equation model of Spalart and Allmaras [1993] in the inner layer is,

$$\frac{1}{\sigma}\left[\frac{\partial}{\partial y}\left(\nu_t \frac{\partial \nu_t}{\partial y}\right) + c_{b2}\left(\frac{\partial \nu_t}{\partial y}\right)^2\right] + c_{b1}|\frac{\partial u}{\partial y}|\nu_t - c_{w1}\left(\frac{\nu_t}{y}\right)^2 = 0. \tag{10}$$

By substituting (3) and (7) into (10), the following relationship for the model coefficients can be obtained:

$$c_{w1} = \frac{c_{b1}}{\kappa^2} + \frac{1 + c_{b2}}{\sigma}. \tag{11}$$

The high Reynolds number forms of the $k - \epsilon$ model in the inner layer are,

$$\frac{\partial}{\partial y}\left[\frac{\nu_t}{\sigma_k}\left(\frac{\partial k}{\partial y}\right)\right] + \nu_t \left(\frac{\partial u}{\partial y}\right)^2 - \epsilon = 0 \tag{12}$$

$$\frac{\partial}{\partial y}\left[\frac{\nu_t}{\sigma_\epsilon}\left(\frac{\partial \epsilon}{\partial y}\right)\right] + c_{\epsilon,1}\nu_t \left(\frac{\partial u}{\partial y}\right)^2 \frac{\epsilon}{k} - c_{\epsilon,2}\frac{\epsilon^2}{k} = 0 \tag{13}$$

and the eddy viscosity can be expressed as:

$$\nu_t = c_\mu \frac{k^2}{\epsilon}. \tag{14}$$

With the assumption that the flow is in local equilibrium, $-\overline{u'v'}/k = \tau_w/k = \sqrt{c_\mu}$ and by substituting (3) and (7) into (12) to (14), the following relationship among the model coefficients must be established:

$$\frac{\sqrt{c_\mu}\sigma_\epsilon}{\kappa^2}(c_{\epsilon,2} - c_{\epsilon,2}) = 1. \tag{15}$$

Finally, we may define a general variable ϕ in terms of k and ϵ,

$$\phi = k^m \epsilon^n, \tag{16}$$

such that a family of two-equation models for k and ϕ can be defined. For example, $m = -1$ and $n = 1$ give the $k - \omega$ model, where $\omega = \epsilon/k$.

The governing equations for the $k - \phi$ family in the inner layer of the boundary layer are:

$$\frac{\partial}{\partial y}\left[\frac{\nu_t}{\sigma_k}\left(\frac{\partial k}{\partial y}\right)\right] + \nu_t\left(\frac{\partial u}{\partial y}\right)^2 - \epsilon = 0 \tag{17}$$

$$\frac{\partial}{\partial y}\left[\frac{\nu_t}{\sigma_\phi}\left(\frac{\partial \phi}{\partial y}\right)\right] + c_{\phi,1}\nu_t\left(\frac{\partial u}{\partial y}\right)^2 \frac{\phi}{k} - c_{\phi,2}\frac{\epsilon}{k}\phi = 0. \tag{18}$$

where $c_{\phi,1}$ and $c_{\phi,2}$ can be related to the coefficients for ϵ transport equation, $c_{\epsilon,1}$ and $c_{\epsilon,2}$ as,

$$c_{\phi,1} = n\, c_{\epsilon,1} + m \tag{19}$$

$$c_{\phi,2} = n\, c_{\epsilon,2} + m. \tag{20}$$

With the assumption that $-\overline{u'v'}/k = \tau_w/k = \sqrt{c_\mu}$ and by substituting (3) and (7) into (17) to (18), the following relationship among the model coefficients must be established:

$$\frac{\sqrt{c_\mu}\sigma_\phi}{n^2\kappa^2}(c_{\phi,2} - c_{\phi,2}) = 1. \tag{21}$$

3. The law of the wall under weak adverse pressure gradients

When a boundary layer flow is subject to a not-too-large adverse pressure gradient, the departure of the velocity profile from the log law is "progressive" rather than "general" as shown in Figure 2 [Galbraith *et al.*, 1977 and

Huang and Bradshaw, 1995]. Such an observation leads to the conclusion that (3) is still valid for the law of the wall under adverse pressure gradients [Huang and Bradshaw, 1995] and hence,

$$\nu_t = \kappa y u_\tau \tau^+ \tag{22}$$

where

$$\tau^+ = \frac{\tau}{\tau_w} \tag{23}$$

with τ being the local shear stress. For flows with adverse pressure gradients, $\tau^+ > 1$ and τ^+ can be approximated to increase linearly with y^+ in the logarithmic layer: $\tau^+ \approx 1 + \beta y^+$.

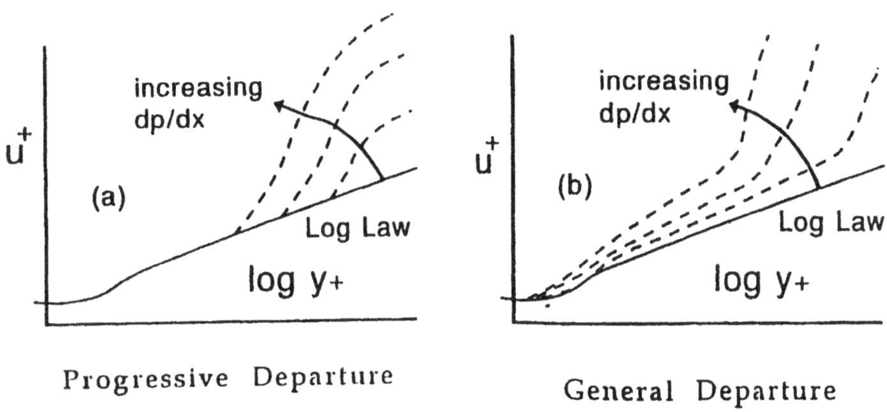

Figure 2. Possible mode of velocity profile departing from the logarithmic law of the wall in adverse pressure gradients.

It should be noted that all turbulence models were adjusted to satisfy the law of the wall for a zero pressure gradient, as described in the previous section, and their extensions to the law of the wall for pressure gradients are not automatically guaranteed. As a result, the models' responses to flows with adverse pressure gradients may vary from model to model. For example, the mixing length model shows that [Huang and Bradshaw, 1995]

$$\frac{\partial u^+}{\partial y^+} = \frac{\sqrt{\tau^+}}{\kappa y^+} \tag{24}$$

while the two-equation $k - \epsilon$ model gives rise to [Huang and Bradshaw, 1995],

$$\frac{\sqrt{c_\mu}\sigma_\epsilon}{\kappa^{*2}}(c_{\epsilon,2} - c_{\epsilon,2}) = \frac{1}{\tau^+}. \tag{25}$$

Since all model coefficients were adjusted according to (15), the κ^* value shown in (25) must be made equal to $\kappa\sqrt{\tau^+}$. This leads to the following law of the wall behavior:

$$\frac{\partial u^+}{\partial y^+} = \frac{1}{\kappa y^+ \sqrt{\tau^+}}. \tag{26}$$

Note that both the mixing-length and the $k-\epsilon$ models incorrectly predict two opposite "general" departing behaviors for flows under adverse pressure gradients; they respectively predict a $\partial u^+/\partial y^+$ that is both too large and too small when comparing with the experimental data.

Finally, the corresponding law of the wall behavior for model coefficients of the $k-\phi$ family becomes [Huang and Bradshaw, 1995],

$$\frac{\sqrt{c_\mu}\sigma_\phi}{n^2\kappa^2}(c_{\phi.2}-c_{\phi.2}) = \tau^+ + \frac{1}{n^2}\left[d_1 y^+ \frac{d\,\tau^+}{d\,y^+} + d_2 y^{+2}\frac{d^2\,\tau^+}{d\,y^{+2}} + d_3\frac{y^{+2}}{\tau^+}\left(\frac{d\,\tau^+}{d\,y^+}\right)^2\right] \tag{27}$$

where $d_1 = m - 2n^2 - 2mn$, $d_2 = m + n$ and $d_3 = (m+n)^2$. As can be seen from (27). different models respond differently to the variation of τ^+ in y^+. It should be noted that if one chooses $m = -1$ and $n = 1$, or the ω ($\equiv \epsilon/k$) variable, the right hand side of (27) is unity. Hence, it is expected that the $k-\omega$ model should perform well for flow with pressure gradients that are not too large.

The important need for the models to satisfy the law of the wall for pressure gradients can be demonstrated from the following two examples.

Figure 3 shows predictions of the experiments of Blackwell *et al.* [1972] for a boundary layer flow with adverse pressure gradients (case 11087). As can be seen from the figure, the $k-\epsilon$ model predicts too small a value of $\partial u^+/\partial y^+$ for y^+ between 20 and 200, leading to too small a value of $u_\infty/u_\tau(\equiv \sqrt{2/c_f})$ at the edge of the boundary layer, whereas the results obtained by the $k-\omega$ model agree well with the data. Thus, the skin friction is over-predicted by the $k-\epsilon$ model.

This observation is further confirmed by the prediction of the adverse pressure gradient experiments of Samuel and Joubert [1975]. As can be seen from Figure 4, The $k-\epsilon$ model predicts a level of skin friction coefficients that is too high while the $k-\omega$ model shows a relatively good agreement with the experimental data.

4. Predictions of flows with strong adverse pressure gradients

Although the above two examples demonstrate the importance of the models to match the law of the wall for flows with adverse pressure gradients,

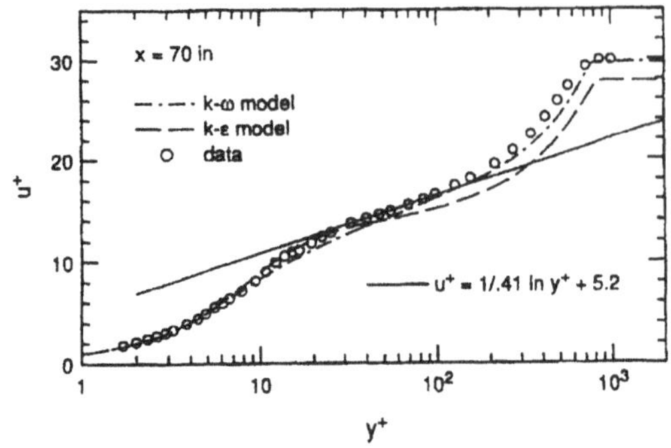

Figure 3. Comparison of the law of the wall for velocity in adverse pressure gradients using data of Blackwell *et al.* [1972].

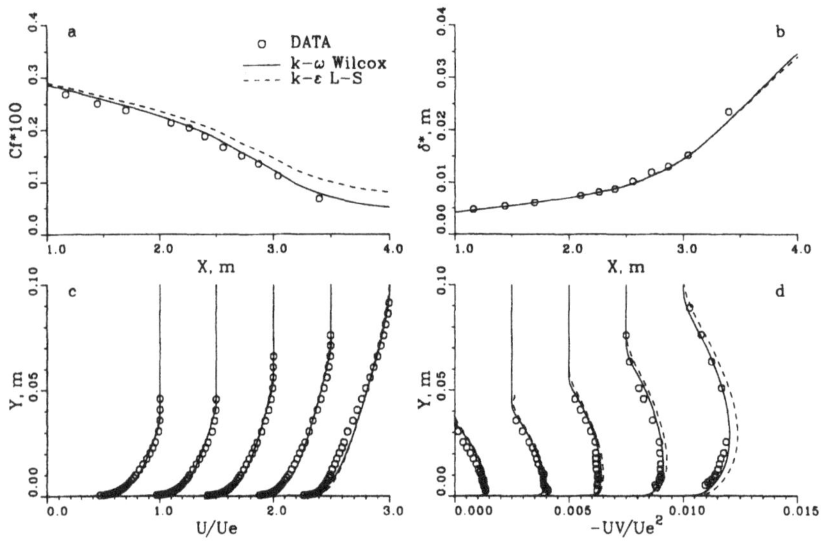

Figure 4. Comparisons of the skin friction, displacement thickness, velocity profiles and shear stress profiles using data of Samuel and Joubert [1975].

the pressure gradients tested were relatively weak. If a strong adverse pressure gradient is encountered such that the flow separation occurs, it has been found that satisfying the law of the wall alone is not sufficient to predict the flow.

For example, Figure 5 shows the failure of the $k-\omega$ model in the predictions of an axisymmetric boundary layer flow under strong adverse pressure gradients by Driver [1991]. As can be seen from the figure, although the $k-\omega$ solution is better than the $k-\epsilon$ solution, it is still not good enough. Both solutions showed a gradual departure away from the

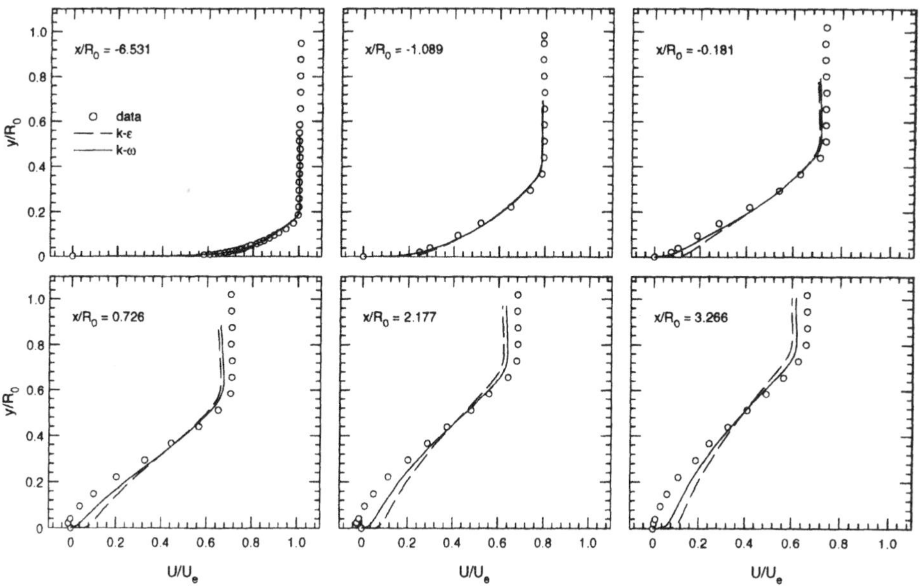

Figure 5. Prediction of the Driver's axisymmetric flow with strong adverse pressure gradients [1991].

experimental data when the flow is approaching the separation zone. The defect has been found to be associated with the failure of the models in predicting the correct level of turbulent shear stress, $\overline{u'v'}$, for flows subject to strong adverse pressure gradients. This over-prediction of the turbulent shear stress is observed outside the inner layer and therefore is not directly related to the improper approximation of the law of the wall. Johnson and King [1985] were the first to report this problem and argued that the cause is a result of the lag between the turbulent stresses and the rate of strain tensor. They proposed an O.D.E. to account for the transport of shear stress to remedy this problem [1985]. Menter [1994] proposed a more direct adjustment of the model by limiting the magnitude of the eddy viscosity according to Bradshaw's model [1972], which assumes that the shear stress in a boundary layer is proportional to the turbulent kinetic energy. Combining the use of a hybrid of the $k-\omega$ model near the wall and the $k-\epsilon$ model in the free-stream, he came up with the SST model [1994]. It should be mentioned that the treatment to limit the level of turbulent shear stress for flows under strong adverse pressure gradients is not unique. For example, Coakley [1997] proposed to make the model constant $c_{\epsilon,1}$ a function of dimensionless pressure gradients such that the growth rate of turbulent shear stress for flows under adverse pressure gradients can be reduced.

As can be seen from Figure 6, the SST results demonstrate that the

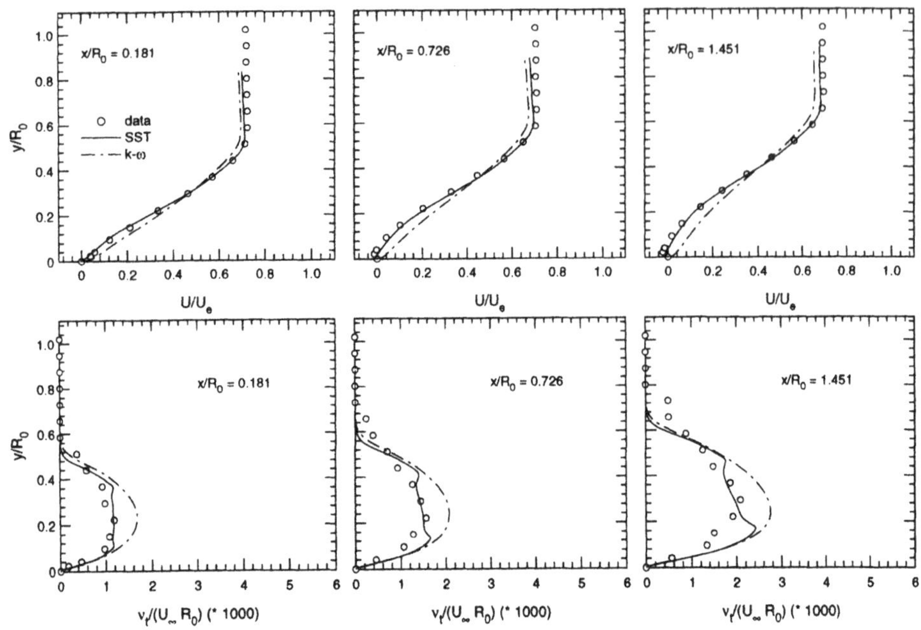

Figure 6. Effects of the limiting the eddy viscosity in the middle of boundary layer. Driver's experiments [1991].

solution can be greatly improved by limiting the value of the eddy viscosity in the middle of the boundary layer. It should be noted that non-linear algebraic turbulence models developed recently have provided a similar adjustment, even though their derivations are considered more rigorous. As can be seen from Figure 7, all improved models show a decrease of c_μ with increasing ratios of the turbulent kinetic energy generation to the energy dissipation rates. Based on this observation, it is believed that these nonlinear models will provide similar improvements; these improvements, however, are not directly related to the inclusion of the non-linear terms.

Although drastic improvements can be made in regions with adverse pressure gradients by limiting the level of the predicted eddy viscosity, predictions of the flows near and in the recovery regions are not at all satisfactory. As shown in Figure 8, the measured turbulent shear stress data in recovery regions of Driver's experiments overshoot the predictions, whether they are corrected or non-corrected ones. Driver's data do not extend long enough in the recovery region. The "slingshot" effect of the shear stress can be better represented by the experiments of Bachalo and Johnson [1986], in which a transonic bump flow was studied.

Before we present the results in the recovery region, Figure 9 shows the improvement of the SST model when c_μ is allowed to decrease in regions

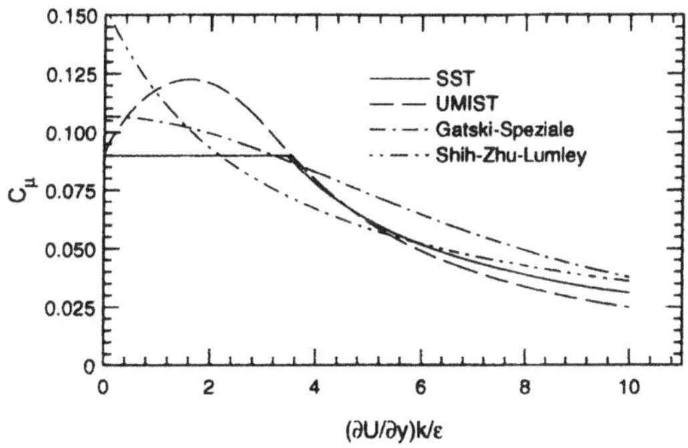

Figure 7. Variation of c_μ vs. $P_k/\rho\epsilon$

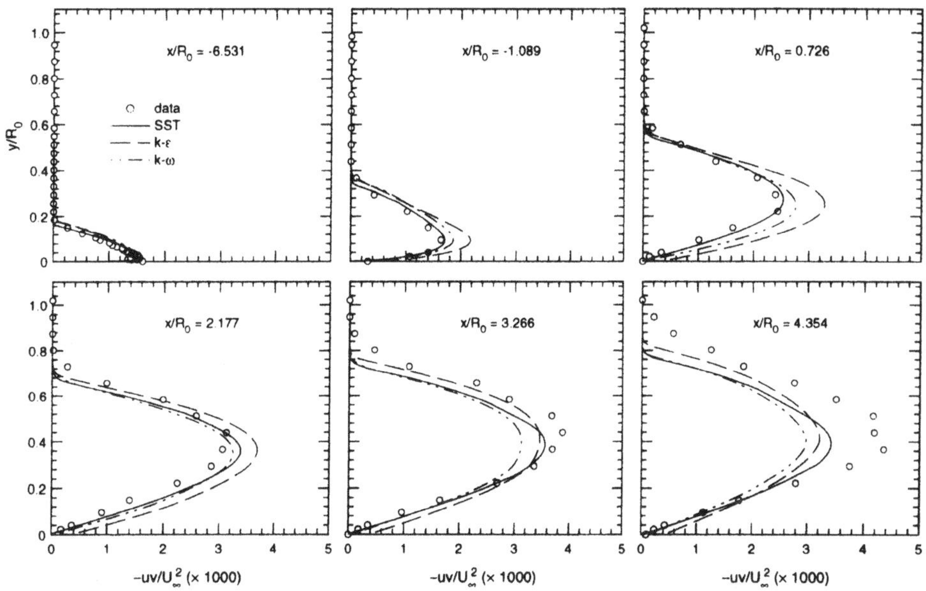

Figure 8. The slingshot effect of the turbulent shear stress observed in Driver's experiments [1991].

where the generation of turbulent kinetic energy is larger than the energy dissipation rate. As can be seen from the figure, when the eddy viscosity is corrected for the adverse pressure gradients, the shock position and the subsequent flow separation are better predicted.

Figure 10 shows the comparison of turbulent shear stress profiles in the recovery region of the Bachalo-Johnson transonic bump flow. The "slingshot" effect of the turbulent shear stress is clearly observed in the experiment. After the flow re-attaches ($x/c > 0.8$), the turbulent shear stress in-

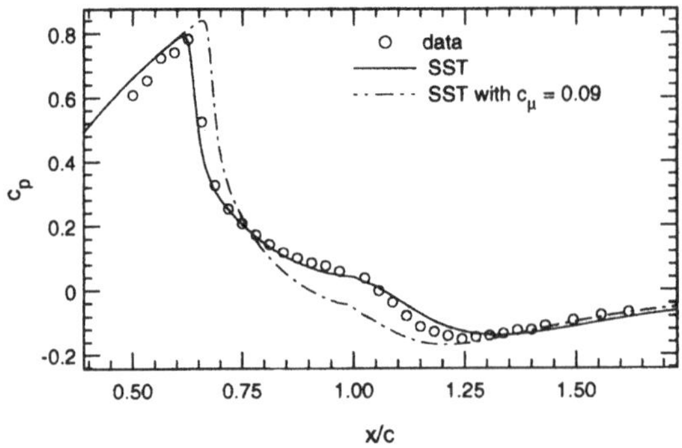

Figure 9. Effects of variable c_μ - Pressure coefficients along the surface of Bachalo-Johnson transonic bump [1986].

creases rapidly and then reduces back to the normal flat-plate values. None of the models used in this paper, whether they are corrected or uncorrected for the adverse pressure gradients, correctly predicted the overshoot of the shear stress in the recovery regions.

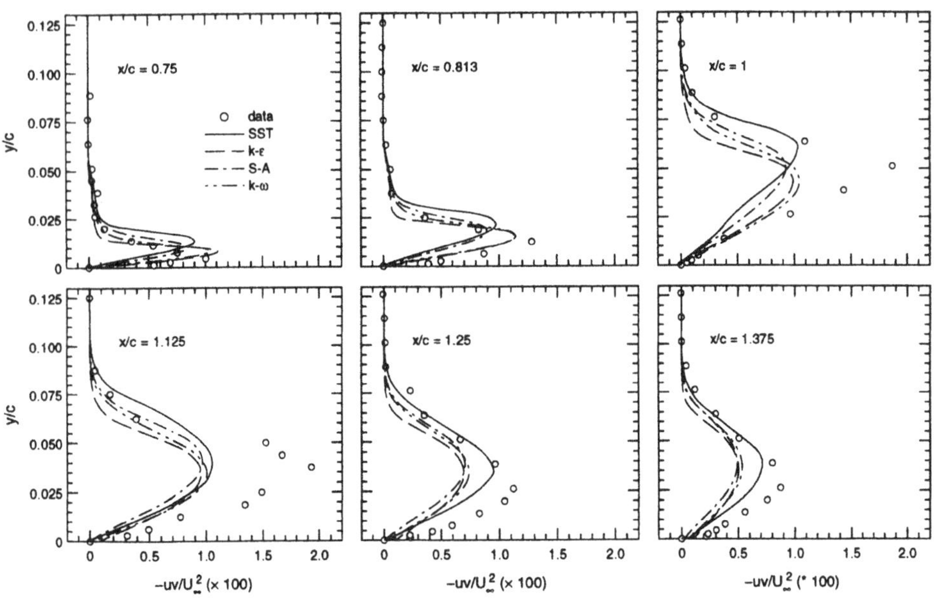

Figure 10. The slingshot effect of the shear stress observed in Bachalo-Johnson transonic bump [1986].

5. Concluding Remarks

The physics of flows under zero and adverse pressure gradients was discussed. Computations based on a number of eddy-viscosity models showed that models that satisfy the law of the wall are in better agreement with experimental data for flows subject to weak adverse pressure gradients. However, when the pressure gradients are large enough to cause flow separation, satisfying the law of the wall alone is not sufficient to predict the flow. The existing two-equation models tend to underpredict the extent of flow separation due to the fact that the turbulent shear stresses were over-predicted in the mid-section of the boundary layer. Remedies to this problem have been proposed which aim to reduce the level of the predicted turbulent shear stresses when the ratios of the generation of turbulent kinetic energy to the energy dissipation rates are large. Models incorporating these modifications have provided significant improvements in predicting flows with strong adverse pressure gradients.

The "slingshot" effect which appeared in flow recovery regions has been discussed. The turbulent shear stress in the flow recovery regions grows rapidly and then reduces back to the flat-plate level. This overshoot of the turbulent shear stress is not predicted by the models, whether they are corrected or non-corrected. The cause of the overshoot is not well-understood but it poses a major challenge for modelers to predict flows with a complete evolution of the pressure-induced separation, reattachment and recovery.

Acknowledgements

This work is supported by a CALSPAN Corp./NASA-Ames contract, NAS2-13605. The author would like to express his gratitude to his former colleagues, Jorge Badina, Peter Bradshaw, Tom Coakley, Dennis Johnson and Joe Marvin for their collaboration during the course of this work. A special thanks goes to Tom Coakley for allowing me to use Figure 4 and to Dennis Johnson for pointing out the slingshot effect to me.

References

Bachalo, W. D. and Johnson, D. A., 1986. "Transonic Turbulent Boundary-Layer Separation Generated on an Axisymmetric Flow Model," *AIAA J.* **24**, pp. 437-443.

Blackwell, B. F., Kays, W. W. and Moffat, R. J., 1972. *The Turbulent Boundary Layer on a Porous Plate: An Experimental Study of Heat Transfer Behavior with Adverse Pressure Gradients,* Thermosciences Div., Dept. of Mechanical Engineering, Stanford University, Rept. HMT-16, Stanford, CA,

Bradshaw P., Ferriss, D. H., and Atwell, N. P., 1967. "Calculation of Boundary Layer Development Using the Turbulent Energy Equation," *Journal of Fluid Mechanics* **28**, Pt. 3, pp. 593-616.

Coakley, T. J., 1997. "Development of Turbulence Models for Aerodynamics Application," *AIAA 97-2009.*

Driver, D. M., 1991. "Reynolds Shear Stress Measurements in a Separated Boundary Layer Flow," *AIAA 91-1781.*

Galbraith, R. A., McD., Sjolander, S., and Head, M. R., 1977. "Mixing Length in the Wall Region of Turbulent Boundary Layers," *Aeronautical Quarterly* **28**, pp. 97-110.

Huang, P. G. and Bradshaw, P., 1995. "Law of the Wall for Turbulent Flows in Pressure Gradients," *AIAA Journal* **33**, No. 4, pp. 624-632.

Johnson, D. A. and King, L. S., 1985. "A Mathematically Simple Turbulence Closure Model for Attached and Separated Turbulent Boundary Layers," *AIAA J.* **23**, pp. 1684-1692.

Menter, F. R., 1994. "Two-Equation Eddy Viscosity Turbulence Models for Engineering Applications," *AIAA J.* **32**, pp. 1299-1310.

Samuel, A. E. and Joubert, P. N., 1975. "A Boundary Layer Developing in an Increasingly Adverse Pressure Gradient," *Journal of Fluid Mechanics* **66**, 481.

Spalart, P. R. and Allmaras, S. R., 1992. "A One-Equation Turbulence Model for Aerodynamic Flows," *AIAA Paper 92-0439.*

COMPUTATIONS OF COMPLEX TURBULENT FLOWS USING THE COMMERCIAL CODE FLUENT

S.-E. KIM, D. CHOUDHURY AND B. PATEL
Fluent Inc.
10 Cavendish Court
Lebanon, New Hampshire

Abstract. The present paper is primarily concerned with practical aspects of modeling complex turbulent flows. The issues of meshing and discretization, which are prerequisite to successful prediction of complex turbulent flows, are discussed. Near-wall treatments are briefly reviewed with the main focus on the wall function approach in view of its practicality in simulating complex industrial flows. Computational results obtained using a selected number of turbulence models, ranging from a simple one-equation model to a differential Reynolds-stress model, are presented and discussed to assess what the models can offer for complex turbulent flows involving strong pressure gradients, separation, crossflow, and shocks.

1. Introduction

The ever-expanding computing power and ongoing developments in solution algorithms and mesh technology offer Computational Fluid Dynamics (CFD) practitioners both opportunities and challenges. We are now in a much better position than a decade ago to minimize numerical errors from various sources. Faster machines and efficient numerical algorithms have brought about a remarkable speed-up of computations, greatly reducing the turnaround time of CFD solutions. All this progress, more than ever, makes turbulence modeling stand out as a major unresolved issue and a pacing factor in modern CFD. Although the same progress has accelerated the development in high-level simulations of turbulence such as direct numerical simulation and large eddy simulation, the progress in engineering turbulence modeling has been rather slow.

M. D. Salas et al. (eds.), Modeling Complex Turbulent Flows, 259–276.

The fidelity of numerical predictions of turbulent flows is dictated by turbulence modeling, and therefore there is every reason to pursue more accurate turbulence models. Yet, there are still other practical issues to consider in the simulation of industrial flows. These issues are associated with the fact that, as the computing power has increased, so has the size and complexity of the problems engineers attempt to solve, and that, as a consequence, memory and turnaround time considerations, despite the tremendous computing power available, still limit the resolution of meshes and the level of turbulence modeling usable in industrial flow simulations.

This paper addresses the issues alluded to above, from a practical perspective with the computation of industrial flows in mind. We will start by looking at some of the implications a new mesh technology has for the computation of complex turbulent flows. The impact of the discretization scheme on the accuracy of numerical solutions for complex turbulent flows is discussed next. A brief review of near-wall treatments is made with its primary focus on the wall function approach, in view of its widespread use in industrial flow simulations. Lastly, computational results obtained using a selected number of popular turbulence models are presented to shed some light on what these models can offer for complex turbulent flows involving strong pressure gradient, separations, vortices and shocks. For the computations, the commercial CFD software FLUENT (Weiss & Smith, 1995; Mathur & Murthy, 1997; Kim *et al.*, 1998) was used.

2. Impacts of Meshes and Numerics

2.1. MESH

Most industrial flows involve highly complex geometries. Complex geometries with all the significant details often defy appropriate meshing, limiting the usability of CFD. In recent years, unstructured mesh technology has attracted a great deal of attention from the CFD community. It allows one to employ computational cells of arbitrary topology including quadrilaterals, hexahedra, triangles, tetrahedra, prisms, and combinations of all these. One immediate advantage offered by unstructured meshes over structured meshes is the flexibility in dealing with complicated flow configurations. Another advantage is that unstructured meshes naturally provide a convenient framework for pursuing solution-adaptive, local mesh refinements. This approach has a good potential to benefit the computation of industrial turbulent flows that have a wide range of length scales to be resolved. Furthermore, the hybrid mesh capability allows one to employ different meshing strategies in different regions, depending on the nature of the flow, taking account of the impact of the chosen mesh topology on numerical accuracy.

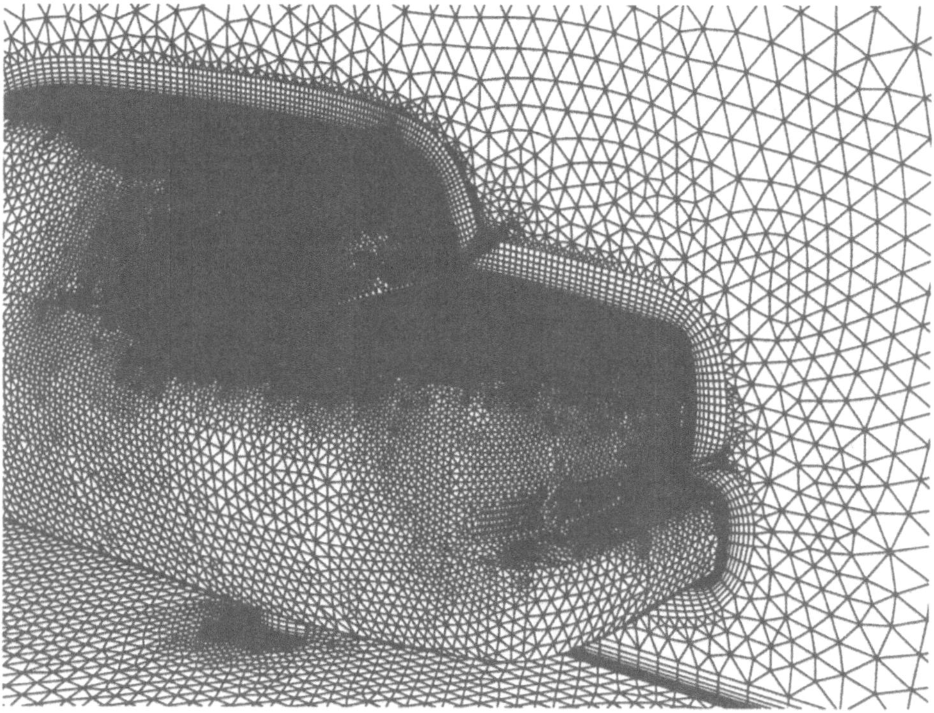

Figure 1. Viscous hybrid mesh for a passenger sedan.

We illustrate here a hybrid unstructured mesh for the flow around a passenger sedan. The car has a smooth surface, but the topology of the body surface and its geometrical details make it difficult to to obtain a quality mesh. The flow has highly complex structures including the three-dimensional flow separation, crossflow and vortices on the rear-body and in the near-wake. An appropriate mesh for this flow, therefore, should properly resolve the near-wake where the flow evolves most rapidly, and the near-wall region as well, where most of the flow features originate. Figure 1 depicts the hybrid mesh which consists of an inner block of prismatic cells grown from the triangular surface mesh on the body, and an outer block of tetrahedral cells wrapped around the inner prism layers. The inner prism layers were extruded several boundary-layer thicknesses from the body, providing a good resolution of the near-wall flow. The mesh resolution can be further enhanced using solution-adaptive mesh refinement to better resolve the near-wake flow.

Under the constraints of computational resources and design cycle in today's industrial setting, the unstructured mesh technology is anticipated to enhance the usability of CFD and the quality of CFD predictions for complex industrial turbulent flows.

2.2. NUMERICS

It is well-known but often overlooked that inaccurate discretization can plague the fidelity of predictions no matter what turbulence models are used. The impact of discretization on the accuracy of numerical solutions is more pronounced for complex three-dimensional flows involving separation, vortices, and secondary flows, all of which are the salient features of industrial flows. Examples showing inappropriateness of overly diffusive low-order discretization schemes are abundant. Here we present such an example. The case considered is the three-dimensional turbulent flow around a model ship hull (HSVA tanker) (Weighardt & Kux, 1980). The Reynolds number based on the ship's length is 5×10^6. Among many important features of the flow, the crossflow and associated axial velocity contours near the ship's stern are of major practical concern. The measured axial velocity contours on a transverse section near the ship's stern are depicted in Figure 2, which shows a unique (hook-like) shape of the contours. The flow was computed on a reasonably fine, $60 \times 40 \times 42$ hexahedral mesh using the Reynolds-stress model (Gibson & Launder, 1978), supposedly one of the more accurate turbulence models. Two different discretization schemes, i.e, the first-order upwind and the QUICK schemes (Leonard, 1979), were employed to discretize the convection term.

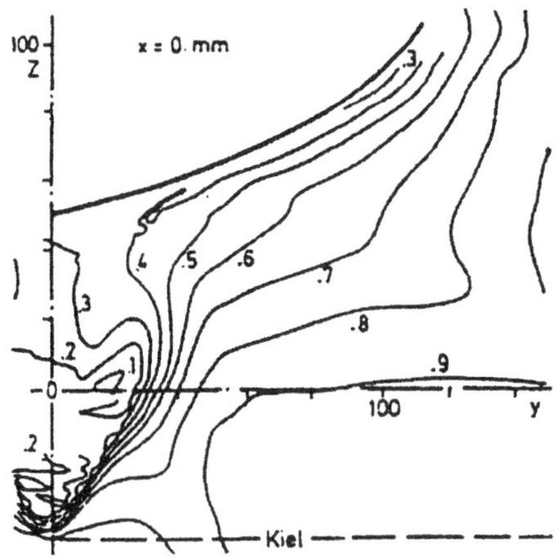

Figure 2. Measured axial velocity contours for the HSVA ship model ($Re_L = 5 \times 10^6$) at $x/L = 0.974$

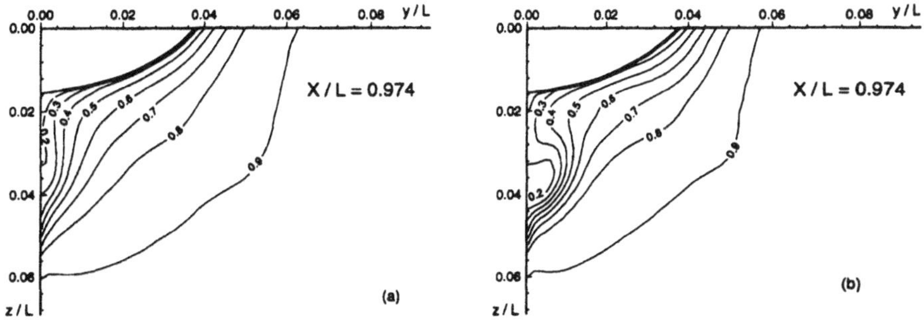

Figure 3. Axial velocity contours at $x/L = 0.974$ for the HSVA ship model ($Re_L = 5 \times 10^6$) predicted using Reynolds-stress model with different discretization schemes – (a) First-order upwind scheme (b)QUICK scheme

Figure 3 shows the predicted contours of the axial velocity at the same section ($x/L = 0.974$) as in Figure 2. A comparison of the predictions with the measured contours indicates that the first-order upwind scheme completely smears out the hook-shaped structure in the contours, but the QUICK scheme is able to predict the hook structure quite well.

This example indicates that it is imperative to employ accurate discretization schemes for reliable predictions of turbulent flows and for a fair evaluation of the performance of turbulence models.

3. Turbulence Modeling

3.1. NEAR-WALL TREATMENTS

For computations of wall-bounded turbulent flows, wall functions based on the law-of-the-wall and related hypotheses have long been used as economical, robust, and reasonably accurate means of treating the near-wall region. The so-called "universal" law-of-the-wall representing the logarithmic mean velocity profile in the fully-turbulent region of the inner layer is given by;

$$\frac{U_P}{u_\tau} = \frac{1}{\kappa} ln \left(E \frac{\rho u_\tau y_P}{\mu} \right) \qquad (1)$$

where $u_\tau \equiv \sqrt{\tau_w/\rho}$, τ_w is the wall shear stress, and κ and E the model constants, whose standard values are $\kappa = 0.42$ and $E = 9.8$.

The log-law in Equation (1) provides the wall boundary condition for momentum equations. In finite-volume implementations, it specifically gives the wall-shear stress ($\tau_w = \rho u_\tau^2$) for the given mean velocity (U_P) at wall-adjacent nodes. When the turbulence equations for k and ε are solved, the usual practice is to specify the turbulence quantities at the wall-adjacent cells in terms of the wall-shear stress using the assumptions of local equilibrium and constant-shear in the logarithmic region, i.e.,

$$k_P = \frac{u_\tau^2}{\sqrt{C_\mu}}, \quad \varepsilon_P = \frac{u_\tau^3}{\kappa y_P} \tag{2}$$

where $C_\mu = 0.09$. Determining the wall-shear with Equation (1) does not require any turbulence quantities. It should be noted that when turbulence equations for k and ε are solved, the turbulent quantities come to be rigidly tied to the wall-shear via Equation (2). One serious predicament encountered when using these wall functions in complex flow situations is that the turbulent velocity scale (u_τ) becomes zero at the points of flow detachment or attachment where the wall-shear vanishes. This has often been found to be responsible for anomalous predictions of the turbulent heat transfer at such points when the corresponding law-of-the-wall for mean temperature is employed. Another problem is that k and ε at these points also vanish according to Equation (2), while experimental evidence indicates quite the contrary.

These problems have perhaps made the wall functions of Launder and Spalding (1974) (called hereafter "standard" wall functions) more popular in the computation of industrial turbulent flows. In this alternative wall function approach, the universal law-of-the-wall for mean velocity, Equation (1), is replaced by;

$$\frac{U_P C_\mu^{1/4} k_P^{1/2}}{\tau_w/\rho} = \frac{1}{\kappa} ln \left(E \frac{\rho C_\mu^{1/4} k_P^{1/2} y_P}{\mu} \right) \tag{3}$$

where k_P is the turbulent kinetic energy at wall-adjacent cells. Equation (3) can be formally derived using the Kolmogorov-Prandtl relation ($\mu_t \sim \rho \sqrt{k}\, y$) and constant-shear assumption in the log-layer ($\tau_t = \tau_w$). k at wall-adjacent cells is obtained by solving its transport equation. The production of k and its dissipation rate at the wall-adjacent cells, which are needed to solve the k-equation, are evaluated using the local equilibrium and constant-shear hypotheses. As a consequence, this wall function approach allows k to be transported in and out of the wall region through convection and diffusion, instead of being rigidly tied to the wall-shear as in the first wall function approach. It is our experience that the Launder and Spalding wall functions outperform the wall functions based on the universal law-of-the-wall for complex wall-bounded turbulent flows.

Efforts have been made to improve the performance of the wall function approaches. In finite-volume implementations adopting the Launder and Spalding standard wall functions, major efforts have been centered upon reformulating the budget of k (i.e., cell-averaged production and dissipation rate of k) at wall-adjacent cells. In short, the new k-budget makes use of a simple two-layer notion that near-wall region consists of a viscosity-dominated region and a fully-turbulent core region, and each region contributes to the cell-averaged production and dissipation in a different manner. The resulting cell-averaged k-budget, i.e., the balance between the production and dissipation, changes from cell to cell depending on the proportions of the two layers at the wall-adjacent cells. This line of effort has been pursued by Chieng and Launder (1980), Amano (1984), Viegas and Rubesin (1985), and Johnson (1988), among others. There have also been some effort to take the effects of pressure gradient into account in the wall functions. Townsend (1961), based on the Prandtl's mixing length formula, extended the log-law, Equation (1), to account for the effects of pressure gradient. Similar approaches have been followed by Viegas and Rubesin (1985) and the present authors (Kim & Choudhury, 1995). The present authors modified the original Launder and Spalding wall functions to include the effects of pressure gradients within the two-layer framework.

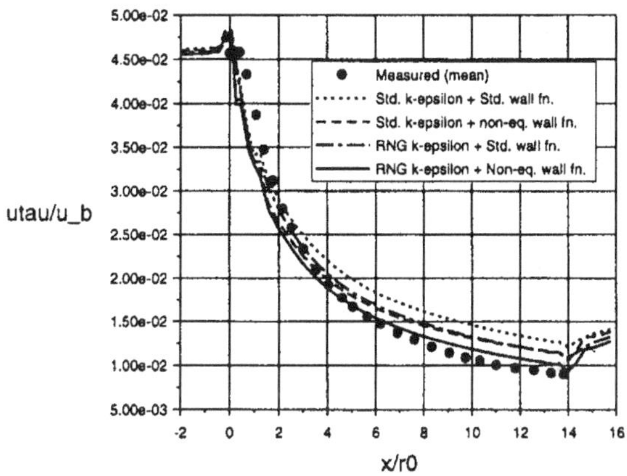

Figure 4. Flow through the conical diffuser of Azad and Kassab - Skin-Friction predictions by two different wall function approaches

We present here several examples where the wall functions (Kim & Choudhury, 1995) sensitized to pressure gradients and nonequilibrium effects improve the accuracy of the predictions for highly nonequilibrium complex flows involving strong pressure gradients. The first example is the incompressible flow through an axisymmetric diffuser whose "included" angle (2θ) is 8° (Azad & Kassab, 1989). The Reynolds number based on the inlet diameter is $Re_D = 1.15 \times 10^5$. We used a relatively coarse mesh appropriate for use of the wall functions, whose y^+ at the wall-neighboring cells is in the range of $y^+ = 30 \sim 60$. Experiments indicate that the flow remains attached throughout the diffuser, but the wall-shear rapidly decreases in the flow direction due to the expansion of the diffuser section. The standard k-ε (Launder & Spalding, 1974) and Renormalization-Group (RNG) based k-ε model (Yakhot *et al.*, 1992) were employed in conjunction with the standard wall functions and the nonequilibrium wall functions. Figure 4 shows the predictions of the dimensionless friction-velocity (u_τ/U_b, where U_b is the mean bulk velocity at the diffuser inlet) along the diffuser wall. The results indicate that the non-equilibrium wall functions incorporating the effects of pressure gradient appreciably improve the wall-shear predictions for both turbulence models.

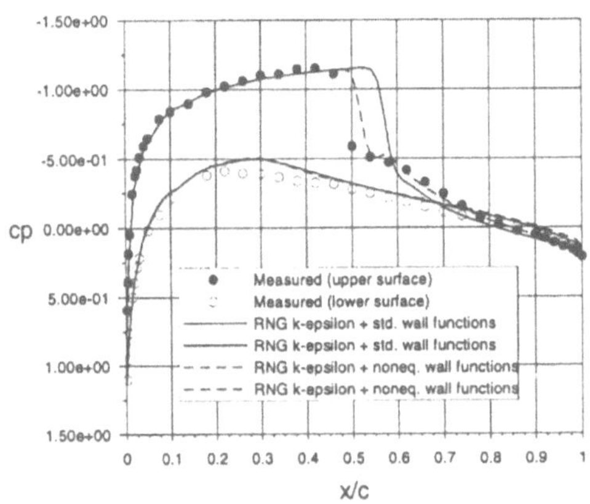

Figure 5. Flow around NACA-0012 airfoil ($M_\infty = 0.799$, $\alpha = 2.26°$, $Re_\infty = 9 \times 10^6$) - Predictions of pressure distributions by two different wall function approaches

The second example is the transonic flow ($M_\infty = 0.799$) around a NACA 0012 airfoil (Thibert *et al.*, 1979) at an incidence angle of $\alpha = 2.26°$ (corrected for the interference of wind tunnel wall), which has been subjected to many numerical studies. This flow has a fairly strong shock near the mid-chord ($x/c = 0.45$) on the suction (upper) side. The computations were made using the RNG k-ε model with the standard and the non-equilibrium wall functions. Again, a coarse mesh warranting the use of wall functions was adopted. Figure 5 shows the results. Both predictions on the pressure (lower) side are seen to be almost identical to each other. However, on the suction (upper) side, the non-equilibrium wall functions are shown to give substantially better predictions of the shock location and the characteristic features of the pressure distribution including the plateau behind the shock.

It seems that these efforts to improve the wall function approaches have not been taken all that seriously by the turbulence modeling community. From the viewpoint of industrial applications, however, the undertaking to improve the wall functions is worthier of pursuit than is usually thought, because the wall function approach is the most widely adopted in the computation of industrial flows, and because, quite often, it is the only practically usable option for complex high-Reynolds number flows. Note that, for some applications involving very high Reynolds numbers such as full-scale hydrodynamic applications whose Reynolds number may well be as high as 10^9, it is practically impossible to resolve the viscous sublayer as required in low-Reynolds number models.

In many of the computational examples to be presented in the next section, where we will compare the performances of various turbulence models, the wall functions are deliberately adopted. This conscious use of wall functions is to show the efficacy of the wall function approaches. However, we should note in passing that wall functions become inadequate in situations where low-Reynolds number effects are pervasive in the flow domain in question, or when the near-wall region needs to be resolved all the way to the wall including the viscous sublayer. Such situations call for use of low-Reynolds number turbulence models that are valid in the viscosity-dominated near-wall region. We will present later some results obtained using a low-Reynolds number near-wall turbulence model.

3.2. CORE TURBULENCE MODELING

Surveying the recent literature shows that there is no shortage of turbulence models. There are, indeed, numerous engineering turbulence models ranging from simple algebraic models to second-moment closure models. Among them, the group of two-equation turbulence models seem to be the most widely used. In fact, the standard k-ε (Launder & Spalding, 1974) model and its variants have become an "industry-standard" over the years for their simplicity, robustness and reasonable accuracy for a wide range of problems. Despite their efficacy, however, the conventional k-ε models have inherent drawbacks associated with the underlying hypotheses including the isotropic eddy-viscosity hypothesis. The conventional k-ε models thus have been found to be less than satisfactory or, often to fail miserably, in situations where the underlying hypotheses break down, *i.e.*, the flows involve strong anisotropic and nonequilibrium turbulence. The last decade has seen active research that attempts to improve the performance of the two-equation models. A number of alternatives have been proposed (Yakhot *et al.*, 1992; Wilcox, 1993; Shih *et al.*, 1993) including several anisotropic k-ε models (Gatski & Speziale, 1993; Craft *et al.*, 1993; Shih *et al.*, 1994)

In the last few years, one increasingly encounters industrial flow simulations employing second-moment closure models where the transport equations for individual Reynolds-stresses are solved. Anisotropy and transport of Reynolds-stresses are all accounted for in these models. Although these more elaborate models find their regular use in specific applications where the conventional two-equation models completely fail, e.g. cyclones, swirl combustors, etc., the second-moment closure models are not yet as widely used as the two-equation models for industrial flow simulations. The main reason for this limited use lies in their considerably higher cost. There are also several open issues yet to be resolved, such as pressure-strain correlations, wall-echo effects, and modeling of ε. Nevertheless, due to their theoretically sound basis, it is anticipated that use of second-moment closures will become more commonplace in the coming years for industrial flow simulations, as computers become more powerful.

With so many choices of turbulence models with varying sophistication and associated computational effort, one would naturally ask what one particular model offers vis-à-vis other models and what level of turbulence modeling is needed for a given flow. To answer these questions requires a comprehensive evaluation of the models for a wide range of flows. This is an impossible task to accomplish in a single study such as this one, in view of the number of the turbulence models and the breadth and complexity of the flow phenomena in industrial flows. Rather, we want to share our experience with some of the contemporary turbulence models through a selected number of computational examples involving strong pressure gradient, separation, and crossflow, all of which are salient features of industrial flows.

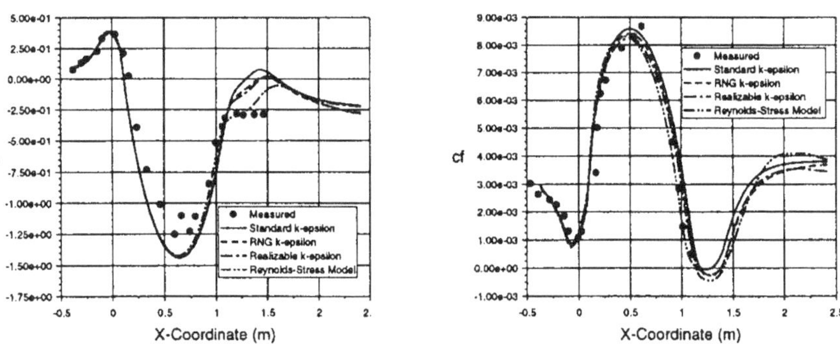

Figure 6. Flow over the curved two-dimensional hill - Predictions using four different turbulence models, left: pressure distribution, right: skin-friction distribution

The first example is the subsonic boundary layer over a two-dimensional curved hill, which was studied experimentally by Baskaran *et al.* (1987). The incoming flow upstream of the hill is a typical flat-plate turbulent boundary layer. As the boundary layer passes the hill, it is subjected to a streamline curvature of alternating signs (convex and concave) and strong favorable and adverse pressure gradients. The experimental observation indicates that there is an incipient separation at about $1.1m$ downstream from the leading-edge on the leeward side of the hill. The freestream Reynolds number per unit length (U_0/ν) is approximately $1.33 \times 10^6\, m^{-1}$. The computations were performed on a 170×90 quadrilateral mesh using three different k-ε models (standard k-ε (Launder & Spalding, 1974), realizable k-ε (Shih *et al.*, 1993), RNG k-ε (Yakhot *et al.*, 1992) models), and a second-moment closure model (Gibson & Launder, 1978). The standard wall functions (Launder & Spalding, 1974) were employed. Figure 6 shows the predicted pressure and skin-friction distributions. Note that $x = 0$ corresponds to the leading-edge, and the trailing-edge is located at about $x = 1.284\ m$. All four turbulence models yield practically the same pressure distribution over the upstream half of the hill. However, the predictions depart from each other downstream of the hill crest. The k-ε models are shown to significantly overpredict the pressure recovery in the recirculation region, although the RNG k-ε model and the realizable k-ε model slightly reduce the discrepancy. The best prediction is given by the second-moment closure model. The same ranking can be made among the turbulence models for the skin-friction distributions as well. The RNG k-ε and the realizable k-ε models are seen to reproduce the onset of the flow reversal $(x = 1.1\ m)$ quite closely. The second-moment closure yields the best result of all.

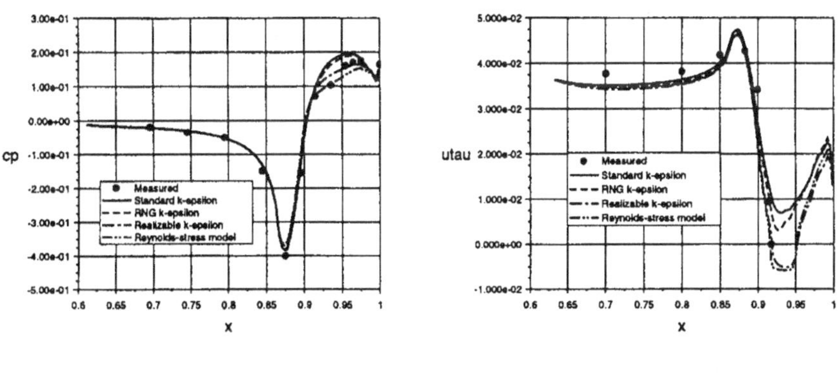

Figure 7. Flow past the axisymmetric underwater body - Predictions using four different turbulence models, left: pressure distribution, right: friction-velocity distribution

For another example, the flow past an axisymmetric under-water body

($Re_L = 5.9 \times 10^6$) was considered. The main feature of this low-subsonic flow is the incipient flow separation experimentally observed at $x/L = 0.92$. The computations were made on a 120×60 quadrilateral mesh. The same turbulence models and wall functions as in the previous example were employed. The results are shown in Figure 7. The discussion for the previous example largely holds true for this flow too. All three k-ε models overpredict the pressure rise toward the tail, although the prediction by the realizable k-ε model is appreciably better than those by the standard k-ε and the RNG k-ε models. The second-moment closure model is again found to best predict the pressure distribution near the tail including the sudden change of the slope just downstream of the onset of separation. Both the standard and the RNG k-ε models are found to fail to predict the onset of flow reversal. The realizable k-ε model and the second-moment closure model are seen to capture the flow reversal quite closely.

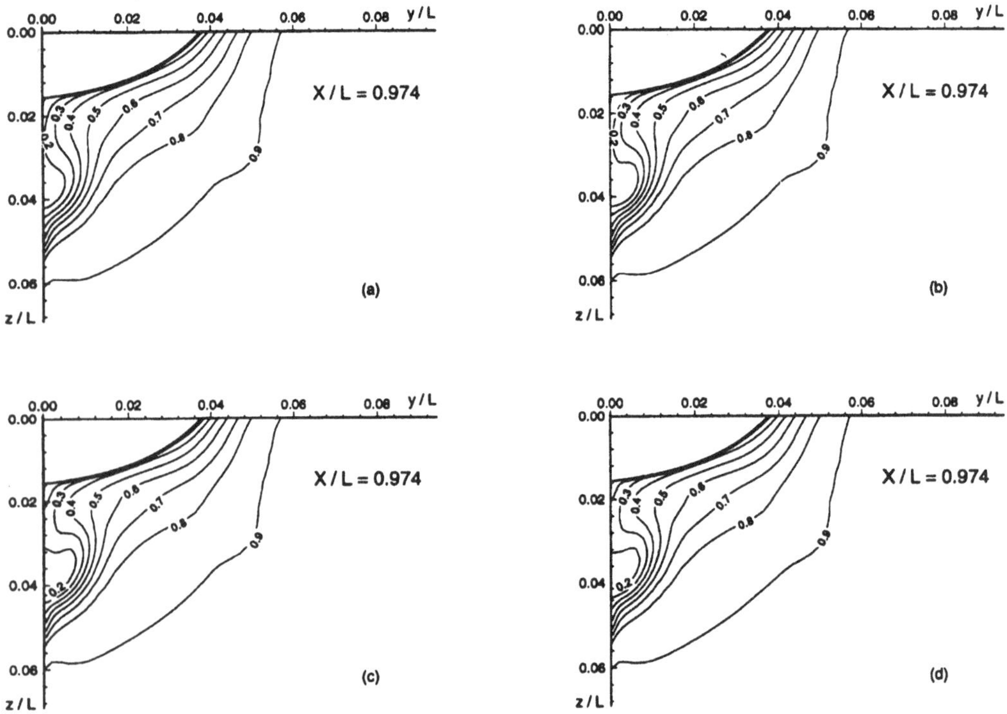

Figure 8. Flow around the HSVA ship model - Axial velocity (U/U_0) contours at $x/L = 0.974$ predicted by four different turbulence models (a) Standard k-ε model (b) RNG k-ε model (c) Realizable k-ε model (d) Reynolds-Stress model

As a three-dimensional example, we revisit the flow around the HSVA ship model discussed earlier in relation to the discussion on the impact of discretization scheme. The flow is characterized by strong pressure gradients (both axial and transverse), three-dimensional boundary layers with strong crossflow and streamline curvature, and streamwise vortices emanating from the boundary layer and evolving into the wake. Once again, the computations were made with the same turbulence models and wall functions as in the previous two examples. As noted earlier, our main interest here is how well the salient features of the axial velocity contours are predicted by the models. Figure 8 shows the predicted axial velocity contours. Comparing these results with the measured contours in Figure 2 indicates that the RNG and the realizable k-ε models and the second-moment closure predict the characteristic shape of the velocity contours quite closely. It is particularly noteworthy that the prediction by the realizable k-ε model is almost indiscernible from that by the second-moment closure model.

While more universal turbulence models are being actively sought among two-equation and higher-order models, one-equation turbulence models such as Baldwin-Barth (1990) model and Spalart-Allmaras (1992) model are getting renewed attention these days. Of these two, the Spalart-Allmaras model seems to enjoy a wider acceptance, probably because it is easier to implement, and also possibly because some recent studies have found that overall it performs better than the Baldwin-Barth model. In the rest of this paper, we present a few more computational examples in which the Spalart-Allmaras model as well as the k-ε models are employed.

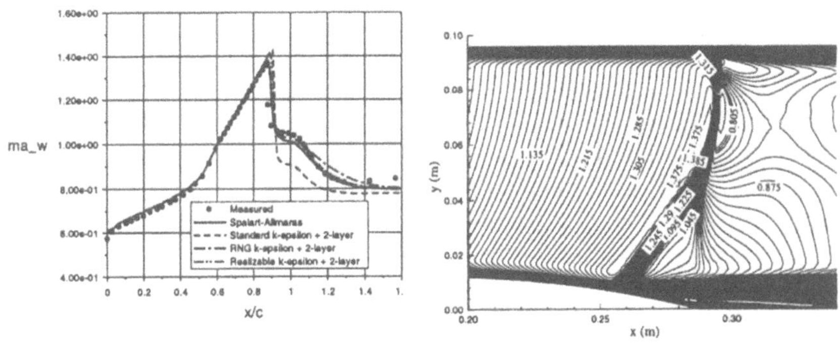

Figure 9. Transonic flow over the two-dimensional bump - Left: Wall Mach number distributions,Right: Mach number contours near the shock

We took the transonic flow over a two-dimensional bump mounted on the bottom wall of a wind tunnel, which was selected as a test case (Case No. 8612) at the 1980-81 AFOSR-HTTM-Stanford conference. The flow involves a fairly strong shock on the rear part of the bump at around $x/c = 0.88$, and a pocket of separation right behind the shock. The maximum Mach number of the flow, which occurs near the shock, was found to be about 1.4. The computations were carried out on a fine mesh ($y^+ \simeq 1$) using the Spalart-Allmaras model and the three k-ε models used in the preceding examples. Unlike the earlier examples, however, we employed here a no-slip condition at the wall boundary, instead of the wall functions. For the k-ε models, which are not valid in the viscosity-dominated near-wall region, the two-layer zonal approach of Chen and Patel (1988) was used. Figure 9 shows the results. The plot on the left shows the wall Mach number predictions. All the models, except the standard k-ε model, are seen to predict the wall Mach number distribution fairly well. Particularly noteworthy is that the prediction by the Spalart-Allmaras model is largely comparable to those of the RNG k-ε and the realizable k-ε models. The plot on the right depicts the Mach number contours in the flow predicted by the Spalart-Allmaras model. We can see that the predicted maximum Mach number (1.39) closely matches the experimental value (1.4).

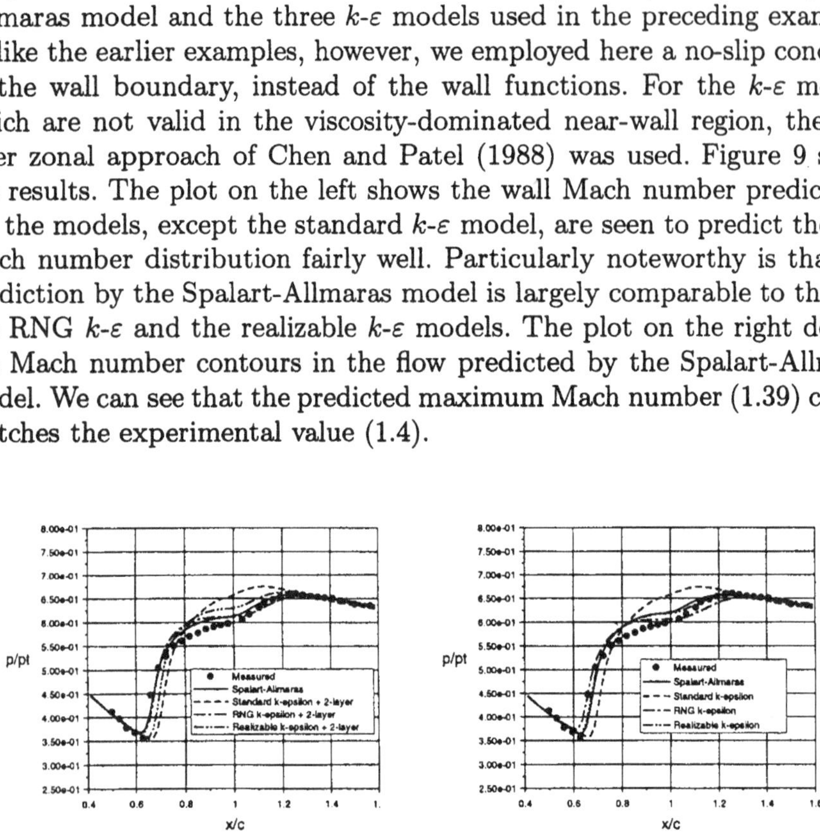

Figure 10. Transonic flow over the axisymmetric bump ($M_\infty = 0.875$, $Re_\infty = 13.2 \times 10^6$ - Surface pressure distributions predicted using different turbulence models, Left:coarse (wall function) mesh solutions, right:coarse (wall function) mesh solutions

For another example, the transonic flow over an axisymmetric bump, measured by Bachalo and Johnson (1986), was considered. This case was also one of test cases for the 1980-81 AFOSR-HTTM-Stanford conference (Case No. 8611), and has the features similar to those of the two-dimensional bump flow. The freestream Mach number is 0.875, and the Reynolds number based on the bump length is 13.2×10^6. The computations were made using the same turbulence models as in the previous example. This time, however, we also employed a coarse mesh, in addition to a fine mesh. For the fine mesh, no-slip was imposed at the wall, with the two-layer zonal model employed for the k-ε models. For the coarse mesh, the wall functions were adopted, *i.e.*, the standard wall functions for the k-ε models, and the universal law-of-the-wall, Equation (1), for the Spalart-Allmaras model.

A few remarks are deemed necessary here regarding the adoption of the law-of-the-wall. The Spalart-Allmaras model was originally proposed to be used on fine meshes with no-slip as the wall boundary condition. However, it can also be employed, without any modification, in conjunction with the law-of-the-wall when the chosen mesh is too coarse to resolve the viscosity-dominated near-wall layer. The combination of the Spalart-Allmaras model and the law-of-the-wall offers a practical alternative in industrial flow simulations where using a fine near-wall mesh that resolves the viscous sublayer is not always feasible.

Figure 10 shows the results. The results for the fine mesh (the plot on the left) show again that the predictions by the Spalart-Allmaras, RNG k-ε and realizable k-ε models all are in good agreements with the experimental data, while the standard k-ε model severely overpredicts the pressure recovery in the separated region. The wall function results, shown on the right, are rather intriguing. As far as the pressure distribution is concerned, the predictions employing the wall functions do not show significant differences from the fine mesh results. This seems to be largely the case for all the models including the Spalart-Allmaras model.

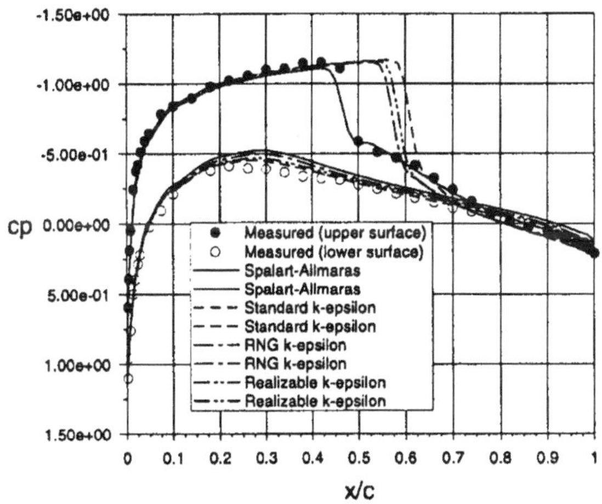

Figure 11. Flow around NACA 0012 airfoil ($M_\infty = 0.799$, $Re_\infty = 9 \times 10^6$, $\alpha = 6°$) - Pressure distributions predicted using different turbulence models

As a last example, we considered the flow around the NACA 0012 airfoil discussed earlier. The computations were carried out on a coarse mesh using the same turbulence models and the same wall functions as in the last example. It is shown in Figure 11 that the Spalart-Allmaras model captures both the shock location and the pressure distribution behind the shock remarkably well. The predictions by the k-ε models are seen to be rather poor, although the results from the RNG k-ε and realizable k-ε models look slightly better.

The last three examples seem to suggest that the Spalart-Allmaras model has a good potential to provide a practical alternative to more sophisticated models, at least for a certain class of flows such as the ones presented in this study. Further testing for a broader range of flows is needed to establish its efficacy as an engineering turbulence model.

4. Closure

Mesh, choice of discretization schemes, and turbulence modeling dictate the successful computation of complex industrial turbulent flows. We anticipate that the unstructured mesh technology will further enhance both the usability of CFD and the accuracy of the predictions for industrial flows which involve highly complex geometries and flow physics. We showed that employing accurate discretization schemes is essential for reliable predictions of complex turbulent flows and proper evaluation of turbulence models. As indicated by many computational examples presented in this paper, wall function approaches provide an effective means of near-wall treatment for complex wall-bounded turbulent flows. From the standpoint of industrial applications of CFD, the efforts to improve the wall function approach deserve more attention. Some of the more recent k-ε turbulence models (RNG k-ε and realizable k-ε models) were shown to bring appreciable improvements over the standard k-ε model for complex flows involving pressure gradient, streamline curvature, crossflow, and separation. More comprehensive investigations are required to evaluate the performance of more recent turbulence models including other models not included in this study. The Spalart-Allmaras one-equation model yields quite encouraging results. Further studies are needed to understand its strengths and weaknesses for a broader class of complex turbulent flows.

5. Acknowledgements

The authors would like to acknowledge the efforts of all the developers and industry business units at Fluent Inc. who contributed to this paper.

References

Amano, R.S., 1984. "Development of a Turbulence Near-Wall Model and Its Application to Separated and Reattached Flows," *Numerical Heat Transfer* **7**, pp. 59–75.

Azad, R.S. and Kassab, S.Z., 1989. "Turbulent Flow in a Conical Diffuser: Overview and Implications," *Physics of Fluids, A* **1**, No. 3, pp. 564-573.

Bachalo, W.D. and Johnson, D.A., 1986. "Transonic Turbulent Boundary-Layer Separation Generated on an Axisymmetric Flow Model," *AIAA J.* **24**, No. 3, pp. 437–443.

Baldwin, B.S. and Barth, T.J., 1990. "A One-Equation Transport Model for High-Reynolds number Wall-Bounded Flows," *NASA TM 102847.*

Baskaran, V., Smits, A.J., and Joubert, P.N., 1987. "A Turbulent Flow Over a Curver Hill - Part 1. Growth of an Internal Boundary Layer," *J. Fluid Mech.* **182**, pp. 47–83.

Chen, H.C. and Patel, V.C., 1988. "Near-Wall Turbulence Models for Complex Flows Including Separation," *AIAA Journal* **26**, No. 6, pp. 641-648.

Chieng, C.C. and Launder B.E., 1980. "On the Calculation of Turbulent Heat Transport Downstream from an Abrupt Pipe Expansion," *Numerical Heat Transfer* **3**, pp. 189–207.

Craft, T.J., Launder, B.E., Suga, K., 1993. "Extending the Applicability of Eddy Viscosity Models through the Use of Deformation Invariants and Non-Linear Elements," *Proceedings of the 5th International Symposium on Refined Flow Modeling and Turbulence Measurements*, Paris, pp. 125-132.

Delery, J. P. and P. Le Diuzet, 1979. "Découlement résultant d'une interaction onde de choc/couche limite turbulente," T.P. No. 1979-146, ONERA.

Gatski, T.B. and Speziale, C.G., 1993. "On Explicit Algebraic Stress Models for Complex Turbulent Flows," *J. Fluid Mech.* **254**, pp. 47–83.

Gibson, M.M. and Launder, B.E., 1978. "Ground Effects on Pressure Fluctuations in the Atmospheric Boundary Layer," *J. Fluid Mechanics* **86**, pp. 491–511.

Huang, T.T., Wang, H.T., Santelli, N., and Groves, N.C., 1976. "Propeller/Stern/Boundary-Layer Interaction on Axisymmetric Bodies,: Theory and Experiment," David W. Taylor Naval Ship Research and Development Center, Bethesda, MD, Rep. 76-0113.

Johnson, R.W., 1988. "Numerical Simulation of Local Nusselt Number for Turbulent Flow in a Square Duct with a 180° Bend," *Numerical Heat Transfer* **13**, pp. 205–228.

Kim, S.-E. and Choudhury, D., 1995. "A Near-Wall Treatment Using Wall Functions Sensitized to Pressure Gradient," *ASME FED-Vol. 217, Separated and Complex Flows*, ASME.

Kim, S.-E., Mathur, S.R., Murthy, J.Y., and Choudhury, D., 1998. "A Reynolds-Averaged Navier-Stokes Solver Using an Unstructured Mesh Based Finite-Volume Scheme," *AIAA-98-0231.*

Launder, B.E. and Spalding, D.B., 1974. "The Numerical Computation of Turbulent Flows," *Comp. Meth. Appl. Mech. Eng.* **3**, pp. 269–289.

Leonard, B.P., 1979. "A stable and accurate convective modeling procedure based on quadratic upstream interpolation," *Computational Methods Appl. Mech. Eng.* **19**, p. 59.

Mathur, S. R. and Murthy, J. Y., 1997. "A Pressure-Based Method for Unstructured Meshes," *Numerical Heat Transfer* **31**, pp. 195–215.

Shih, T.-H., Liou, W.W., Shabbir, A., and Zhu, J., 1993. "A New k-ε Eddy-Viscosity Model for High Reynolds Number Turbulent Flows - Model Development and Validation," *Computers & Fluids* **24**, No. 3, pp. 227–238.

Shih, T.-H., Zhu, J., and Lumley, J. L., 1994. "A New Reynolds Stress Algebraic Equation Model," *NASA Technical Memorandum 106644*, ICOMP-94-15; CMOTT-94-8.

Spalart, P. and Allmaras, S., 1992. "A One-Equation Turbulence Model for Aerodynamic Flows," *AIAA-92-0439.*

Thibert, J.J., Granjacques, M., and Ohman, L.H., 1979. "NACA 0012 Airfoil, Experimental Data Base for Computer Program Assessment," AGARD Advisory Rept. No. 138.

Townsend, A.A., 1961. "Equilibrium Layers and Wall Turbulence," *J. Fluid Mech.* **11**, pp. 97–120.
Viegas, J.R. and Rubesin, M.W., 1985. "On the Use of Wall Functions as Boundary Conditions for Two-dimensional Separated Compressible Flows," *AIAA-85-0180.*
Weiss, J.M. and Smith, W.A., 1995. "Preconditioning Applied to Variable and Constant Density Flows," *AIAA J.* **33**, No. 11.
Wieghardt, K. and Kux, J., 1980. "Nominal Wakes Based on Wind Tunnel Tests," (in German), *Jahrbuch des Schiffbautechnischen Gesellschaft* (STG), Springer-Verlag, pp. 303–318.
Wilcox, D.C., 1993. "Comparison of Two-Equation Turbulence Models for Boundary Layers in Pressure Gradient," *AIAA J.* **31**, No. 8, pp. 1414–1424.
Yakhot, V., Orszag, S.A., Thangam, S., Gatski, T.B. and Speziale, C.G., 1992. "Development of Turbulence models for Shear Flows by a Double Expansion Technique," *Physics of Fluids, A* **4**, No. 7, pp. 1510–1520.

SIMULATION OF SHOCK WAVE-TURBULENT BOUNDARY LAYER INTERACTIONS USING THE REYNOLDS-AVERAGED NAVIER-STOKES EQUATIONS

DOYLE D. KNIGHT
Department of Mechanical and Aerospace Engineering
Rutgers University - The State University of New Jersey
New Brunswick, New Jersey

Abstract. The paper examines the capability for numerical simulation of three-dimensional shock wave turbulent boundary layer interactions using the Reynolds-averaged Navier-Stokes equations. Two configurations are considered, namely, the single and double fin interactions. Numerical simulations were performed by an international group of researchers as part of an AGARD-sponsored study of capabilities for simulation of high speed flight. A broad range of turbulence models were employed including zero-, one-, two-equation and full Reynolds stress equation models. The paper presents a comparison of the numerical simulations with experiment and summarizes the results.

1. Introduction

Accurate simulation of the three dimensional flowfields arising from the interaction of shock waves with turbulent boundary layers is essential for prediction of the aerodynamic performance of high speed aircraft and missiles. For example, the deflection of a control surface (*e.g.*, vertical stabilizer or aileron) in supersonic flight generates a shock wave whose interaction with the boundary layer determines the pressure distribution on the control surface and hence the aerodynamic forces and moments.

There are several extensive reviews of shock wave-turbulent boundary layer interaction focusing on the physics of these flows. Examples include Greene (Greene, 1970), Korkegi (Korkegi, 1971), Peake and Tobak (Peake *et al.*, 1980), Delery and Marvin (Delery *et al.*, 1986), Settles and Dolling (Settles *et al.*, 1986; Settles *et al.*, 1990), Stollery (Stollery, 1989), Degrez

M. D. Salas et al. (eds.), Modeling Complex Turbulent Flows, 277–296.

et al. (Degrez, 1993), Delery and Panaras (Delery *et al.*, 1996) and Zheltovodov (Zheltovodov, 1996).

The objective of this paper is to examine the capability for numerical simulation of 3-D shock wave turbulent boundary layer interactions. Two specific configurations are examined, namely, the single fin (Fig. 1) and double fin interactions (Fig. 2). The focus of the investigation is the interaction of the shock wave(s) generated by the fin(s) with the supersonic turbulent boundary layer on the flat plate. The fin heights are assumed semi-infinite. These configurations represent geometrical simplifications of the corresponding engineering configuration, *i.e.*, the single fin represents a vertical stabilizer-fuselage juncture, and the double-fin represents a hypersonic (Trexler-type) inlet.

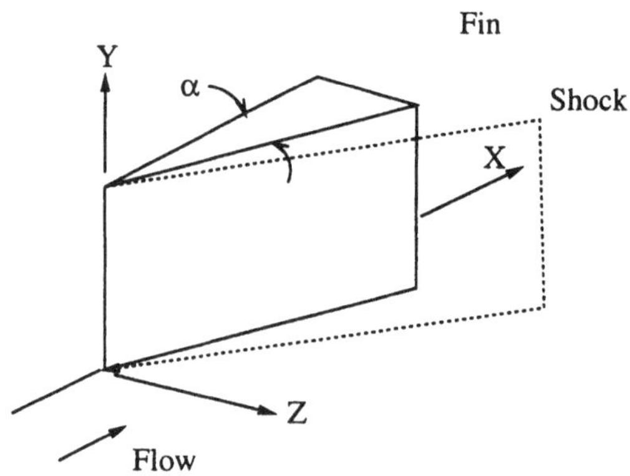

Figure 1. Single fin geometry

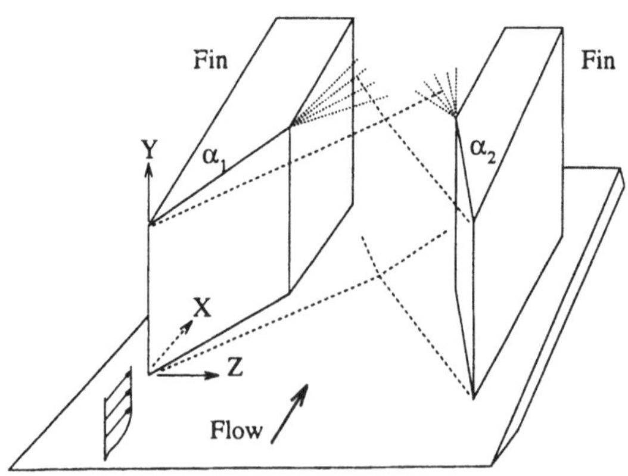

Figure 2. Double fin geometry

The results presented in this paper were obtained by an international group of researchers (Table 1) who participated in an extensive study of the capability for numerical simulation of high speed flows involving shock wave-turbulent boundary layer interaction. The results of that study, organized by the AGARD Fluid Dynamics Panel Working Group 18, was presented in an AGARD report (Knight *et al.*, 1997). This paper summarizes the results for one specific case each of the single and double fin configurations. These two cases are typical of the results obtained for the entire test matrix described in the full report (Knight *et al.*, 1997) which included seven cases for the single fin, four cases for the double fin, and two cases for a hollow cylinder flare interaction.

TABLE 1. Participants

Participant	*Organization*
Greg Alexopoulos	North Carolina State Univ
J.-M. Bousquet	ONERA
R. Bur	ONERA
Gérard Degrez	Von Karman Institute
Jack Edwards	North Carolina State Univ
Datta Gaitonde	Wright Labs, WPAFB, OH
Marianna Gnedin	Rutgers University
F. Grasso	Università di Roma
Hassan Hassan	North Carolina State Univ
C. C. Horstman	NASA Ames
Doyle Knight	Rutgers University
John Moss	NASA Langley
Natraj Narayanswami	Rutgers University
Argyris Panaras	Hellenic AF Academy
Patrick Rodi	Univ of Texas-Austin
Balu Sekar	Wright Labs, WPAFB, OH
Edwin Van der Weide	Von Karman Institute
Gecheng Zha	Rutgers University

2. Single Fin

The single fin geometry (Fig. 1) is defined by the fin angle α. The fin is attached normal to the flat plate on which an equilibrium compressible turbulent boundary layer has developed. The fin generates an oblique shock wave which interacts with the boundary layer on the flat plate. The flow parameters are the Mach number M_∞, Reynolds number Re_{δ_∞} based on the upstream boundary layer thickness δ_∞, fin angle α and wall temperature ratio T_w/T_∞.

The structure of the single fin flowfield is well understood. Alvi and Settles (Alvi *et al.*, 1991) and Zheltovodov and Shilein (Zheltovodov *et al.*, 1986) provide a detailed description. When the shock strength is sufficient to cause separation of the boundary layer on the flat plate, then the wave structure and mean streamline pattern behave in an approximately conical manner[1] outside an initial "inception zone" near the juncture of the fin leading edge and flat plate. Fig. 3 (from (Alvi *et al.*, 1991)) shows the conical flowfield structure.[2] The oblique shock bifurcates to form a λ-shock whose triple point is the origin of a slip line (Courant *et al.*, 1948). The adverse pressure gradient forces a separation of the boundary layer which forms a vortex whose center is approximately beneath the primary inviscid shock. Depending on the shock strength, additional features may appear including secondary separation of the boundary layer beneath the vortex, a normal shock in the impinging jet which turns back over the vortex, and supersonic reversed flow.

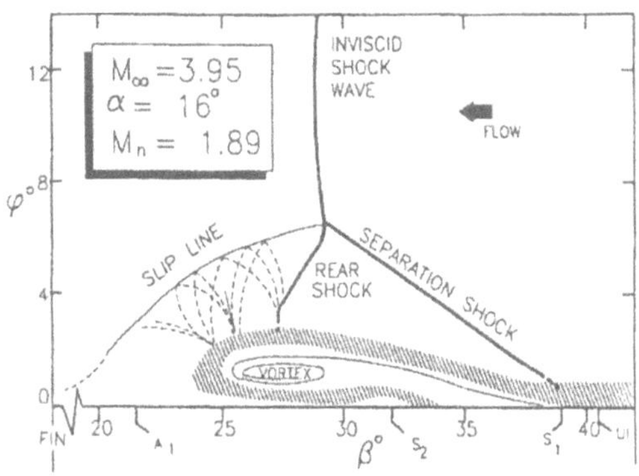

Figure 3. Flowfield structure for single fin (from Alvi & Settles)

The flow conditions for the single fin case are $M_\infty = 4.0$, $\alpha = 20°$, $Re_{\delta_\infty} = 2.1 \times 10^5$ and $T_w/T_{aw} = 1.06$. The experiments were performed by Kim *et al.* (Kim *et al.*, 1990). Computations were performed using several different turbulence models as indicated in Table 2.

The surface pressure is shown in Fig. 4. The Baldwin-Lomax-Panaras and Spalart-Allmaras-Edwards models show closest agreement with experiment. These two models predict the surface pressure in the plateau region

[1] Conical flow is defined in Appendix A.

[2] This particular case corresponds to $M_\infty = 4$, $\alpha = 16°$ (Kim *et al.*, 1990).

TABLE 2. Turbulence Models for Single Fin

Participant	*Model*	*Ref*
J. Edwards	Spalart-Allmaras-Edwards	(Edwards *et al.*, 1996)
C. C. Horstman	$k-\epsilon$	(Jones *et al.*, 1972)
D. Knight	Baldwin-Lomax	(Baldwin *et al.*, 1978)
A. Panaras	Baldwin-Lomax-Panaras	(Panaras, 1996)

($36° \leq \beta \leq 47°$) within 5% to 10% of the experimental data. Also, both models display a pressure trough at $\beta = 32°$, in agreement with experiment, although their predictions differ from the experiment by 30%. In the vicinity of the corner, both models overestimate the peak pressure by 11%. The Baldwin-Lomax and $k-\epsilon$ models exhibit the general trends of the experiment, but show greater disagreement with experiment.

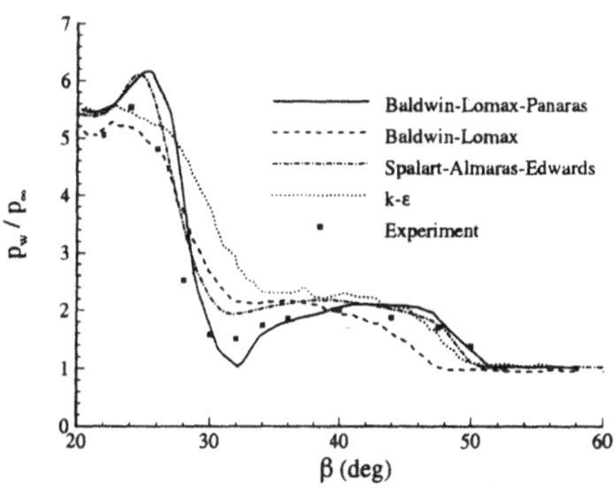

Figure 4. Surface pressure for single fin

The surface streamline angle Φ on the flat plate[3] is displayed in Fig. 5. The Baldwin-Lomax-Panaras is the most accurate, with the Spalart-Allmaras-Edwards model providing nearly the same results. The two models differ principally in the region of the secondary separation at $\beta = 40°$. Again, the Baldwin-Lomax and $k-\epsilon$ models show general agreement with experiment, but are less accurate.

[3]The ordinate of Fig. 5 is $\Phi - \beta$, where Φ is the angle of the surface streamline on the flat plate measured relative to the direction of the approaching flow (x-direction in Fig. 1). The angle β is the azimuthal angle measured relative to the x-axis (see Appendix A). For conical flow, the separation and attachment lines are defined by $\Phi - \beta = 0$, and thus the choice of $\Phi - \beta$ as the ordinate allows easy identification of these features.

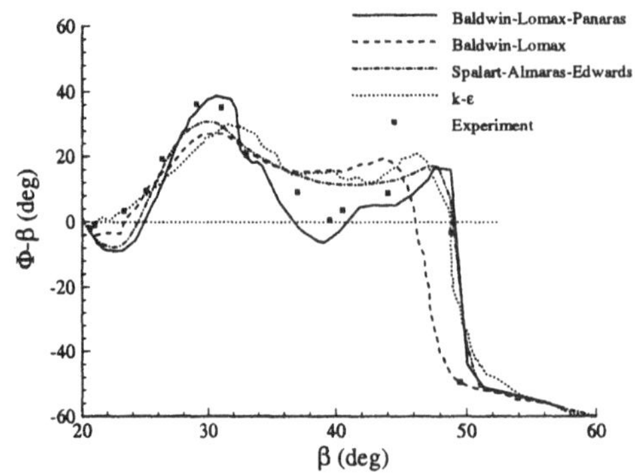

Figure 5. Surface streamline angle for single fin

The skin friction coefficient is shown in Fig. 6. The Baldwin-Lomax-Panaras and Spalart-Allmaras-Edwards models predict a peak in the vicinity of the corner which is not seen in the experimental data.[4] Their computed values at the experimental location $\beta = 26.5°$ are significantly above the experiment. Additional measurements in the region $22° < \beta < 26°$ would be useful to ascertain whether a peak appears[5]. Elsewhere, all four models provide generally good agreement with experiment.

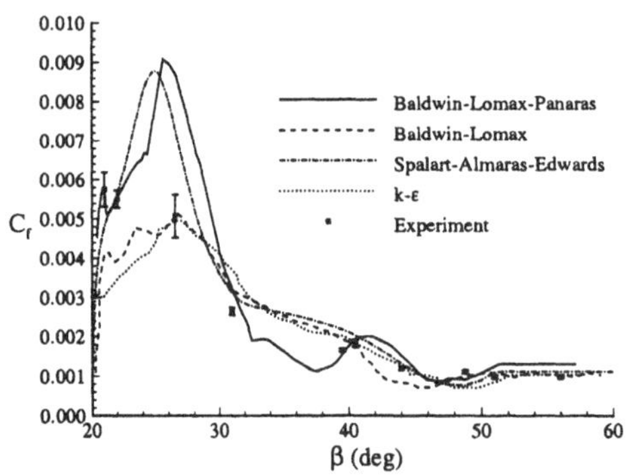

Figure 6. Skin friction coefficient for single fin

[4]Corrected data for $\beta = 22°$ and $26.5°$, provided by Prof. G. Settles, is included in Fig. 6.

[5]Note that the attachment line is $\beta = 26°$ (Alvi *et al.*, 1991).

3. Double Fin

The double fin geometry is defined by the two fins of angles α_1 and α_2. The fins are attached normal to a flat plate (Fig. 2) on which an equilibrium turbulent boundary layer has developed. The fins generate intersecting oblique shock waves which interact with the turbulent boundary layers on the flat plate and inner fin surfaces. The flow parameters are the Mach number M_∞, Reynolds number Re_{δ_∞}, fin angles α_1 and α_2, contraction ratio L_2/L_1, throat middle line offset L_3/L_1, boundary layer to throat width ratio δ_∞/L_2, and wall temperature ratio T_w/T_{aw}. The fins are assumed semi-infinite in height. For the symmetric double fin, $L_3 = 0$.

The flowfield structure of the double fin interaction is not fully understood. The interaction of the incident λ-shocks, generated by the initial single fin interactions, forms a complex wave system which is described, for example, in Garrison *et al.* (Garrison *et al.*, 1992) and Gaitonde *et al.* (Gaitonde *et al.*, 1995) for the symmetric case, and in Knight *et al.* (Knight *et al.*, 1995) for the asymmetric case. A pair of counter-rotating vortices, generated by the initial single fin interactions, merge into a vortex pair which is associated with a region of low total pressure. A detailed discussion of the streamline structure for the symmetric case is presented, for example, in Narayanswami *et al.* (Narayanswami *et al.*, 1992), and Gaitonde *et al.* (Gaitonde *et al.*, 1995) for the symmetric case, and for an asymmetric case in Knight *et al.* (Knight *et al.*, 1995). Fig. 7 shows an example of the streamline structure.

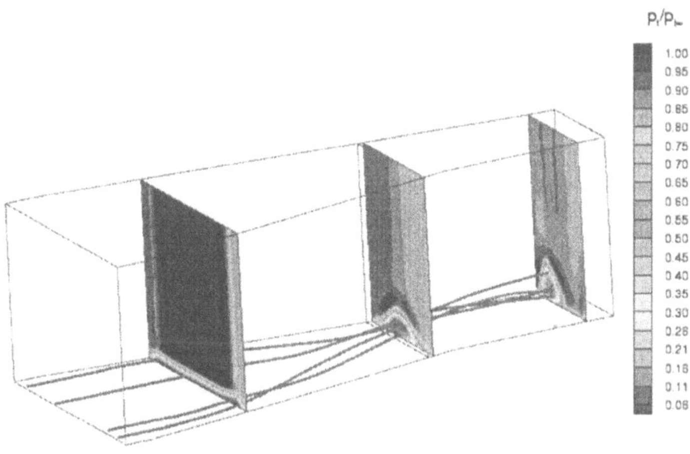

Figure 7. Computed streamlines and total pressure contours using $k-\epsilon$ Chien model

The flow conditions for the double fin case are $M_\infty = 4.0$, $\alpha_1 = 7°$, $\alpha_2 = 11°$, $Re_{\delta_\infty} = 2.0 \times 10^5$, $L_2/L_1 = 0.44$, $L_3/L_1 = 0.014$, $\delta_\infty/L_2 = 0.11$ and $T_w/T_{aw} = 1.11$. The experiments were performed by Zheltovodov *et al.* (Zheltovodov *et al.*, 1994). Computations were performed using several different turbulence models as shown in Table 3.

TABLE 3. Turbulence Models for Double Fin

Participant	*Model*	*Ref*
J. Edwards	Spalart-Allmaras-Edwards	(Edwards *et al.*, 1996)
M. Gnedin	$k-\epsilon$ Knight	(Becht *et al.*, 1995)
H. Hassan	$k-\omega$	(Alexopoulos *et al.*, 1997)
D. Knight	$k-\epsilon$ Chien	(Chien, 1982)
A. Panaras	Baldwin-Lomax-Panaras	(Panaras, 1996)
G. Zha	Reynolds Stress Equation	(Zha *et al.*, 1996)

The computed surface skin friction lines using the $k-\epsilon$ Chien model and $k-\epsilon$ Knight model are shown in Figs. 8 and 9, respectively, and the experimental surface visualization in Fig. 10. The incident separation lines, originating from the fin leading edges (**1** and **2**), are accurately predicted using both turbulence models. The computed separation line angles, measured relative to the x-axis, are within 10% of the experiment. The $k-\epsilon$ Chien model displays a coalescence of the incident separation lines into a narrow band (**3**) offset to the left side, in agreement with experiment.[6] The $k-\epsilon$ Chien results also show a second line of coalescence form alongside on the right and farther downstream (**4**) associated with a secondary separation underneath the left side of the right vortex (Knight *et al.*, 1995), and a line of divergence alongside the right fin (**5**). A similar line of divergence (unmarked) is near the left fin. The $k-\epsilon$ Knight model (Fig. 9) does not show a coalescence of the incident separation lines. Instead, the incident separation lines continue further downstream almost in parallel until they converge at $x \approx 110$ mm to form a narrow band of skin friction lines (**3**), which is offset to the left side of the channel. Lines of divergence are seen near the right fin (**4**) and left fin (**5**). These are attributable to the incident single fin interaction. The second line of coalescence observed in the $k-\epsilon$ Chien results (**4** in Fig. 8) is not present in this computation. Consequently, the $k-\epsilon$ Knight model does not predict a secondary separation underneath the left side of the right vortex. This difference is due

[6]This line is the image of the boundary between the left and right vortices associated with the initial single fin interactions.

to variation in the predictions of the pressure distribution in the spanwise direction, obtained with each turbulence model as described below.

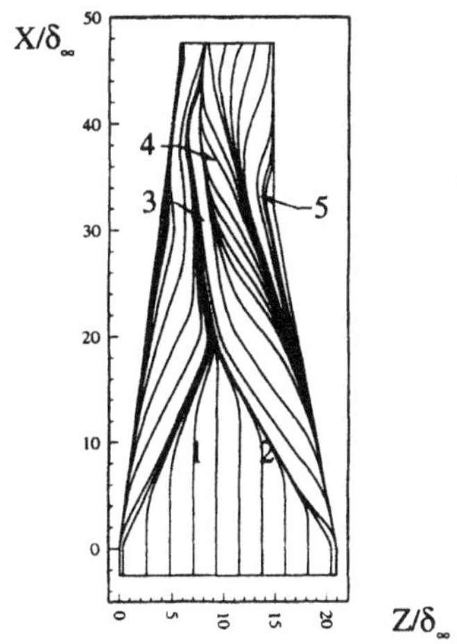

Figure 8. Computed skin friction lines, $k-\epsilon$ Chien model
1 Left incident separation line
2 Right incident separation line
3 Left downstream coalescence line
4 Right downstream coalescence line
5 Line of divergence (similar line near left fin)

The surface pressure along the Throat Middle Line[7] is shown Figs. 11 and 12. All turbulence models display close agreement with experiment for $x < 135$ mm, although the computations slightly underestimate the extent of the upstream influence.[8]

The spanwise behavior of the computed and experimental surface pressure at $x = 46$ and 79 mm is displayed in Figs. 13 to 16. The abscissa $z - z_{TML}$ represents the spanwise distance measured from the TML. The predictions of all models agree closely with experiment.

The spanwise variation of the surface pressure at $x = 112$ mm (the farthest station downstream of the experimental data) is shown in Figs. 17 and 18. The computations using the $k-\epsilon$ Chien, $k-\epsilon$ Knight ("Present $k-\epsilon$ ") and RSE models are in general agreement with experiment. The Spalart-Allmaras-Edwards, Baldwin-Lomax and $k-\omega$ models overpredict the pressure by 16% to 21%. The $k-\epsilon$ Chien model predicts a local adverse pressure

[7]The Throat Middle Line (TML) is the line which bisects the channel, formed by the two fins, at the location of minimum cross section.

[8]The computations do not accurately predict the pressure rise associated with the shock reflection from the 7° fin (beginning at $x = 145$ mm), since the computations omit the boundary layers on the fin surfaces.

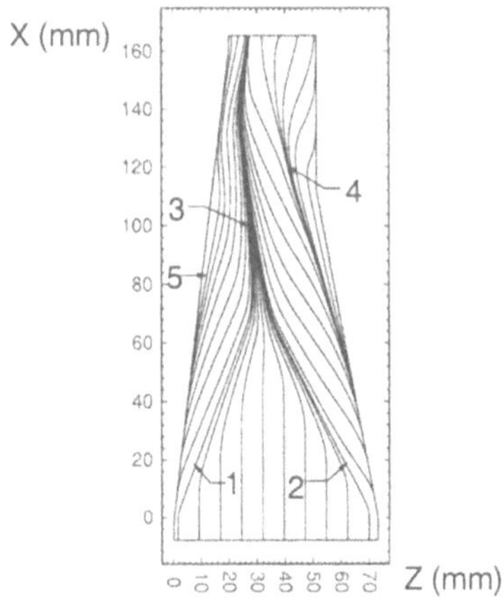

Figure 9. Computed skin friction lines, $k-\epsilon$ Knight model
1 Left incident separation line
2 Right incident separation line
3 Left downstream coalescence line
4,5 Lines of divergence

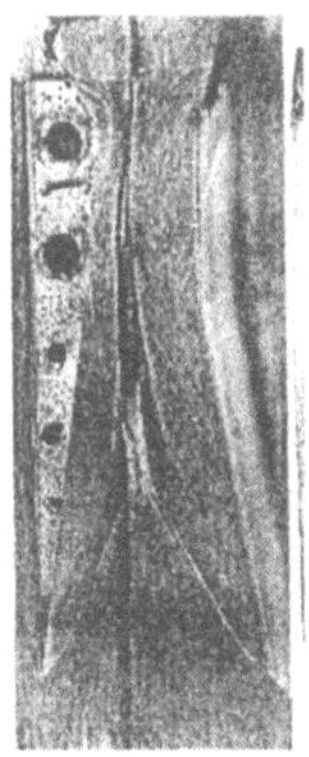

Figure 10. Experimental surface flow for $7^\circ \times 11^\circ$

gradient in spanwise direction in the region $-10\,\mathrm{mm} < z - z_{TML} < -4\,\mathrm{mm}$. Since the flow near the surface at this location is moving towards the left fin, this adverse pressure gradient causes the secondary separation and the appearance of the right downstream coalescence line (**4** in Fig. 8). The $k-\epsilon$ Knight model does not predict a significant adverse pressure gradient in this region, and hence a secondary separation line does not appear.

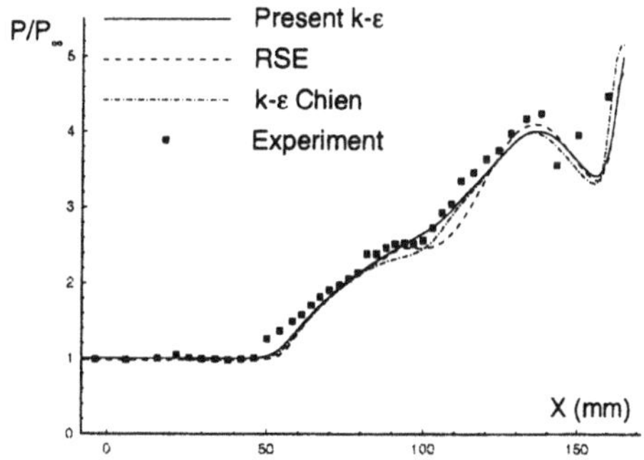

Figure 11. Wall pressure on TML

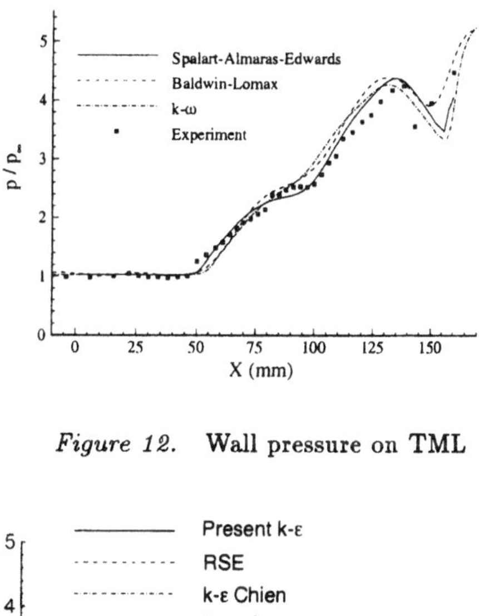

Figure 12. Wall pressure on TML

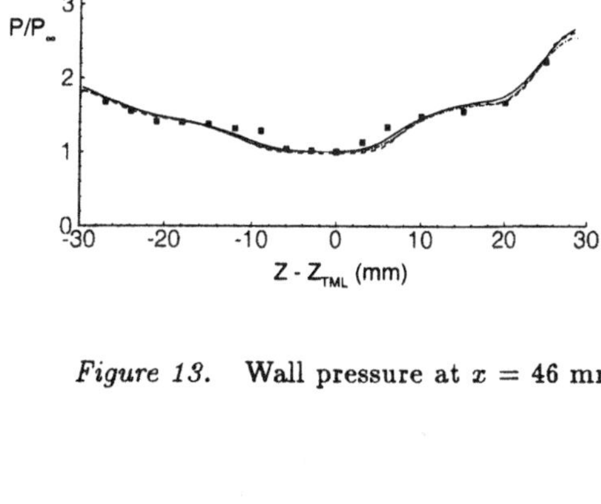

Figure 13. Wall pressure at $x = 46$ mm

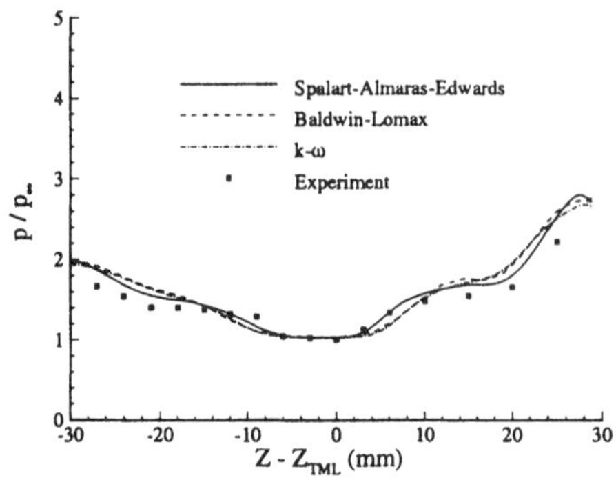

Figure 14. Wall pressure at $x = 46$ mm

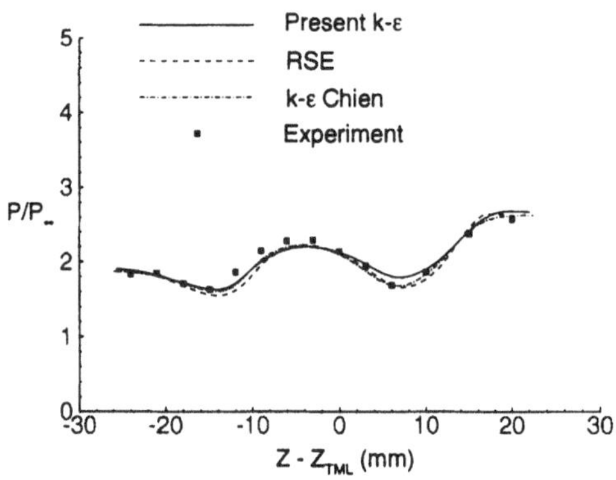

Figure 15. Wall pressure at $x = 79$ mm

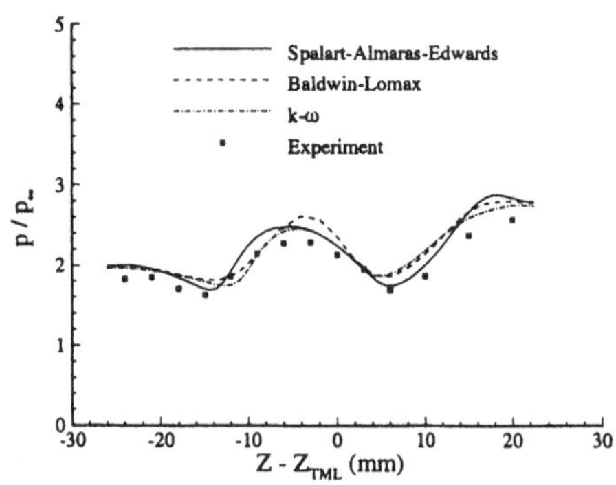

Figure 16. Wall pressure at $x = 79$ mm

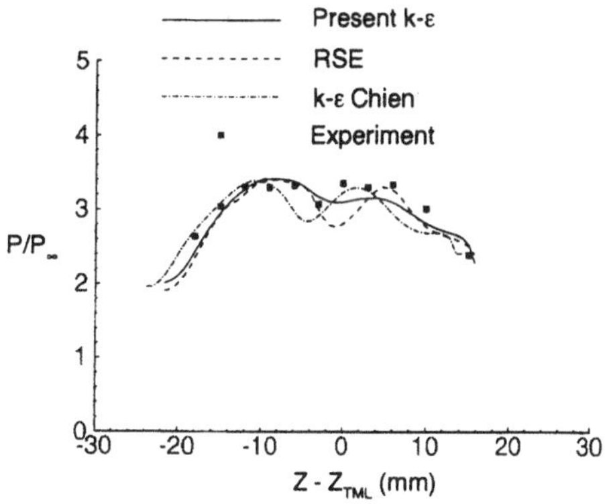

Figure 17. Wall pressure at $x = 112$ mm

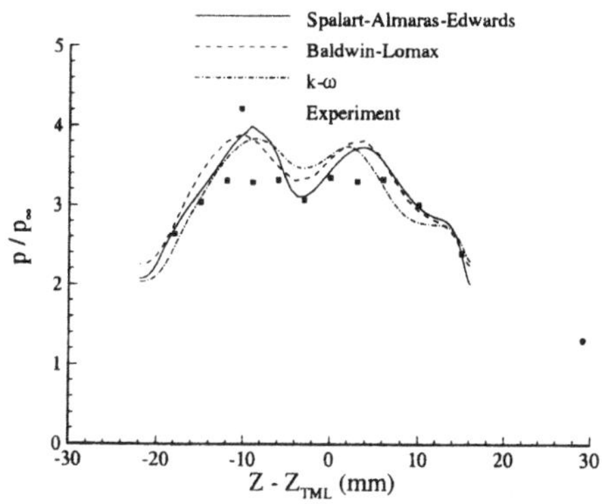

Figure 18. Wall pressure at $x = 112$ mm

The surface heat transfer coefficient[9] C_h is displayed in Figs. 19 to 22. On the TML (Figs. 19 and 20), all turbulence models overestimate C_h by approximately a factor of two downstream of the intersection of the shocks (*i.e.*, $x = 93.7$ mm). The predicted spanwise variation of C_h at $x = 112$ mm (Figs. 21 and 22), located within the strongly three-dimensional region of the flow, shows significant disagreement with experiment. The overprediction in C_h is actually an overestimate in q_w, since a series of studies (Zha *et al.*, 1996; Gnedin, 1996) demonstrated that the computed q_w

[9]The heat transfer coefficient is defined by

$$C_h = \frac{q_w(x,z)}{\rho_\infty U_\infty c_p (T_w(x,z) - T_{aw}(x,z))}$$

is proportional to the computed $T_w - T_{aw}$. A possible explanation (Gnedin *et al.*, 1996) is that all of the turbulence models overestimate the production of turbulence by the shock system, and thus generate too much turbulence kinetic energy, thereby leading to a overprediction of the turbulent eddy viscosity.

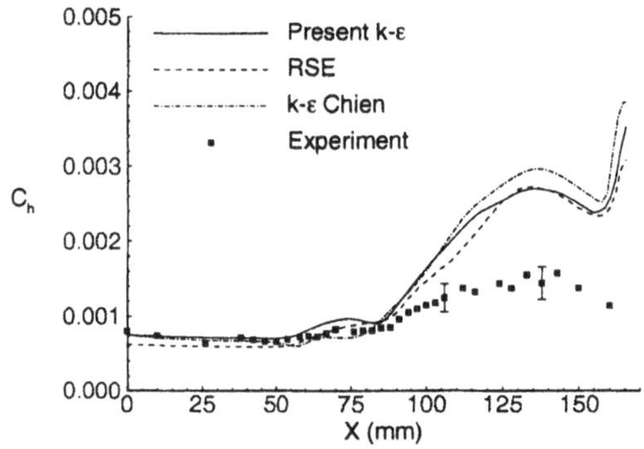

Figure 19. C_h on TML

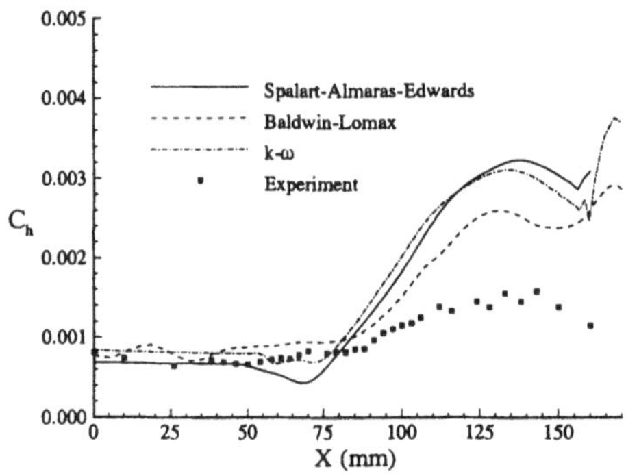

Figure 20. C_h on TML

The adiabatic wall temperature T_{aw}/T_∞ on the TML is shown in Figs. 23 and 24. The $k-\epsilon$ Knight and $k-\omega$ models show best agreement with experiment. The computed results for spanwise variation of T_{aw} at $x = 46, 79$ and 112 mm show comparable good agreement.

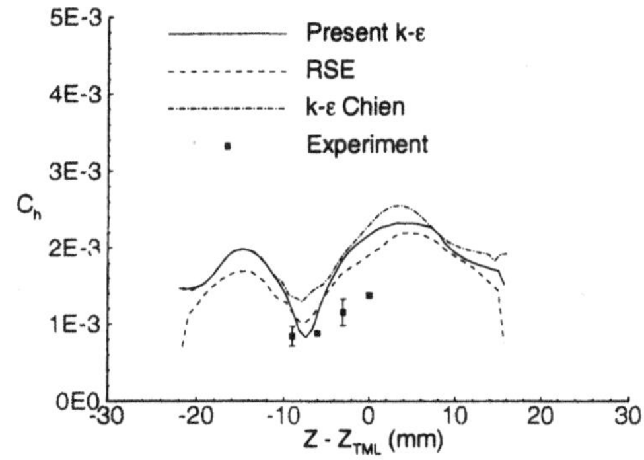

Figure 21. C_h at $x = 112$ mm

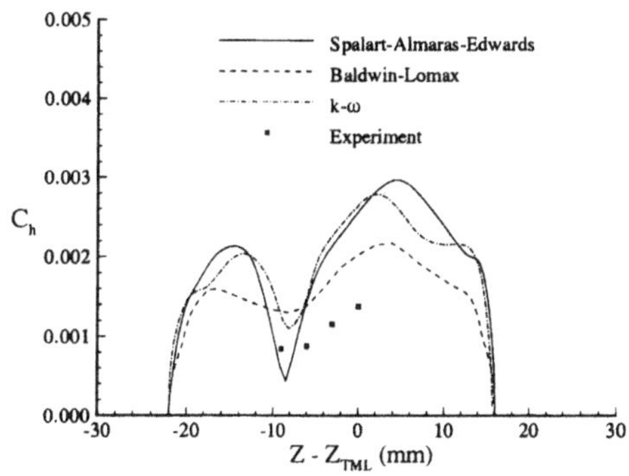

Figure 22. C_h at $x = 112$ mm

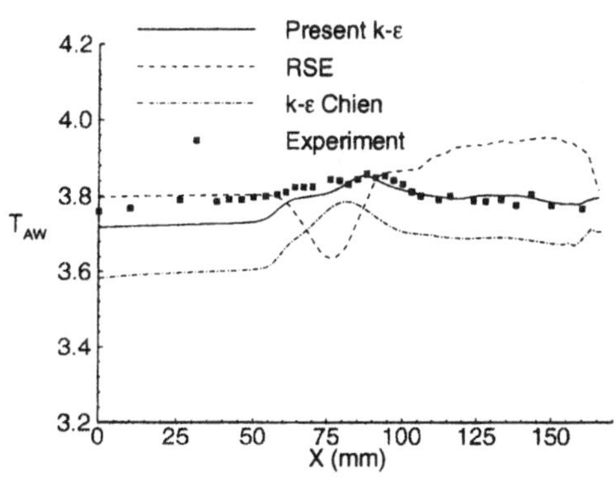

Figure 23. T_{aw} on TML

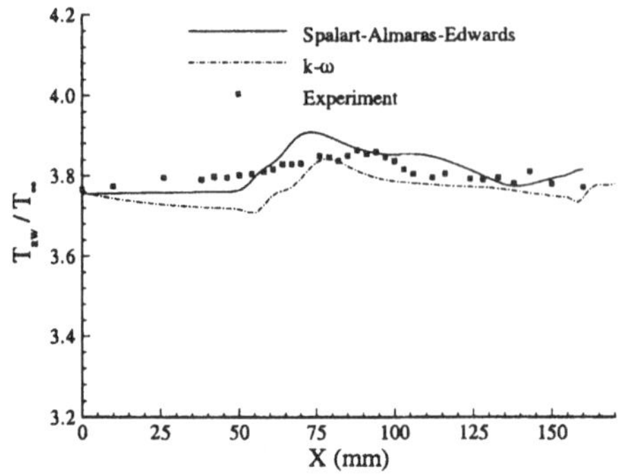

Figure 24. T_{aw} on TML

4. Conclusions

The paper assessed the capability of prediction of three-dimensional shock wave-turbulent boundary layer interactions using the Reynolds-averaged Navier-Stokes equations. Two configurations – the single and double fin – were considered. Computations were performed by an international group of researchers as part of an AGARD-sponsored study focusing on assessment of computational capability for high speed flight. A variety of turbulence models were considered ranging from zero-equation to full Reynolds stress equation models.

The mean pressure distributions are generally predicted satisfactorily, with modest variations observed between different turbulence models. The similarity of the computations may be due to an approximate triple-deck structure of the flowfield (Knight, 1993). The triple-deck, developed in the early work of Stewartson (see, for example, (Stewartson, 1981)) and extended to non-separated two- and three-dimensional shock wave turbulent boundary layer interactions (see, for example, Inger (Inger, 1985; Inger, 1987)), is defined by three layers. In the first layer ("deck"), located immediately adjacent to the surface, the fluid motion is determined by both viscous (*i.e.*, laminar and turbulent stresses and heat transfer) and inviscid effects. In the second layer, immediately above the first deck, the flow is approximately rotational and inviscid, *i.e.*, turbulent stresses and heat transfer have a small effect. This region encompasses most of the boundary layer in the interaction. The third region is the inviscid, irrotational flow outside the boundary layer. To a first approximation, the surface pressure distribution is determined by the interaction of the second and third layers, and therefore does not depend strongly on the choice of turbulence model. The principal contribution of the turbulence model is to provide the correct vorticity distribution in the incoming boundary layer (*i.e.*, the correct inflow mean velocity profile), but does not otherwise significantly affect the predicted surface pressure. Since all turbulence models employed provide an accurate inflow profile, there is little difference in the prediction of surface pressure.

The primary separation location is accurately predicted. However, secondary features (*e.g.*, secondary lines of separation) are sensitive to the turbulence model and not always predicted accurately.

The heat transfer coefficient is poorly predicted (except for weak interactions (Knight *et al.*, 1997)). Computed values exceed experimental measurements by as much as 100%. No single reason can be given for the discrepancy. For two equation turbulence models, for example, the model equations predict a net production of turbulence kinetic energy k in the vicinity of a shock wave which behaves like Δ^{-1} where Δ is the grid spacing normal to the shock wave.[10] This can lead to an unphysical increase in k in the vicinity of a shock, and hence an overprediction of turbulent eddy viscosity μ_t and heat transfer.

Efforts are needed to identify promising new avenues for research in simulation of shock wave turbulent boundary layer interaction. One possible approach is Large Eddy Simulation, which has been extensively applied to incompressible flows but rarely applied to compressible flows.

5. Acknowledgments

The author is grateful to the AGARD WG18 participants whose contributions formed the basis for this paper. The study has been supported by the Air Force Office of Scientific Research under grant F49620-93-1-0005 (monitored by Dr. Len Sakell). The experimental data for the double fin was provided by Dr. Alexander Zheltovodov and his colleagues who were supported by the Russian Foundation for Basic Research (Project Code 96-01-01777) and United Technologies Research Center (monitored by Dr. Michael Werle).

[10] Consider the $k-\epsilon$ model. For a normal shock wave, the integrated production of turbulence kinetic energy k across a normal shock wave for a fluid particle is

$$\int P_k \, dt = \int 2\mu_t \left(\frac{\partial u}{\partial x}\right)^2 dt \approx 2\mu_t \left(\frac{\Delta u}{\Delta x}\right)^2 \frac{\Delta x}{u} \sim \frac{1}{\Delta x}$$

since Δu is finite across the shock. This implies that the integrated production of k for a fluid particle becomes infinite as $\Delta x \to 0$ which is unphysical. Of course, it is possible to derive physically correct shock jump conditions for k which avoid this behavior; however, they require determination of the location of the shocks which can be complex in 3 -D shock boundary layer interactions.

References

Alexopoulos, G., and Hassan, H. (1997) Computation of Crossing Shock Flows Using the k-enstrophy Turbulence Model, AIAA Paper No. 97-0206.

Alvi, F., and Settles, G. (1991), Physical Model of the Swept Shock / Boundary Layer Interaction Flowfield, *AIAA Journal*, **Vol. 30**, pp. 2252–2258.

Baldwin, B., and Lomax, H. (1978), Thin Layer Approximation and Algebraic Model Separated Flows, AIAA Paper 78-257.

Becht, C., and Knight, D. (1995), A Simple Low Reynolds Number Modification for the Compressible $k-\epsilon$ Model. Part I. Boundary Layer Flows, AIAA Paper 95-2218.

Chien, K.-Y. (1982), Predictions of Channel and Boundary Layer Flows with a Low Reynolds Number Turbulence Model, *AIAA Journal*, **Vol. 20**, pp. 33–38.

Courant, R., and Friedrichs, K. (1948), *Supersonic Flow and Shock Waves*, Springer-Verlag, New York.

Degrez, G., ed. (1993) AGARD Special Course on Shock-Wave/Turbulent Boundary-Layer Interactions in Supersonic and Hypersonic Flow, AGARD Report 792, Advisory Group for Aerospace Research and Development.

Delery, J., and Marvin, J. (1986), Shock-Wave Boundary-Layer Interactions, AGARDograph No. 280.

Delery, J., and Panaras, A. (1996), Shock-Wave Boundary-Layer Interactions in Hih Mach Number Flows, AGARD AR-319, Advisory Group for Aerospace Research and Development, Vol. 1.

Edwards, J., and Chandra, S. (1996), Comparison of Eddy Viscosity-Transport Turbulence Models for Three-Dimensional, Shock-Separated Flowfields, *AIAA Journal*, **Vol. 34**, pp. 756-763.

Gaitonde, D., Shang, J., Visbal, M. (1995), Structure of a Double-Fin Interaction at High Speed, *AIAA Journal* **Vol. 33**, pp. 193–200.

Garrison, T., Settles, G., Narayanswami, N., and Knight, D. (1993), Structure of Crossing-Shock Wave/Turbulent Boundary Layer Interactions, *AIAA Journal*, **Vol. 31**, pp. 2204-2211.

Gnedin, M. (1996), Numerical Simulation of 3-D Shock Wave Turbulent Boundary Layer Interaction Using a Two Equation Model of Turbulence, PhD Thesis, Dept of Mech & Aero Engr, Rutgers University.

Gnedin, M., Knight, D., Zheltovodov, A., Maksimov, A., Shevchenko, A., and Vorontsov, S. (1996), 3-D Crossing Shock Wave Turbulent Boundary Layer Interaction, AIAA Paper No. 96-2001.

Greene, J. E. (1970), Interactions between Shock Waves and Turbulent Boundary Layers, *Progress in Aerospace Science*, Pergammon Press, **Vol. 11**, pp. 235–340.

Inger, G. (1985), Theoretical and Experimental Study of Nonadiabatic Transonic Shock Boundary-Layer Interaction, *AIAA Journal*, **Vol. 23**, pp. 1476–1482.

Inger, G. (1987), Spanwise Propagation of Upstream Influence in Conical Swept Shock / Boundary Layer Interactions, *AIAA Journal*, **Vol. 25**, pp. 287–293.

Jones, W., and Launder, B. (1972), The Prediction of Laminarization with a Two-Equation Model of Turbulence, *International Journal of Heat and Mass Transfer* **Vol. 15**, pp. 301–304.

Kim, K., Lee, Y., Alvi, F., Settles, G., and Horstman, C. (1990), Laser Skin Friction Measurements and CFD Comparison of Weak-to-Strong Swept Shock / Boundary Layer Interactions, AIAA Paper No. 90-0378.

Knight, D. (1993), Numerical Simulation of 3-D Shock Wave Turbulent Boundary Layer Interactions, in *AGARD/VKI Special Course on Shock-Wave Boundary-Layer Interactions in Supersonic and Hypersonic Flows*, Von Karman Institute for Fluid Dynamics, AGARD R-792, pp. 3-1 to 3-32.

Knight, D., Garrison, T., Settles, G., Zheltovodov, A., Maksimov, A., Shevchenko, A., and Vorontsov, S. (1995), Asymmetric Crossing-Shock Wave / Turbulent Boundary Layer Interaction, *AIAA Journal*, **Vol. 33**, pp. 2241–2249.

Knight, D., and Degrez, G. (1997), Shock Wave Boundary Layer Interactions in High Mach Number Flows – A Critical Survey of Current CFD Prediction Capabilities, AGARD AR-319, Advisory Group for Aerospace Research and Development, Vol. 2.

Korkegi, R. (1971), Survey of Viscous Interactions Associated with High Mach Number Flight, *AIAA Journal*, **Vol. 9**, pp. 771–784.

Narayanswami, N., Knight, D., Bogdonoff, S., and Horstman, C. (1992), Interaction Between Crossing Oblique Shocks and a Turbulent Boundary Layer, *AIAA Journal*, **Vol. 30**, pp. 1945-1952.

Panaras, A. (1996) Algebraic Turbulence Modelling for Swept Shock Wave / Turbulent Boundary Layer Interactions, Deutsche Forschungsanstalt für Luft- und Raumfahrt e.V., Report No. IB 223-96 A 22.

Peake, D. and Tobak, M. (1980), Three-Dimensional Interactions and Vortical Flows with Emphasis on High Speed, NASA TM 81169.

Settles, G., and Dolling, D. (1986), Swept Shock Wave Boundary Layer Interactions, *Tactical Missile Aerodynamics*, AIAA, pp. 297–379.

Settles, G., and Dolling, D. (1990), Swept Shock / Boundary-Layer Interactions - Tutorial and Update, AIAA Paper 90-0375.

Stewartson, K. (1981), Some Recent Studies in Triple-Deck Theory, in *Numerical and Physical Aspects of Aerodynamic Flows*, T. Cebeci, ed., Springer-Verlag, pp. 142.

Stollery, J. (1989), AGARD Special Course on Three-Dimensional Supersonic and Hypersonic Flows Including Separation, Advisory Group for Aerospace Research and Development.

Zha, G., and Knight, D. (1996), Three-Dimensional Shock / Boundary-Layer Interaction Using Reynolds Stress Equation Turbulence Model, *AIAA Journal*, **Vol. 34**, pp. 1313–1320.

Zheltovodov, A., and Shilein, E. (1986), 3-D Swept Shock Waves / Turbulent Boundary Layer Interaction in Angles Configuration, Institute of Theoretical and Applied Mechanics, Academy of Sciences, Siberian Division, No. 34-86.

Zheltovodov, A., Maksimov, A., Shevchenko, Vorontsov, S., and Knight, D. (1994), Experimental Study and Computational Comparison of Crossing Shock Wave - Turbulent Boundary Layer Interaction, Proceedings of the International Conference on Methods of Aerophysical Research - Part 1, Russian Academy of Sciences, Siberian Division, pp. 221–230.

Zheltovodov, A. (1996), Shock Waves / Turbulent Boundary Layer Interactions - Fundamental Studies and Applications, AIAA Paper No. 96-1977.

A. Conical Flow

A conical flow is a steady flowfield whose Cartesian velocity components u_i, static pressure p and static temperature T are invariant with radial distance from a common vertex (Courant *et al.*, 1948). Consider the spherical polar coordinate system (R, β, ϕ) shown in Fig. 25. Thus,

$$\frac{\partial u_i}{\partial R} = 0$$
$$\frac{\partial p}{\partial R} = 0$$
$$\frac{\partial T}{\partial R} = 0$$

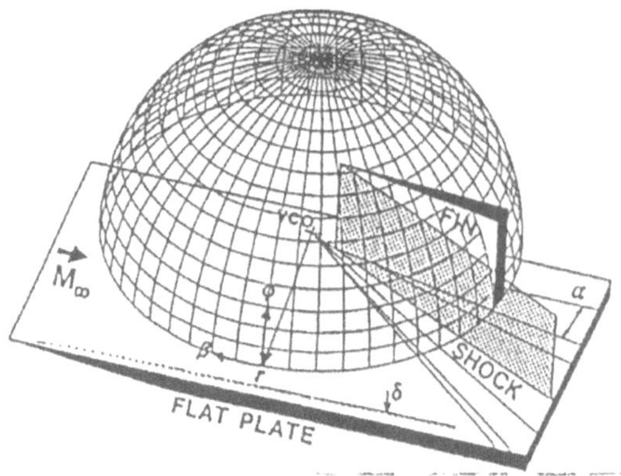

Figure 25. Spherical polar coordinates (from Alvi & Settles)

where R is the spherical polar radius

$$R = \sqrt{(x-x_0)^2 + (y-y_0)^2 + (z-z_0)^2}$$

where (x_0, y_0, z_0) is the Virtual Conical Origin. For the single fin, the VCO is close to the intersection of the fin with the flat plate. The velocity, pressure and temperature are functions of the spherical polar coordinates

$$\beta = \tan^{-1}\left((z-z_0) / (x-x_0)\right)$$
$$\phi = \tan^{-1}\left((y-y_0) / \sqrt{(x-x_0)^2 + (z-z_0)^2}\right)$$

SOME RESULTS RELEVANT TO STATISTICAL CLOSURES FOR COMPRESSIBLE TURBULENCE

J.R. RISTORCELLI
Institute for Computer Applications in Science and Engineering
NASA Langley Research Center, Hampton, Virginia

Abstract. For weakly compressible turbulent fluctuations there exists a small parameter, the square of the fluctuating Mach number, that allows an investigation using a perturbative treatment. The consequences of such a perturbative analysis in three different subject areas are described: 1) initial conditions in direct numerical simulations, 2) an explanation for the oscillations seen in the compressible pressure in the direct numerical simulations of homogeneous shear, and 3) for turbulence closures accounting for the compressibility of velocity fluctuations.

1) Initial conditions consistent with small turbulent Mach number asymptotics are constructed. The importance of consistent initial conditions in the direct numerical simulation of compressible turbulence is dramatically illustrated: spurious oscillations associated with inconsistent initial conditions are avoided, and the fluctuating dilatational field is some two orders of magnitude smaller for a compressible isotropic turbulence. For the isotropic decay it is shown that the choice of initial conditions can change the scaling law for the compressible dissipation.

2) A two-time expansion of the Navier-Stokes equations is used to distinguish compressible acoustic and compressible advective modes. A simple conceptual model for weakly compressible turbulence — a forced linear oscillator — is described. It is shown that the evolution equations for the compressible portions of turbulence can be understood as a forced wave equation with refraction. Acoustic modes of the flow can be amplified by refraction and are able to manifest themselves in large fluctuations of the compressible pressure.

3) The consequences of a small turbulent Mach number expansion for the closure of two covariances appearing in the kinetic energy equation, the pressure-dilatation and the dilatational dissipation, are investigated. Comparisons with different models and a discussion of the results in the context of the homogeneous shear is given. In agreement with observations

M. D. Salas et al. (eds.), Modeling Complex Turbulent Flows, 297–327.

of DNS of compressible turbulence, the dilatational covariances can not account for the large reductions in the growth of compressible shear layers.

1. Introduction

The term 'phrase weakly compressible' describes the turbulence relevant to many aerodynamic applications. Several interesting insights into the nature of compressible turbulence can be found by investigations of the weakly compressible limit: the weakly compressible limit allows a perturbative treatment. Such weakly compressible investigations are relevant to compressible shear layers encountered in practical applications.

For weakly compressible fluctuations the turbulent Mach number squared is small, $M_t^2 \ll 1$, and serves as a small parameter. Here $M_t = u_c/c$; u_c is a fluctuating velocity scale and c is the local mean speed of sound. The turbulent Mach number reflects the weakly compressible nature of the turbulence — the mean flow itself may be highly supersonic. An analytical development for the covariances of the fluctuating dilatation suitable for shear flows has been given in Ristorcelli (1997). This article describes the implications of such an analysis for the initial conditions in the direct numerical simulation of turbulence as discussed in Ristorcelli and Blaisdell (1997). This article also reports an extension of that analysis that can explain the peculiar oscillations seen in the DNS of homogeneous shear. In the process a simple conceptual model - the forced linear oscillator - for compressible turbulence is derived. This article also explores, in the context of the DNS, representations for terms requiring closure in the kinetic energy equation.

The primary subject of this article, given that it is to appear in the proceedings of a conference devoted to turbulence modeling, is a summary of recent advances made in the understanding of the effects of compressibility as related to turbulence models for engineering flows. To this end a substantial portion of space is spent understanding numerically generated data bases which are used to develop turbulence models. This is done to insure that the physics of the numerically simulated flow is consistent with the engineering problem and thus constitutes a relevant data base for statistical closures relevant to engineering flows.

This article is divided into three separate sections.

Section 2 treats initial conditions for the DNS of compressible turbulence. It is shown that as a consequence of a finite turbulent Mach number there are finite non-zero density, temperature and dilatational fluctuations. These fluctuations are, to lowest order, specified by the incompressible field.

These results are relevant to initial conditions on thermodynamic quantities employed in DNS of compressible turbulence that are started using *incompressible* fluctuating velocity fields. Use of these initial conditions gives rise to a smooth development of the flow, in contrast to cases in which these fields are specified arbitrarily (or set to zero). For the isotropic decay it is shown that the choice of the initial conditions appears to change the scaling law for the compressible dissipation. It will be seen that, for the isotropic decay, the well accepted M_t^2 scaling for the compressible dissipation appears to be an artifact of an evanescent wave field set up by the initial conditions.

In Section 3 a simple conceptual model for weakly compressible turbulence is described: the evolution equations for the compressible portions of turbulence in the presence of a homogeneous shear are seen to obey a forced wave equation with refraction. The forcing and refraction effects are a function of the shear rate as indicated by a gradient Mach number.

In Section 4 the results of the modeling of Ristorcelli (1997) are explored. It is shown that M_t^4 scaling for the compressible dissipation is consistent with the DNS of compressible turbulence results with *the appropriate initial conditions*. A scaling of M_t^2 for the compressible dissipation is consistent with the DNS if "inconsistent" initial conditions are used. A verification of the scalings and comparison with other models for the pressure-dilatation is also conducted.

2. Initial Conditions

For turbulence with finite turbulent Mach number there is a finite effect of compressibility. A methodology, consistent with finite Mach number asymptotics, for generating initial conditions for the fluctuating pressure, density, temperature, and dilatational velocity fields is described. Relationships between diverse thermodynamic quantities appropriate to weakly compressible turbulence are derived. The phrase *consistent initial conditions* is used to denote initial conditions that are suggested by low turbulent Mach number asymptotics.

2.1. SINGLE TIME PERTURBATION ANALYSIS

Ristorcelli (1997) has conducted a small Mach number expansion of the compressible Navier-Stokes equations. Some of those results, as they are relevant to the present subject, are now briefly sketched. The problem of compressible turbulence can be viewed as a singular perturbation. The inner problem is related to the acoustic source problem with time and length scales based on the turbulence time and length scales, ℓ/u_c and ℓ. Here u_c is the turbulence velocity scale and ℓ is the turbulence integral length scale.

The outer problem is the acoustic propagation problem, which in the low M_t^2 limit has time and length scales $\lambda/c = \ell/u_c$ and $\lambda = \ell/M_t$. The inner expansion is relevant to the compressible turbulence modeling problem of acoustically compact flows, Ristorcelli (1997). The outer problem is not relevant to the problem of compressible turbulence for a compact fields, Ristorcelli (1997).

The compressible Navier-Stokes equations in a form convenient for weakly compressible high Reynolds number turbulence are

$$\rho_{,t} + u_k^* \rho_{,k} = -(1+\rho) u_{p,p}^* \tag{1}$$

$$(1+\rho)[u_{i,t}^* + u_k^* u_{i,k}^*] + \frac{c_\infty^2}{\gamma} p_{,i} = 0 \tag{2}$$

$$p - \gamma\rho = \frac{1}{2}\gamma(\gamma-1)\rho^2 \tag{3}$$

$$\rho_{,tt} - \frac{c_\infty^2}{\gamma} p_{,jj} = ((1+\rho)u_i^* u_j^*)_{,ij}\,. \tag{4}$$

The last equation, a wave equation, is derived from continuity and momentum. The velocity field is decomposed according to $u_i = v_i + \epsilon^2 w_i + \ldots$, where v_i represents the solenoidal velocity field and the the small parameter is $\epsilon^2 = \gamma M_t^2$, where $M_t = u_c/c$. Thus $d = u_{i,i} = \epsilon^2 w_{i,i} + \ldots$. The thermodynamic variables are decomposed according to a perturbation about a mean state, $(P, \bar{\rho}, T)$, thus $p^* = P(1+p)$, $\rho^* = \bar{\rho}(1+\rho)$ and $T^* = T(1+\theta)$. Perturbation series of the form $p = \epsilon^2\,[p_1 + \epsilon^2 p_2 + \ldots]$ are assumed. To lowest order

$$v_{i,t} + v_p v_{i,p} + p_{1,i} = 0 \tag{5}$$

$$v_{i,i} = 0 \tag{6}$$

$$p_{1,jj} = -(v_i v_j)_{,ij} \tag{7}$$

$$\rho_1 = \frac{1}{\gamma} p_1 \tag{8}$$

$$\theta_1 = \frac{\gamma-1}{\gamma}\, p_1. \tag{9}$$

The zeroth-order problem is the incompressible problem. The fluctuating pressure is not an independent variable but is set by the solenoidal velocity fluctuations and which also produces the density and temperature fluctuations. On the inner scales the incompressible pressure fluctuations dominate the pressure field. This pressure is sometimes called the "pseudo-pressure" in contradistinction to the propagating pressure associated with the sound field. The last two equations come from the ideal gas law. In light of the homogeneous shear flow simulations of Blaisdell *et al.* (1991, 1993) (the fluctuations were seen to follow a polytropic gas law with coefficient close to the adiabatic value), the adiabatic case is treated.

The next order expansion for the continuity equation produces a diagnostic relation for the fluctuating dilatation,

$$-\gamma d = p_{1,t} + v_k p_{1,k} \,. \tag{10}$$

The dimensional equivalents of the above equations, using $\rho_1 = \rho'/\bar{\rho}$, $\theta_1 = \theta'/T$, $p_1 = p'/P$ are

$$\begin{aligned} \rho' &= \frac{1}{c^2}\, p' \\ \theta' &= \frac{\gamma - 1}{\gamma R \bar{\rho}}\, p' \\ -\gamma P d' &= p'_{,t} + v_k p'_{,k} \,, \end{aligned} \tag{11}$$

where p' is the solenoidal pressure fluctuation. For finite turbulent Mach number there are *unique specified finite* fluctuations of density, temperature, dilatation and pressure.

Obtaining an initial condition for the compressible portion of the velocity, w_i, which reflects the relation (11) is possible, Ristorcelli and Blaisdell (1997). The fluctuating pressure p_1 needed in (11) is obtained from the Poisson equation (7), and its time derivative can be found by taking the time derivative of (7) and substituting for the time derivative of the velocity from (5). This gives a Poisson equation for $p_{1,t}$,

$$(p_{1,t})_{,jj} = 2\left[(v_k v_{i,k} + p_{1,i})v_j\right]_{,ij} \,. \tag{12}$$

Solving the two Poisson equations (7) and (12), the dilatation is found from (10). For homogeneous turbulence the dilatational velocity can be found from the dilatation by working in Fourier space. As the dilatational velocity, w_i, is irrotational the Fourier coefficients of the velocity are aligned with the wavevector, $\vec{k}$, and they can be found from

$$\widehat{w}_j = -i\frac{k_j}{k^2}\widehat{d}. \tag{13}$$

The dilatational velocity w_i is then combined with the solenoidal velocity v_i to obtain the full initial velocity field. This set of initial conditions, (11) and(13), will called the consistent (or pseudo-sound) initial conditions.

2.2. SIMULATIONS AND RESULTS

In order to determine the effectiveness of the new initial condition method, direct numerical simulations of decaying isotropic turbulence are carried out with three different types of initial conditions, Ristorcelli and Blaisdell (1997). Simulations were performed with initial turbulent Mach numbers $M_t = 0.231$, 0.115, and 0.058. The initial turbulent Reynolds number $Re_T = \bar{\rho} q^4/\varepsilon\nu = 200$ where $q^2 = \langle \rho u_i u_i \rangle / \bar{\rho}$, and ε is the dissipation rate of turbulent kinetic energy per unit volume. The mean density gradient is zero. For the present analysis and simulations the Reynolds and Favre averages are the same. Details about the simulation method can be found in Blaisdell *et al.* (1991).

1. The first type of initial conditions, IC1, has zero density and pressure fluctuations. The velocity field is solenoidal.
2. The second type of initial conditions, IC2, has zero density fluctuations; the pressure is found by solving the Poisson equation (7). The velocity field is again solenoidal.
3. The third type of initial conditions, IC3, the fluctuating pressure field is found by solving the Poisson equation (7). The relations (11) determine the density and dilatational fluctuations from (7). The dilatation is used to determine the dilatational velocity from (13).

2.2.1. *Magnitude of the dilatation*

Data from the $M_t = 0.231$ run has been chosen to illustrate the effect of the different initial conditions. Figure 1 shows the development of the variance of the dilatation, $\langle dd \rangle$. It is clear from the figure that there are sizeable oscillations near the origin for the first two initial conditions as the flow seeks to adjust to "inconsistent" initial conditions. This is not the case for the pseudo-sound initial conditions, IC3. Also interesting to note is the fact that the variance of the dilatation for IC1 is one order of magnitude larger than IC3. In general, at lower M_t this difference is more pronounced; at $M_t = 0.058$ the difference is two orders of magnitude. This excess dilatation is associated with an evanescent wave field, see §3, set up by an inappropriate choice of initial conditions and is not the dilatation due to the compressible turbulence field.

An intuitive argument can be given to explain the behavior seen in Figure 1. One might speculate that there is a wave field generated by the inconsistency between the pressure, density, temperature, and dilatational fields whose decay rate is far slower than the turbulence decay. This gives rise to a background evanescent acoustic radiation, for a homogeneous flow, that lasts the course of the simulation. For IC1 the fluctuating pressure field is set to zero. This can be viewed as a combination of the incompressible pressure satisfying (7) plus an acoustic pressure field which exactly cancels the incompressible pressure field – so that the initial field is zero. As time evolves the two pressure fields become decorrelated, giving rise to large acoustic pressure and dilatation fluctuations. For IC2 the pressure field is correct; however, there is no dilatation field. Again this can be viewed as a combination of the dilatation found from (7) plus an acoustic dilatation field that exactly cancels this. As the flow evolves away from the initial conditions, the two dilatation fields become decorrelated so that one is left with the correct dilatation plus acoustic fluctuations of dilatation and pressure associated with the slowly decaying evanescent field. The evolution of the pressure variance and $\langle pd \rangle$ (not shown here, Blaisdell (1996)) corroborates this picture.

2.2.2. *Scaling of the dilatation*

Figure 2, provided by Professor G.A. Blaisdell, is a very rich figure with many implications. Figure 2 is a graphic indication of the importance of initial conditions; it indicates the scaling of the dissipation with the turbulent Mach number.

The figure describes the isotropic decay conducted for three different initial turbulent Mach numbers. The three different initial conditions are seen to produce two different scalings for the compressible dissipation or, equivalently, the variance of the dilatation. For the consistent initial conditions the dilatation is seen to have M_t^4 scaling. This is indicated by the upper solid line. For the two sets of "inconsistent" initial conditions an M_t^2 scaling, as indicated by the lower dashed lines, is seen. The M_t^2 scaling is the scaling usually agreed on for some models of the compressible dissipation – it is a scaling that is arrived at from observations of DNS with inconsistent initial conditions.

2.3. CONCLUSIONS

It has been shown that as a consequence of a finite turbulent Mach number there are finite non-zero density, temperature and dilatational fluctuations. These fluctuations are, to lowest order, specified by the incompressible field. These results are relevant to initial conditions on thermodynamic quantities employed in DNS of compressible turbulence that are started using *incompressible* fluctuating velocity fields. A potential practical consequence of the present results is a reduction in the amount of computational effort spent adjusting to transients associated with the relaxation from arbitrary initial conditions.

The major point is that the initial conditions on the fluctuating thermodynamic variables of density, temperature, pressure, and dilatation should not be arbitrarily specified in the DNS of compressible turbulence. To lowest order these fluctuations, whose nature is connected to the underlying fluctuating vortical turbulence field, are generated by the pressure field associated with the divergence-free portion of the vortical motions. The methodology for generating initial conditions presented here allows the flow to develop more naturally.

The significance of the initial conditions is likely to depend on the type of flow considered; for example the effects described above are likely to be much less evident in a homogeneous shear in as much as the final M_t substantial exceeds the initial M_t. Nonetheless, one is led naturally to speculate about the nature and relevance of simulations that start from arbitrary initial conditions. There are surely a large number of interesting studies in which one can study the relaxation from diverse arbitrary initial conditions. It is also possible to argue that they have practical value. However, the strongest argument, given the current engineering problems and the lack of knowledge regarding the effects of compressibility, can be made for the initial conditions in which the density, temperature, pressure, and dilatational fields are related to the underlying local turbulence field that is the source of the fluctuations. This seems better than an arbitrary guess at initial conditions that may or may not have physical relevance or meaning.

3. Two-timing compressible turbulence

The present discussion is limited to a physical understanding of the effects of compressibility as arrived at mathematically. A two-time expansion of the compressible Navier-Stokes equations is used to investigate the compressible modes of a turbulent flow. Such a two-time scale analysis sheds substantial insight on the nature of compressibility.

The present treatment of the problem contrasts to that given by Ristorcelli (1997). That treatment, in which the the small parameter was also the turbulent Mach numbers, used a single time expansion. The single time expansion was a consequence of nondimensionalizing time with a time scale characteristic of the turbulence, the eddy turnover time. Such an analysis is suitable for statistical closure that evolves on a nonlinear advective time scale. Here the investigation is directed towards distinguishing quantities that evolve on a fast time scale from those evolving on a slow time scale. It will be seen that the quantities evolving on the fast time, as can be assessed by their evolution equations, cannot be understood as turbulence. The fast time equations suggest that it is useful to interpret the fast time solution as an evanescent wave field.

It is appropriate to indicate the observations that motivate the present analysis. The experimental data that suggested this investigation is the DNS of the homogeneous compressible shear as might be seen in Blaisdell *et al.* (1991, 1993), Sarkar, Erlebacher and Hussaini (1991a), or Sarkar (1992). Figure 3 is the pressure-dilatation from Blaisdell's recent simulations; the instantaneous ensemble averaged as well as the time averaged pressure-dilatation time history is shown. The initial conditions follow those outlined in Ristorcelli and Blaisdell (1997). Even with the initial conditions suggested by a single-time small turbulent Mach number expansion there is a build up of rapidly varying fluctuations in the pressure-dilatation. This is also seen in the earlier simulations by Sarkar (1992); in Sarkar's (1992) simulations the oscillations occur earlier and this is consistent with the choice of initial conditions. The oscillations in the instantaneous pressure-dilatation exhibits a time scale characteristic of the sound crossing time over an eddy, ℓ/c, as can be estimated from the DNS. The evolution of the time averaged trace evolves on a slower time scale, the eddy turnover time, $\ell/u_c \sim k/\varepsilon$. The appearance of these two disparate time scales suggest that a two-time analysis of the compressible DNS may be appropriate to further understand the nature of the physics. Consider a characteristic fluctuating velocity, u_c, say related to the kinetic energy of the turbulence k, $u_c^2 = \frac{2}{3}k$. The ratio of these two time scales is the fluctuating Mach number:

$$M_t = \frac{u_c}{c}. \tag{14}$$

If it is assumed that the ratio of the eddy crossing time to the turnover time is small, $\frac{\ell}{u_c} >> \frac{\ell}{c}$, (or equivalently the Mach number is small) a multi-scale expansion of the Navier-Stokes equations will produce a useful model. The small parameter in the two-time problem is now $\epsilon = M_t$. In the single time problem the appropriate small parameter was M_t^2. In the present development, as will become clear, the fluctuating Mach number plays two unique roles: It indicates the relative time scales of advection versus propagation – in this role it appears to the first power. It is also a measure of the dynamical effects of the compressibility of the fluctuating field – in this role it appears to the second power.

3.1. TWO-TIMING: ANALYSIS

A set of equations describing and distinguishing the relevant aspects of the flow physics is now derived. The procedure is straightforward and standard for those familiar with perturbation methods. Perturbing about a reference state, (p_∞, ρ_∞), the dimensional pressure, density and velocity are taken as

$$\begin{aligned} p^* &= p_\infty(1+p) && (15)\\ \rho^* &= \rho_\infty(1+\rho). && (16)\\ u_i^* &= u_c(v_i + \phi_w w_i) && (17) \end{aligned}$$

where

$$\begin{aligned} p &= \phi_0(p_s + \phi_1 p_c) && (18)\\ \rho &= \phi_0(\rho_s + \phi_1 \rho_c). && (19) \end{aligned}$$

The dependent variables are now all nondimensional: $[v_i, w_i, p_s, p_c, \rho_s, \rho_c]$ are in units of $[u_c, p_\infty, \rho_\infty]$. These forms of the dimensional variables are then inserted into the compressible equations, (1), (2), (3) of §2. The leading order problem, as in §2.1, is described by equations used for a solenoidal velocity field with characteristic time, length and velocity scales $[\frac{\ell}{u_c}, \ell, u_c]$,

$$
\begin{aligned}
v_{p,p} &= 0 && (20)\\
v_{i,t} + v_k v_{i,k} + p_{s,i} &= 0 && (21)\\
p_{s,jj} &= -(v_i v_j)_{,ij} && (22)\\
\gamma \rho_s &= p_s. && (23)
\end{aligned}
$$

Note that the leading order density fluctuations are given by p_s. The subscript "s" is understood to indicate solenoidal as the field v_i has zero divergence. The gauge function $\phi_0 = \gamma M_t^2$ in order to obtain a meaningful balance with pressure.

The compressible portion of the problem, $[p_c, \rho_c, w_i]$, is described by the following set of equations:

$$
\begin{aligned}
[1 + \phi_0(\rho_s + \phi_1 \rho_c)][w_{i,t} &+ v_k w_{i,k} + w_k v_{i,k} + \phi_w w_k w_{i,k}]\phi_w \\
&+ \frac{c_\infty^2}{\gamma}\phi_0 \phi_1 p_{c,i} \\
&= -\phi_0(\rho_s + \phi_1 \rho_c)[v_{i,t} + v_k v_{i,k}] && (24)\\
\phi_0 \phi_1 [\rho_{c,t} + v_k \rho_{c,k} + \phi_w w_k \rho_{c,k}] &+ [1 + \phi_0(\rho_s + \phi_1 \rho_c)] w_{k,k}\, \phi_w \\
&= -\phi_0 [\rho_{s,t} + v_k \rho_{s,k}] && (25)\\
\phi_o \phi_1 (p_c - \gamma \rho_c) &= \frac{1}{2}\gamma(\gamma - 1)\phi_0^2 (\rho_s + \phi_1 \rho_c)^2. && (26)
\end{aligned}
$$

It is important to note that the variable time is dimensional; length has been rescaled with ℓ so that the convective operators have dimensions of inverse time. The scalings for length and time will determine ϕ_1 and ϕ_w. The method of dominant balance, after choosing the acoustic length and time scales, indicates that the only self-consistent choices for the scale functions are $\phi_w = M_t^2$ and $\phi_1 = M_t$. Additional details relevant to a similar procedure, though in a different context, can be found in Ristorcelli (1997a).

The compressible problem evolves on two time scales. The fast time scales with ℓ/c. The slow time scales with ℓ/u_c. Following the usual multi-scale *ansatz* the original time variable is replaced with two independent time variables, $[t] \Rightarrow [t_0, t_1]$; t_0 is the fast time scale and t_1 is the slow time scale. The multi-time-scale *ansatz* for a dependent variable is $f(x,t) = f(x, t_0, t_1)$. The time derivative, *following a fluid particle* of $f(x,t)$ is then written

$$
\frac{D}{Dt} f(x,t) = \frac{D}{Dt_0} f(x, t_0, t_1) + \epsilon \frac{D}{Dt_1} f(x, t_0, t_1), \qquad (27)
$$

where the small parameter is now understood to be $\epsilon = M_t$. The equations are made nondimensional with the integral length scale, ℓ, and the fast time scale, ℓ/c. The compressible dependent variables are now expanded in a regular perturbation expansion, $f = f_0 + \epsilon f_1 + \dots$. A Reynolds decomposition of the velocity field is used: the mean flow is characterized by a shear of magnitude S. The mean flow is assumed, for present purposes, to be homogeneous with $V_{i,i} = 0$.

3.2. THE LEADING ORDER PRIMITIVE EQUATIONS

The leading order compressible problem satisfies the following momentum and continuity equations:

$$\frac{D}{Dt_0} w_{0i} + [\frac{S\ell}{c}]\, w_{0k} V_{i,k} + p_{0,i} = -p_s\, [\frac{S\ell}{c}]\, v_k V_{i,k} \tag{28}$$

$$\frac{D}{Dt_0} p_0 + w_{0k,k} = -[\frac{D}{Dt_1} p_s + v_k p_{s,k}]. \tag{29}$$

The leading order equation of state, $\gamma p_0 = \rho_0$ has been used to eliminate the density in favor of pressure. There are several items worth noting:

1. The so-called gradient Mach number $\frac{S\ell}{c}$ appears as an independent parameter. It appears in what is called a production term (also sometimes called a refraction term).
2. The forcing terms are all functions of the slow time: $p_s = p_s(x, t_1)$, $v_i = v_i(x, t_1)$.
3. The equations are linear: there is no nonlinear compressible-compressible mode coupling in the low Mach number limit. Nonlinearity is only a characteristic, as one might expect in a perturbative treatment, of the incompressible modes.
4. The solution is written as a sum of homogeneous and particular solutions which, to leading order, have a different dependence on the fast (t_0) and slow (t_1) time variables, Ristorcelli (1997a):

$$w_{0i} = w_{0iE}(x, t_0) + w_{0iP}(x, t_1)$$
$$p_0 = p_{0E}(x, t_0) + p_{0P}(x, t_1).$$

5. The initial value problem, $[w_{0iE}(x, t_0), p_{0E}(x, t_0)]$, is the solution of the homogeneous problem:

$$\frac{D}{Dt_0} w_{0i} + [\frac{S\ell}{c}]\, w_{0k} V_{i,k} + p_{0,i} = 0 \tag{30}$$

$$\frac{D}{Dt_0} p_0 + w_{0k,k} = 0. \tag{31}$$

These are the equations of an evanescent wave field. The solution is a function of, to leading order, of only the fast time.

3.3. A CONCOMITANT WAVE EQUATION

If one takes the divergence of the momentum equation, (28), and the Lagrangian derivative of the continuity equation, (29), one obtains the following third-order wave equation with source:

$$\frac{D}{Dt_0}\left[\frac{D^2}{Dt_0^2}\, p_0 - \nabla^2 p_0\right] + 2\left[\frac{S\ell}{c}\right] V_{i,k}\, p_{0,ik} - 4\left[\frac{S\ell}{c}\right]^2 V_{k,i}\, V_{i,j}\, w_{0j,k} = 2\left[\frac{S\ell}{c}\right] V_{j,i}\, f_{i,j} \quad (32)$$

On the right hand side $f_i = -p_s\,[\frac{S\ell}{c}]\, v_k V_{i,k}$, which is known in terms the solenoidal field. The problem can be viewed as a linear fourth order forced oscillator, whose natural time scale, t_0, is fast compared to the forcing time scale, t_1. The homogeneous solution satisfies the initial value problem and only varies on the fast time scale; the particular solution varies on the slow time scale. If the initial conditions are homogeneous, the evanescent wave field with the fast eigenmodes is not stimulated. In such a case the solution for the compressible modes follows the slow time forcing with a phase lag related to the damping.

In the absence of forcing the homogeneous problem satisfies the following wave equation

$$\frac{D}{Dt_0}\left[\frac{D^2}{Dt_0^2}\, p_0 - \nabla^2 p_0\right] + 2\left[\frac{S\ell}{c}\right] V_{i,k}\, p_{0,ik} = 4\left[\frac{S\ell}{c}\right]^2 V_{k,i}\, V_{i,j}\, w_{0j,k}\,. \quad (33)$$

For a parallel unidirectional shear the right hand side is zero,

$$\frac{D}{Dt_0}\left[\frac{D^2}{Dt_0^2}\, p_0 - \nabla^2 p_0\right] + 2\left[\frac{S\ell}{c}\right] V_{i,k}\, p_{0,ik} = 0, \quad (34)$$

and the order of the system is reduced. A equation similar in form has been deduced by Lilley (1996). Apart from some numerical factors the major difference from Lilley's result is that, as the equation is expressed on the fast time scale, it describes specifically the fast modes of the flow. This is the wave equations in a moving medium with refraction due to the mean shear. The mean shear alters the magnitude and direction of the wave field.

For a homogeneous medium this equation describes an evanescent wave field determined by the initial conditions. If the initial conditions are zero the solution remains zero; arbitrary non-zero initial conditions will stimulate the eigenmodes of this equation. These are waves with phase speed equal to speed of sound (unity in the present equations). The production (refraction) mechanism can amplify these modes. The production mechanism consequently leads to secular behavior; the growing amplified modes, extracting energy from the mean shear, will grow on a time scale t_0 such that the oscillations will no longer be small on a scale of the turbulent Mach number. (The above equations will also no longer be valid). This refraction mechanism is likely to be the source of the very large compressible pressure fluctuations seen in the compressible DNS, Blaisdell *et al.* (1991), Sarkar (1992). This possibility is the topic of research now in progress. Blaisdell *et al.* (1993) have also reported on a tilting of dilatational surfaces as the DNS proceeds; a refraction mechanism as indicated above would explain this behavior.

3.4. THE CONCOMITANT VORTICITY EQUATION

If one takes the curl of the momentum equation one obtains the vorticity equation:

$$\frac{D}{Dt_0}\,\omega_{0q} + [\frac{S\ell}{c}]\,\varepsilon_{qij}[w_{0k,j}\,V_{i,k} + w_{0i,k}\,V_{k,j}] \;=\; -[\frac{S\ell}{c}]\,\varepsilon_{qij}(p_s v_k)_{,j}\,V_{i,k}\,. \tag{35}$$

The compressible fast modes are not irrotational (even if they start irrotational). It appears that it is inappropriate to call the field acoustic: there are rotational modes that propagate with acoustic phase speed which are stimulated by the initial conditions. Like the leading order compressible pressure field the compressible vortical modes can be amplified by extracting energy from the mean shear.

3.5. SUMMARY OF TWO-TIMING RESULTS

A simple two-time analysis has indicated some very interesting and simple models for the compressible portions of a turbulent flow. The fast compressible modes are governed by a forced convective wave equation with production (refraction) terms; this can be understood as a linear forced oscillator model for compressible effects. In a medium undergoing a simple homogeneous shear the following linear wave equation describing the evolution of the initial conditions determining an evanescent compressible wave field with acoustic phase speed:

$$\frac{D}{Dt_0}\,[\frac{D^2}{Dt_0^2}\,p_0 - \nabla^2 p_0] \;+\; 2\,[\frac{S\ell}{c}]\,V_{i,k}\,p_{0,ik} = 0. \tag{36}$$

This is essentially a third order linear oscillator, *modulo* mean convection effects. The equation contains the production (refraction) term, $V_{i,k}\,p_{0,ik}$, which will produce secular growth that is a possible mechanism for the large oscillations seen in the compressible DNS in the presence of shear.

4. Compressible turbulence modeling

An investigation, using recent DNS data, Blaisdell (1996), of the scalings obtained for the pressure-dilatation and the dilatational dissipation by Ristorcelli (1997) is the primary subject of this section. The representations given in Ristorcelli (1997) were obtained using simple scaling arguments regarding the effects of compressibility, a singular perturbation idea and the methods of statistical fluid mechanics. While the results are expressed in the context of a statistical turbulence closure, they provide, with few phenomenological assumptions, an interesting and clear physical model for the scalar effects of compressibility. The Ristorcelli (1997) analysis is relevant to shear flows with negligible bulk dilatation and small M_t^2. These restrictions are also met in a wide number of engineering flows ranging from simple shear layers of fundamental and practical interest, Papamoschou and Roshko (1988), as well as the complex shear layers associated with supersonic mixing enhancement, Gutmark *et al.* (1995). In most of these supersonic shear layers a Mach number based on the fluctuating velocity of the fluid particle is small. For example, in a Mach 4 mean flow with a turbulence intensity of 8 *per cent* has a turbulent Mach number of $M_t = 0.32$. The square of this turbulent Mach number, the appropriate perturbation expansion parameter arising from the Navier-Stokes equations, $M_t^2 \sim 0.1$, is small. These restrictions are also met in the DNS of homogeneous shear; the homogeneous shear is our laboratory experiment.

4.0.1. *A pseudo-sound analysis for the effects of compressibility*

The existence of a small parameter allows some analytical results. In Ristorcelli (1997) a low turbulent Mach number expansion for the compressible Navier Stokes equations produced a diagnostic relationship for the dilatation. To leading order, the density fluctuations are given, in nondimensional units, by the solenoidal pressure fluctuations, $\gamma \rho_s = p_s$ and the continuity equation becomes a *diagnostic* relation for the fluctuating dilatation,

$$-\gamma d \quad = \quad p_{s,t} + v_k p_{s,k} \,. \tag{37}$$

One does not need to obtain a solution to the evolution equation for the compressible velocity field, w_i, in order to obtain, to leading order, its dilatation, $d = w_{i,i}$. This is a useful result. The dilatation is determined by *local* fluctuations of the pressure and velocity; it is the rate of change of the solenoidal pressure field $p_{s,jj} = (v_i v_j)_{,ij}$, following a fluid particle.

In a two equation turbulence closure an equation for the kinetic energy of the fluctuations, k, is carried. In the kinetic energy equation two new covariances with the fluctuating dilatation appear: the pressure-dilatation and the dilatational dissipation. Constitutive relations for the pressure-dilatation and the dilatational dissipation can be found by taking the appropriate moments of the diagnostic relation for the dilatation to produce, to leading order in M_t^2,

$$-2\gamma \langle pd\rangle = \frac{D}{Dt}\langle p_s p_s\rangle. \tag{38}$$

$$\gamma^2 \langle dd\rangle = \langle \dot{p}_s \dot{p}_s\rangle + 2\langle \dot{p}_s v_k p_{s,k}\rangle + \langle v_k p_{s,k}\, v_q p_{s,q}\rangle. \tag{39}$$

The substituting of the solutions to $p_{s,jj} = (v_i v_j)_{,ij}$, into the right hand side of the above constitutive relations has been used to obtain representations for $\langle pd\rangle$ and ε_c, Ristorcelli (1997).

4.1. THE DILATATIONAL DISSIPATION

The local dilatational dissipation is comprised of a slow and a rapid part: $\varepsilon_c = \varepsilon_c^r + \varepsilon_c^s$, where

$$\varepsilon_c^s = \frac{16}{3\alpha^2}\frac{M_t^4}{R_t}\,\varepsilon_s\,[I_2^s + 6 I_1^s I_3^s] \tag{40}$$

$$\varepsilon_c^r = (\frac{2}{3})^5\,\frac{M_t^4\hat{S}^2}{R_t}\,\varepsilon_s\,[3+5R^2][\frac{3}{5}I_3^r + (\frac{\hat{S}}{15})^2[13+15R^2]\,\alpha^2\, I_1^r]. \tag{41}$$

Here $R^2 = W^2/S^2$ is the mean rotation to strain ratio; $R = 1$ for a pure shear. The nondimensional strain and rotation rates are given by: $\hat{S}^2 = (Sk/\varepsilon_s)^2$, $\hat{W}^2 = (Wk/\varepsilon_s)^2$ where $S = \sqrt{S_{ij}S_{ij}}$ and $W = \sqrt{W_{ij}W_{ij}}$. The strain and rotation tensors are defined in analogy with the incompressible case, *i.e.* traceless, $S_{ij} = \frac{1}{2}[V_{i,j} + V_{j,i} - \frac{2}{3}D\delta_{ij}]$, $W_{ij} = \frac{1}{2}[V_{i,j} - V_{j,i}]$. M_t is the turbulent Mach number, $M_t^2 = \frac{2}{3}k/c^2$, where $c^2 = \gamma < p > / < \rho >$ is the local sound speed. The traditional definition of the Mach number is used, $M_t = u_c/c$, $u_c^2 = \frac{2}{3}k$. The turbulent Reynolds number is given by $R_t = \frac{u_c \ell}{\nu} = \frac{4k^2}{9\varepsilon\nu}$ using the well known Kolmogorov scaling $\varepsilon_s = \alpha u_c^3/\ell$ with $\alpha \approx 1$.

The constants, denoted by I_i, in these expressions are not adjustable empirical calibration constant obtained by matching calculations to obtain the trends in experimental data; they are *fully specified and measurable quantities.* They are defined in terms of diverse integrals of the longitudinal correlation. The reader is referred to Ristorcelli (1997) for additional details.

The behavior that is most easily explored — in as much as the isotropic decay is the cheapest of DNS to conduct — is the scaling with turbulent Mach number. The dilatational dissipation scales as

$$\varepsilon_c^s \sim M_t^4 \varepsilon_s. \tag{42}$$

Figure 2 used in the discussion on initial conditions in §2 is relevant. With the "consistent" initial conditions derived from finite Mach number asymptotics, a M_t^4 scaling is clearly seen. A similar M_t^4 scaling has been seen in the EDQNM simulations of the Bertoglio group at Lyon. The Reynolds number dependence is not so easily explored as, in DNS, the Reynolds number cannot be varied over a substantial range. In the EDQMN simulations of the Bertoglio group a dependence on the Reynolds number is seen.

4.2. THE PRESSURE-DILATATION

In the asymptotic analysis of Ristorcelli (1997) the following representation for the pressure-dilatation was, to leading order, obtained:

$$< pd >= -\chi_{pd} M_t^2 \, [\bar{\rho} P_k - \bar{\rho}\varepsilon + T_k](P_T + \bar{\rho}\varepsilon + T_T)] \; - \bar{\rho} k \;\; M_t^2 \, \chi_{pd}^r \, \frac{D}{Dt} \mathcal{T},$$

where

$$\begin{aligned} \chi_{pd} &= \frac{2I_{pd}}{1 + 2I_{pd}M_t^2 + \frac{3}{2}I_{pd}M_t^4\gamma(\gamma-1)} \\ \chi_{pd}^r &= \frac{I_{pd}^r}{1 + 2I_{pd}M_t^2 + \frac{3}{2}I_{pd}M_t^4\gamma(\gamma-1)} \\ I_{pd} &= \frac{2}{3}I_1^s + I_{pd}^r \, [3\hat{S}^2 + 5\hat{W}^2] \\ I_{pd}^r &= \frac{1}{30}(\frac{2}{3})^3 \, \alpha^2 I_1^r. \end{aligned} \tag{43}$$

Here $\mathcal{T} = [3\hat{S}^2 + 5\hat{W}^2]$. The constants, denoted by I_i, in these expressions are fully specified by the analysis. For the present article, concerned as it is with the scalings of the compressible covariances, their exact values are not relevant.

4.2.1. *Pressure-dilatation in the isotropic decay*

For the isotropic decay the expression for the pressure-dilatation becomes

$$< pd > = \chi_{pd} M_t^2 \, \varepsilon_s \, . \tag{44}$$

Here $\bar{\rho}$ has been set to unity. The sign of $< pd >$ is positive indicating a net transfer of energy from potential to kinetic modes. Rearranging, to isolate the scaling, produces

$$\frac{< pd >}{M_t^2 \, \varepsilon_s} = \chi_{pd} = \frac{\frac{4}{3}I_1^s}{1 + \frac{4}{3}I_1^s M_t^2}. \tag{45}$$

Terms of order M_t^4 have been dropped. Earlier estimates given in Ristorcelli (1995), shown above, indicate $I_1^s = 0.5 - 0.3$. The theory therefore predicts an asymptote for χ_{pd} as the turbulent Mach number vanishes:

$$\chi_{pd} \rightarrow 0.666 - 0.40 \qquad as \qquad M_t^2 \rightarrow 0. \tag{46}$$

The DNS results, shown in Figure 4, were provided by Blaisdell for three different initial turbulent Mach numbers. As a service to the reader, the figure identifies two definitions of the turbulent Mach number: that used by Blaisdell *et al.* (1993) in their simulations, M_t^b, and that which comes from the traditional definition of the Mach number. The present compressible DNS reflect a consistent set of initial conditions as described above and in Ristorcelli and Blaisdell (1997).

The agreement with the DNS, shown in Figure 4, is very good. The analysis is consistent with the DNS without *a posteriori* adjustment of constants. The actual values of the constant could in principle be calculated from the DNS. As they are expected to be weakly dependent on initial conditions, this course is not followed further - what has been presented is sufficient for verification. Moreover, the slow portion of the pressure-dilatation is nominal compared to the rapid portion which is the most important contributor in the shear flows of interest.

4.2.2. *Homogeneous shear*

The pressure-dilatation in the homogeneous shear is now investigated. The instantaneous pressure-dilatation is seen in Figure 5. Also shown are its averaged values following the procedure of Sarkar (1992). Here the time integral of the pressure-dilatation has been taken: the vertical axis being $-\frac{1}{St}\int \frac{<pd>}{\varepsilon_s} d(St)$. The oscillations in the pressure-dilatation associated with the relaxation from initial conditions are not seen. There is, nonetheless, a build up of the oscillations which has been linked to the compressible component of the pressure field, Sarkar (1992), Blaisdell and Sarkar (1993). A plausible physical mechanism for this behavior was given in §3.

For homogeneous shear, the expression for the pressure-dilatation can be simplified. For simple shear $\mathcal{T} = 8\hat{S}^2$ one obtains

$$< pd > = -\chi_{pd} M_t^2 \,[P_k - \varepsilon_s] \; - k \;\, M_t^2 \, \chi_{pd}^r \, 8\frac{D}{Dt}\hat{S}^2. \tag{47}$$

Here $\hat{S} = \frac{Sk}{\varepsilon_s}$. For Blaisdell's homogeneous shear, in which $S = const.$, the expression can be rearranged

$$< pd > = - \; \chi_{pd} M_t^2 \; \varepsilon_s [\frac{P_k}{\varepsilon_s} - 1] \; - \; 16 \chi_{pd}^r \; M_t^2 \varepsilon_s \, (\frac{Sk}{\varepsilon_s})^2 \frac{D}{D(St)} \, \frac{Sk}{\varepsilon_s}. \tag{48}$$

Note that the coefficient of the first term, χ_{pd}, ignoring the small slow pressure contribution, scales as $\chi_{pd} \sim (\frac{Sk}{\varepsilon_s})^2$; thus accounting for the definitions of the χ's, the pressure-dilatation can be rewritten as

$$< pd > \sim -[\alpha^2 (\frac{Sk}{\varepsilon_s})^2 M_t^2 \, \varepsilon_s (\frac{P_k}{\varepsilon_s} - 1)] \, I_1^r \, [1 + \frac{1}{\frac{P_k}{\varepsilon_s} - 1} \frac{D}{D(St)} \frac{Sk}{\varepsilon_s}]. \tag{49}$$

As the flow evolves, it is expected that $I_1^r \to const$ and $\frac{D}{D(St)} \frac{Sk}{\varepsilon_s} \to 0$ thus $[1 + \frac{1}{\frac{P_k}{\varepsilon_s} - 1} \frac{D}{D(St)} \frac{Sk}{\varepsilon_s}] \to 1$. The scaling of $< pd >$ with the term in the first set of brackets will be investigated.

4.2.3. *Pressure-dilatation scalings*

The appropriately scaled integrals of $< pd >$ will now be taken. In this way one can establish whether the scalings predicted by the model are useful. In the latter portions of Blaisdell's DNS, $St > 10$, about three to four eddy turnovers past initialization, a structural equilibrium is approached. In this region the scaling, given by the analysis above, indicates

$$< pd > \sim -[\alpha^2 (\frac{Sk}{\varepsilon_s})^2 M_t^2 \, \varepsilon_s (\frac{P_k}{\varepsilon_s} - 1)] \, I_1^r \tag{50}$$

is investigated. The time integral $\frac{1}{St - St_0} \int (\,) d(St) \equiv \int_{ST}$ of diverse normalizations of will be taken. The integrals

$$\begin{aligned}
I_0 &= \int_{ST} < pd > \\
I_1 &= \int_{ST} \frac{< pd >}{\varepsilon_s} \\
I_2 &= \int_{ST} \frac{< pd >}{\varepsilon_s [\frac{P_k}{\varepsilon} - 1]} \\
I_3 &= \int_{ST} \frac{< pd >}{M_t^2 \varepsilon_s [\frac{P_k}{\varepsilon} - 1] (\frac{Sk}{\varepsilon})^2} \\
I_4 &= \int_{ST} \frac{< pd >}{\alpha^2 M_t^2 \varepsilon_s [\frac{P_k}{\varepsilon} - 1] (\frac{Sk}{\varepsilon})^2}
\end{aligned}$$

will be computed from the data. The integrals are shown in Figure 5. The integration starts at $St = 9$; the integrals are normalized by their values at $St = 10$. The integration is started at $St > 9$ so that non-equilibrium effects associated with the development of the turbulence from the initial conditions has faded. During such transient periods integrals in the pressure-dilatation model will not be constant. For $St > 9$ the quantity

$\frac{1}{St}\int I_1^r d(St) \sim I_1^r$ and $\frac{D}{Dt}\hat{S} \rightarrow 0$ and the validity of the scaling can be satisfactorily assessed. If the scaling suggested by the analysis of Ristorcelli (1995, 1997) is correct the last integral, I_4, will be approximately constant once an average is performed over a few of the oscillations. The period of time $10 < St < 16$ corresponds to about one eddy turnover time.

4.2.4. *The gradient Mach number*

The largest relative collapse of the normalized $< pd >$ integrals is the collapse from I_2 to I_3. The I_3 includes the quantities $M_t^2(\frac{Sk}{\varepsilon})^2$. The quantity $M_t(\frac{Sk}{\varepsilon})$ can be understood as a gradient Mach number. The pressure-dilatation is a strong function of the gradient Mach number. Sarkar (1995) has defined a gradient Mach number as $M_g = S\ell/c$; the transverse two-point correlation of the longitudinal velocity is used as the length scale, ℓ. In this article ℓ will be taken as the traditionally defined longitudinal length scale that occurs in the Kolmogorov scaling: $\ell = \alpha(2k/3)^{3/2}/\varepsilon_s$. In which case a mean strain Mach number is defined:

$$M_S = \frac{S\ell}{c} = \alpha\frac{2}{3}\frac{Sk}{\varepsilon_s}M_t = \alpha\frac{2}{3}\hat{S}M_t \simeq \frac{2}{3}\hat{S}M_t. \tag{51}$$

The approximation $\alpha \approx 1$ has been made. In fact, the curve I_3 overshoots the optimum collapse (a horizontal line): the gradient Mach number is increasing faster than $< pd >$. The curve, I_4, which accounts for the Kolmogorov coefficient in addition to the scalings of I_3, is a further improvement.

4.2.5. *The Kolmogorov scaling coefficient*

A new feature associated with compressibility for this class of turbulence closures, is the appearance of the Kolmogorov scaling coefficient, α. M_t and $\frac{Sk}{\varepsilon_s}$ are not new quantities for describing turbulence in single point closures; α, however, is new. The quantities M_t and $\frac{Sk}{\varepsilon_s}$ are computed as part of a turbulence model simulation; α however is not. This distinction is made in order to recognize α as a new independent quantity. The value of α can be thought of as describing large scale structural aspects of the flow: after all, it relates the kinetic energy, its cascade rate, and the two-point correlations.

The Kolmogorov parameter α appears because it is has been necessary to evaluate spatial two-point integrals that were made nondimensional using ℓ, Ristorcelli (1997). The dependence on the Kolmogorov scaling coefficient appears when the integral length scale, ℓ, is eliminated in favor of k and ε which are carried in standard turbulence closures. Such a procedure does not appear in developments for incompressible closures – the incompressible pressure-strain is independent of length scale.

The values used for α come from Blaisdell's DNS: the longitudinal integral length scale, ℓ, k and ε_s are calculated from the DNS and then the Kolmogorov relationship, $\varepsilon = \alpha(\frac{2}{3}k)^{3/2}/\ell$ is used to find α.

The Kolmogorov constant is thought to be a universal constant for high Reynolds number isotropic turbulence; for nonideal finite Reynolds number anisotropic turbulence it is *a flow specific quantity.* It is this fact that makes creating a single point turbulence model difficult; a choice for α must be made and for any given flow the choice is not, *a priori*, known.

4.3. COMPARISON WITH SELECT PRESSURE-DILATATION MODELS

The analysis of Ristorcelli (1997) has treated the case in which dilatational fluctuations come from the turbulence and not an evanescent wave field due to initial conditions. A good clean set of DNS calculations, uncontaminated by arbitrary initial conditions, now exists for the homogeneous shear. Blaisdell's new calculations allow a study of the effects of compressibility due to the non-zero Mach number of the vortical fluctuations. This is the simplest flow possible and such dilatational fluctuations will be directly related to turbulence present in engineering flows of interest. Comparison with published work on the pressure-dilatation covariance is now made.

Zeman's (1991) model is given by

$$< pd > = -\frac{1}{2\bar{\rho}c^2}\frac{D}{Dt} < pp > = \frac{1}{2\bar{\rho}c^2}\frac{< pp > - < p_e p_e >}{\tau_a}, \tag{52}$$

where $\tau_a = 0.13(2k/\varepsilon)M_t^b$. The variance $< p_e p_e >$, which represents an equilibrium level of pressure variance; it is given phenomenologically:

$$< p_e p_e > = \frac{\alpha M_t^2 + M_t^4}{1 + \alpha M_t^2 + M_t^4} \,. \tag{53}$$

Zeman's modeling is not consistent with low M_t^2 asymptotics: as $M_t^2 \to 0$ Zeman's phenomenology indicates $\langle pp \rangle \to \langle p_e p_e \rangle \to 0$. The pressure variance must, however, asymptote to the incompressible pressure variance, $\langle pp \rangle \to \langle p_s p_s \rangle$, which will be a function of the energy of the turbulence. Zeman (1991) apparently treats the model problem in which there are arbitrary pressure fluctuations due to initial conditions - the initial value problem. The compressible effects due to the compressible nature of the turbulence itself is missed. Over a time scale of the turbulence τ_a/M_t, the pressure variance predicted by this model will have equilibrated to $\langle p_e p_e \rangle$ and the pressure-dilatation will be zero. This is not seen in the DNS of either the isotropic decay or the homogeneous shear. For this reason the Zeman modeling is not further investigated.

Aupoix *et al.* (1990) have, apparently, a similar problem in mind; their resolution is, however, more general. Their value of the pressure-dilatation is obtained from an evolution equation:

$$\frac{D}{Dt} < pd >= -C_1 \frac{M_t^2}{\tau_a} \frac{D}{Dt} k - \frac{C_2}{\tau_a} < pd > . \tag{54}$$

Again τ_a is a fast "acoustic" time scale. If one inserts the evolution equation for k, $\frac{D}{Dt}k = P_k - \varepsilon + < pd >$ one obtains:

$$\frac{D}{Dt} < pd >= -[(C_1 M_t^2 + C_2) < pd > + C_1 M_t^2 (P_k - \varepsilon_s)]/\tau_a. \tag{55}$$

The density has been set to unity. On the slow manifold, characteristic of the decay, the fixed point solution yields

$$< pd >= - \frac{C_1 M_t^2}{C_2 + C_1 M_t^2} [P_k - \varepsilon_s], \tag{56}$$

which has a form similar to the present model.

Sarkar's (1992) pressure-dilatation model for flows without bulk dilatation, is given by

$$< pd >= -0.15 M_t^2 [\frac{P_k}{M_t} - \frac{4}{3}\varepsilon_s]. \tag{57}$$

This is the same form of the model as used in the simulations of Abid (1994). In case of misunderstanding it should be pointed out that there are no singularities at low turbulent Mach number in Sarkar's (1992) model.

4.3.1. *Model comparison - the isotropic decay*

For the isotropic decay the Aupoix *et al.* (1990) model, the Sarkar (1992) model and the present model given above are all, in the low M_t limit, topologically similar. The three models are, respectively,

$$< pd > = \frac{C_1 M_t^2}{C_2 + C_1 M_t^2} \varepsilon, \tag{58}$$

$$< pd > = 0.2 M_t^2 \varepsilon, \tag{59}$$

$$< pd > = \chi_{pd} M_t^2 \varepsilon_s . \tag{60}$$

The coefficients for Sarkar and Aupoix are determined empirically by matching with compressible simulations; for the present model the coefficient, χ_{pd}, is related to a two-point integral and is obtained in a way not related to any compressible experiments. The coefficients in the models show

$\frac{<pd>}{M_t^2 \varepsilon} \sim 1.25$, 0.2 and .6 for the Aupoix, Sarkar and Ristorcelli models. The DNS of Blaisdell, Figure 2, indicates a value ≈ 0.6 is the better value for the coefficient.

4.3.2. *Model comparison - the homogeneous shear*

For the isotropic decay the Aupoix *et al.* (1990) model, the Sarkar (1992) model and the present model are, in the low M_t limit, topologically similar. For the homogeneous shear differences between the models become apparent.

$$< pd > \; = \; - \frac{C_1 M_t^2}{C_2 + C_1 M_t^2} \varepsilon_s \, [\frac{P_k}{\varepsilon_s} - 1], \tag{61}$$

$$< pd > \; = \; - 0.15 M_t^2 \varepsilon_s [\frac{P_k}{\varepsilon_s M_t} - \frac{4}{3}], \tag{62}$$

$$< pd > \; = \; - \chi_{pd} M_t^2 \, \varepsilon_s [\frac{P_k}{\varepsilon_s} - 1]. \tag{63}$$

The models have been arranged to display some of their similar features; certain groupings of parameters have been emphasized. In case of misunderstanding it should be pointed out that there are no singularities at low turbulent Mach number in Sarkar's (1992) model. The grouping in the models highlights an M_t^2 dependence; the term in the brackets can be understood as representing a (weighted) departure from equilibrium. Note that all three models allow for a change in sign of $\langle pd \rangle$. For the Sarkar model the change in sign is dependent on the turbulent Mach number. For Aupoix and Ristorcelli models the change in sign is simply dependent on the whether production exceeds dissipation.

The present model, (63), features an additional dependence on the Kolmogorov scaling parameter, α, and the relative strain $\frac{Sk}{\varepsilon_s}$. This occurs through the definition of χ_{pd}; recall that $\chi_{pd} \sim \alpha^2 (\frac{Sk}{\varepsilon_s})^2$. Furthermore, the combination $M_t \frac{Sk}{\varepsilon_s}$ can be understood as a gradient Mach number; the combination $\chi_{pd} M_t^2$ indicates a proportionality of the pressure-dilatation to the square of the gradient Mach number.

A comparison of the different models with DNS is indicated in Figure 6. The simulation includes a considerable "non-equilibrium" portion of the DNS. The nonconstant values of the integrals I_i have been included in the present form of the model: the values of the required I_i are calculated from the DNS from their definition. At present we are primarily interested in the scalings – all curves have been normalized by their value at the end of the simulation.

Zeman's modeling, Zeman and Coleman (1991), is also not appropriate for the homogeneous shear with consistent initial conditions for the reasons given above. Durbin and Zeman (1992) have made several improvements to Zeman's original ideas – the vortical fluctuations are recognized as producing dilatational fluctuations, and thus Durbin and Zeman (1992) is consistent with low Mach number asymptotics. Their closure is more complex; requiring, in addition to the solution an evolution equation the pressure variance, but also an equation for a length scale associated with the pressure field. Their development treats rapidly distorted flows and their modeling and validation focuses on flows with large mean dilatations – a flow in which the $\langle pd \rangle$ is important. The present development is focussed on compressible shear flows and a first step in the development of a compressible closure that might explain and predict the growth rate suppression seen in many mixing layer experiments. The mixing layer is not a case of a rapidly distorted flow.

4.4. TURBULENCE MODELING CONCLUSIONS

A simple low Mach number analysis of the Navier-Stokes equations has related the fluctuating dilatation to the solenoidal pressure field, Ristorcelli (1997). This is used to produce constitutive relations for the pressure-dilatation and the dilatational dissipation. The scalings of the these expressions with turbulent Mach number, relative strain rate, gradient Mach number and Kolmogorov scaling coefficient have been shown to collapse the DNS data.

The DNS — with consistent initial conditions — indicates that the dilatational dissipation scales as M_t^4. As such, the dilatational dissipation does not appear to be an item of any concern in low turbulent Mach number flows of interest to high speed (supersonic) aerodynamics. The dilatational dissipation with its M_t^4 dependence certainly cannot explain the reduced growth rate of the compressible mixing layer, Thangam *et al.* (1996a, 1996b). This is consistent with recent simulations, Sarkar (1995), Vreman *et al.* (1996), Simone *et al.* (1997).

The pressure-dilatation is found to be a nonequilibrium phenomenon. It scales as $\alpha^2 M_t^2 (\frac{Sk}{\varepsilon_s})^2 [P_k/\varepsilon_s - 1]$. For it to be important requires <u>both</u> 1) the square of the gradient Mach number, M_S^2, to be substantial <u>and</u> 2) for the flow to be out of equilibrium $P_k \neq \varepsilon$. The pressure-dilatation has been observed and is predicted to be either positive or negative. Its dependence on the non-equilibrium nature of the flow, $P_k \neq \varepsilon$, indicates that the pressure-dilatation can be either stabilizing or destabilizing. These trends appear consistent with the DNS of Simone *et al.* (1997), in which such a dependence on P_k/ε, as related to and implied by the evolution of the anisotropy, $b_{12} = \frac{\langle uv \rangle}{2k}$, from isotropic initial conditions, can be inferred.

The homogeneous shear is arguably the most non-equilibrium flow of the the diverse benchmark flows. Accordingly the pressure-dilatation, for this class of flows, will be at its most important. In as much as many aerodynamic flows are close to equilibrium, it also appears that the pressure-dilatation with $\alpha \approx 1$ cannot explain the dramatically reduced growth rate of the mixing layer, Thangam *et al.* (1996a, 1996b). A value of $\alpha \approx 1$ is a reasonable guess for an equilibrium turbulence; an $\alpha \approx 1$ appears to account for about one fifth of the reduction of the growth rate. A choice of $\alpha \approx 2.5$ would produce the observed reduction of the mixing layer growth rate; this, however, would be no more accurate a representation of the physics than when the dilatation dissipation was argued to account for the reduction of the growth rate of the mixing layer.

The present discussion treats only the "scalar" effects of compressibility – the changes in the turbulence energy, k, due to the dilatational covariances in the energy budget. The dilatational covariances cannot account for the reduction in the shear anisotropy, b_{12}, or the normal anisotropy, b_{11}, so important in the production of the shear stress, $\langle v_1 v_2 \rangle$. To account for these more substantial structural effects appears to require a compressible *pressure-strain* representation accounting for the effects of compressibility in the evolution equations for the Reynold stresses.

In the pseudo-sound analysis the pressure-dilatation is found to be a function of the Kolmogorov scaling coefficient and this is anticipated to be an important (and potentially difficult) feature in single point closures for the effects of compressibility. The Kolmogorov coefficient, a reflection of an equilibrium cascade and an inertial subrange is, in general, a flow dependent quantity: there is little known about its dependence in non-ideal — anisotropic, strained, inhomogeneous — flow situations. The appearance of the Kolmogorov coefficient, in as much as it links the energy, the spectral flux, and a two-point length scale, is an indication of dependence on large scale structure.

5. Summary

This article has described the consequences of low Mach number expansions of the Navier Stokes equations for three different problems. In §2, time was spent developing physically relevant initial conditions for DNS of compressible turbulence. From Blaisdell's DNS it was shown that initial conditions can have a substantial effect on the scalings of the dilatational dissipation. It is clear that the DNS of compressible *homogeneous* turbulence – in which the medium can sustain a wave field in addition to the nonlinear advective field – needs careful consideration.

In §3, a two-timing perturbation method indicated that the fast compressible modes of a turbulent field are, to lowest order, described by a forced linear wave equation with refraction. Compressibility effects can be understood, as indicated by the two-time perturbation analysis, using a forced linear oscillator analog.

Several points relevant to understanding initial conditions for DNS of compressible turbulence stand out:

1. Inconsistent initial conditions stimulate modes with acoustic phase speed.
2. Consistent initial conditions eliminate the oscillations associated with the initial value problem.
3. Modes with acoustic phase speeds are solutions to the initial value problem and cannot be referred to as compressible turbulence which would have an advective phase speed.
4. Turbulence models for acoustic modes are as arbitrary as the initial conditions.
5. The initial value problem associated with inconsistent initial conditions which indicates an M_t^2 scaling for the dilatational dissipation. This is a scaling that is consistent with the presence of an evanescent wave field due to the initial conditions. The dilatational dissipation due to the compressible nature of the turbulence has an M_t^4 scaling.

A single time small Mach number analysis was used to produce models for the covariances with the dilatation. The analysis indicated the importance and role of several nondimensional parameters characterizing the compressibility of the turbulence. The nondimensional parameters that arise from a pseudo-sound analysis of weakly compressible turbulence are

1. The square of the turbulent Mach number: M_t^2.
2. The gradient Mach number: $M_S = \frac{2}{3}\frac{Sk}{\varepsilon}M_t$.
3. The Kolmogorov scaling coefficient: $\alpha = \varepsilon_s \ell / u_c^3$.
4. The departure from equilibrium: $\frac{P_k}{\varepsilon_s} - 1$.

Some of these parameters have been identified in previous works.

From the turbulence modeling point of view the results of the pseudo-sound analysis have indicated that: 1.) Compressible dissipation will not be an important phenomenon for low turbulent Mach numbers as occurs in many aerodynamic applications; 2.) The pressure-dilatation will only be important to the degree that the flow is *out of equilibrium*. It might also be mentioned that the pseudo-sound analysis predicts, contrary to expectations, a dilatational dissipation that depends on the Reynolds number. This is a precautionary statement: it is as example of a situation in which extrapolating a model for the effects of turbulence based on DNS data would be inappropriate for high Reynolds number flows. Presently DNS can not explore a wide enough Reynolds number range and, as a consequence, Reynolds number dependent phenomena can potentially be missed.

The low Mach number pseudo-sound analysis, in as much as it is supported by the DNS, appears to be a useful tool: it appears to have successfully

- Eliminated the fast modes associated with the initial value problem.
- Predicted the M_t^4 scalings of the dilatational dissipation and its consequent neglectability.
- Predicted the scalings of the pressure-dilatation.

With the successes of such a rational procedure incorporating the more complex effects of compressibility can be done with more confidence. Of particular interest is the very difficult problem of the pressure-strain covariance. The importance the pressure-strain has been indicated in Blaisdell and Sarkar (1993), Vreman *et al.* (1996), Sarkar (1995), Simone *et al.* (1997), Freund *et al.* (1997).

6. Acknowledgements

A substantial debt is owed to Professor G. Blaisdell for providing the results of his new DNS calculations with "consistent initial conditions."

References

Abid, R., 1994. "On the prediction of equilibrium states in homogeneous compressible turbulence," NASA Contractor Report 4570, Langley Research Center.

Aupoix, B., Blaisdell, G.A., Reynolds, W.C., and Zeman, O., 1990. "Modeling the turbulent kinetic energy equation for compressible turbulence," Proceedings of the Summer Program, Center for Turbulence Research, Stanford Univ., Stanford, CA.

Blaisdell, G.A., 1996. Personal communication.

Blaisdell, G.A., Mansour, N.N., and Reynolds, W.C., 1991. "Compressibility effects on the growth and structure of homogeneous turbulent shear flow," Report TF-50, Department of Mechanical Engineering, Stanford Univ., Stanford, CA.

Blaisdell, G.A., Mansour, N.N., and Reynolds, W.C., 1993. "Compressibility effects on the growth and structure of homogeneous turbulent shear flow," *Journal of Fluid Mechanics* **256**, p. 443.

Blaisdell, G.A. and Sarkar, S.S., 1993. "Investigation of the pressure-strain correlation in compressible homogeneous turbulent shear flow," Transitional and turbulent compressible flows, Proceedings of the 1993 ASME Fluids Engineering Conference, Washington DC, June 21-24, 1993, FED-Vol. 151, pp. 133-138.

Bradshaw, P., 1977. "Compressible turbulent shear layers," *Ann. Rev. Fluid Mech.* **9**, p. 33.

Durbin, P.A. and Zeman, O., 1992. "Rapid distortion theory for homogeneous compressed turbulence," *J. Fluid Mech.* **242**, pp. 349.

Freund, J.B., Lele, S.K., and Moin, P., 1997. "Direct simulation of a supersonic round turbulent shear layer," AIAA-97-0760, 35th Aerospace Sciences Meeting and Exhibit, Reno, NV.

Freund, J.B., Moin, P., and Lele, S.K., 1997a. "Compressibility effects in a turbulent annular mixing layer," 1997 Report TF-72, Stanford Univ., Mechanical Engineering.

Gutmark, E.J., Schadow, K.C., and Yu, K.H., 1995. "Mixing enhancement in supersonic free shear layers," *Ann. Rev. Fluid Mech.* **27**, p. 375.

Lele, S.K., 1994. "Compressibility effects on turbulence," *Ann. Rev. Fluid Mech.* **26**, p. 211.

Lilley, G., 1996. Personal communication.

Papamoschou, D. and Roshko, A., 1988. "The compressible turbulent shear layer: An experimental study," *Journal of Fluid Mechanics* **197**, p. 453.

Ristorcelli, J.R., 1997. "A pseudo-sound constitutive relationship and closure for the dilatational covariances in compressible turbulence: An analytical theory," ICASE Report 95-22, *Journal of Fluid Mechanics* **347**, p. 37.

Ristorcelli, J.R., 1997a. "A closure for the compressibility of the source terms in Lighthill's acoustic analogy," ICASE Report 97-44, submitted to *Journal of Fluid Mechanics*.

Ristorcelli, J.R., 1997b. "The pressure-dilatation covariance in compressible turbulence: Validation of theory and modeling," ICASE Report in progress.

Ristorcelli, J.R. and Blaisdell, G.A., 1997. "Consistent initial conditions for the DNS of compressible turbulence," ICASE Report 96-49, *Physics of Fluids* **9**, p. 4.

Ristorcelli, J.R., Lumley, J.L., and Abid, R., 1995. "A rapid-pressure covariance representation consistent with the Taylor-Proudman theorem materially-frame-indifferent in the 2D limit," ICASE Report 94-01, *J. Fluid Mechanics* **292**, pp. 111-152.

Sarkar, S., 1992. "The pressure-dilatation correlation in compressible flows," *Physics of Fluids A* **12**, p. 2674.

Sarkar, S., 1995. "The stabilizing effect of compressibility on turbulent shear flow," *Journal of Fluid Mechanics* **282**, p. 163.

Sarkar, S., Erlebacher, G., and Hussaini, M.Y., 1991a. "Direct simulation of compressible turbulence in a shear flow," *Theoret. Comput. Fluid Dyn.* **2**(291).

Simone, A., Coleman, G.N., and Cambon, C., 1997. "The effect of compressibility on turbulent shear flow: A rapid-distortion-theory and direct-numerical-simulation study," *Journal of Fluid Mechanics* **330**, p. 307.

Spina, E.F., Smits, A.J., and Robinson, S.K., 1994. "The physics of supersonic boundary layers," *Ann. Rev. Fluids Mech.* **26**, p. 287.

Thangam, S., Zhou, Y, and Ristorcelli, J.R., 1996a. "Analysis of compressible mixing layers using dilatational covariances model," AIAA-96-3076. 32nd AIAA/SAE/ASME/ASEE Joint Propulsion Conference, Lake Buena Vista, FL.

Thangam, S., Ristorcelli, J.R., and Zhou, Y., 1996b. "Development and application of a dilatational covariances model for compressible mixing layers," AIAA-97-0723. 35th Aerospace Sciences Meeting and Exhibit, Reno, NV.

Vreman, W.A., Sandham, N.D., and Luo, K.H., 1996. "Compressible mixing layer growth rate and turbulence characteristics," *Journal of Fluid Mechanics* **320**, p. 235.

Zeman, O., 1991. "On the decay of compressible isotropic turbulence," *Phys. Fluids A* **3**, pp. 951.

Zeman, O., and Coleman, G.N., 1991. "Compressible turbulence subjected to shear and rapid compression," *Eight Turbulent Shear Flow Symposium*, Springer Verlag.

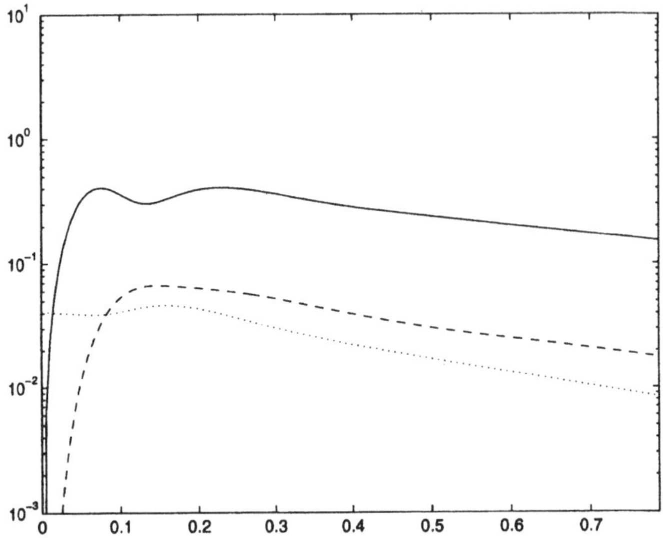

Figure 1. Variance of the dilatation, $\langle dd \rangle$, for the isotropic decay as a function of time — three different initial conditions. IC1 - solid line; IC2 - dashed line; IC3 - dotted line.

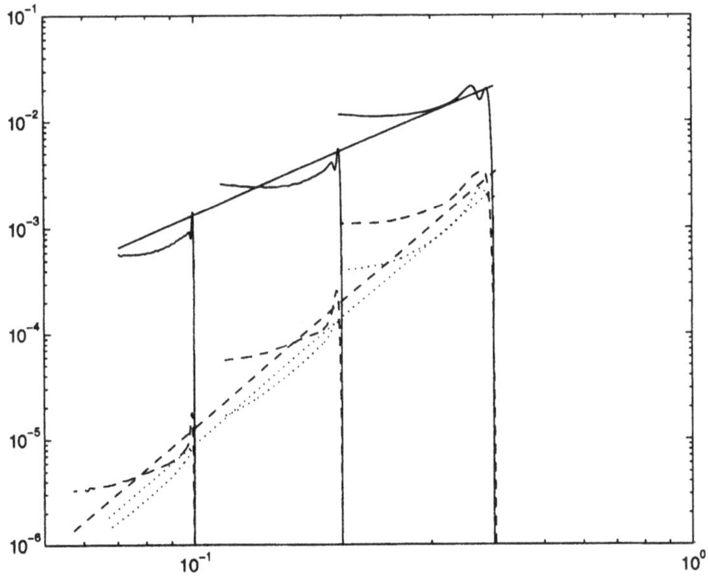

Figure 2. Variance of the dilatation, $\langle dd \rangle$, versus M_t; three different Mach numbers — three different initial conditions. IC1 - solid line; IC2 - dashed line; IC3 - dotted line.

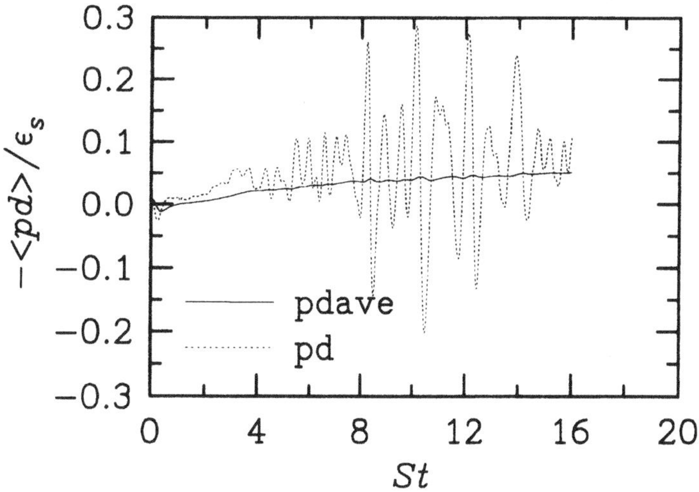

Figure 3. The instantaneous and averaged pressure-dilatation - Blaisdell's rjr_s1 DNS.

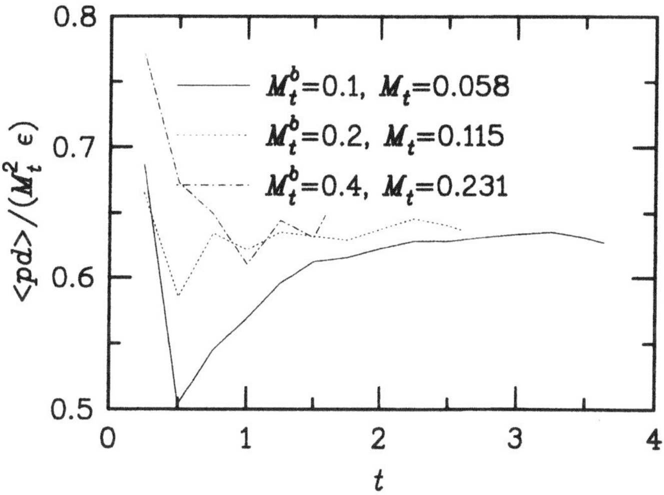

Figure 4. Pressure-dilatation coefficient for the isotropic decay.

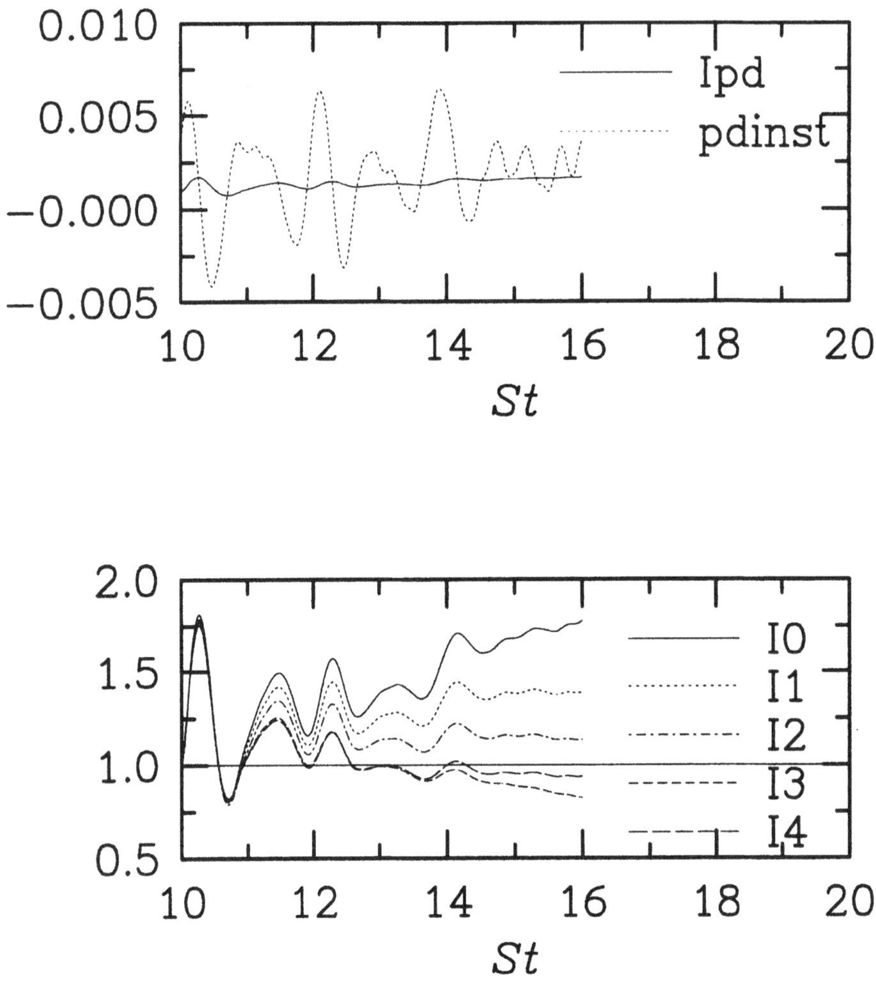

Figure 5. **Integral scalings for the pressure-dilatation in homogeneous shear.**

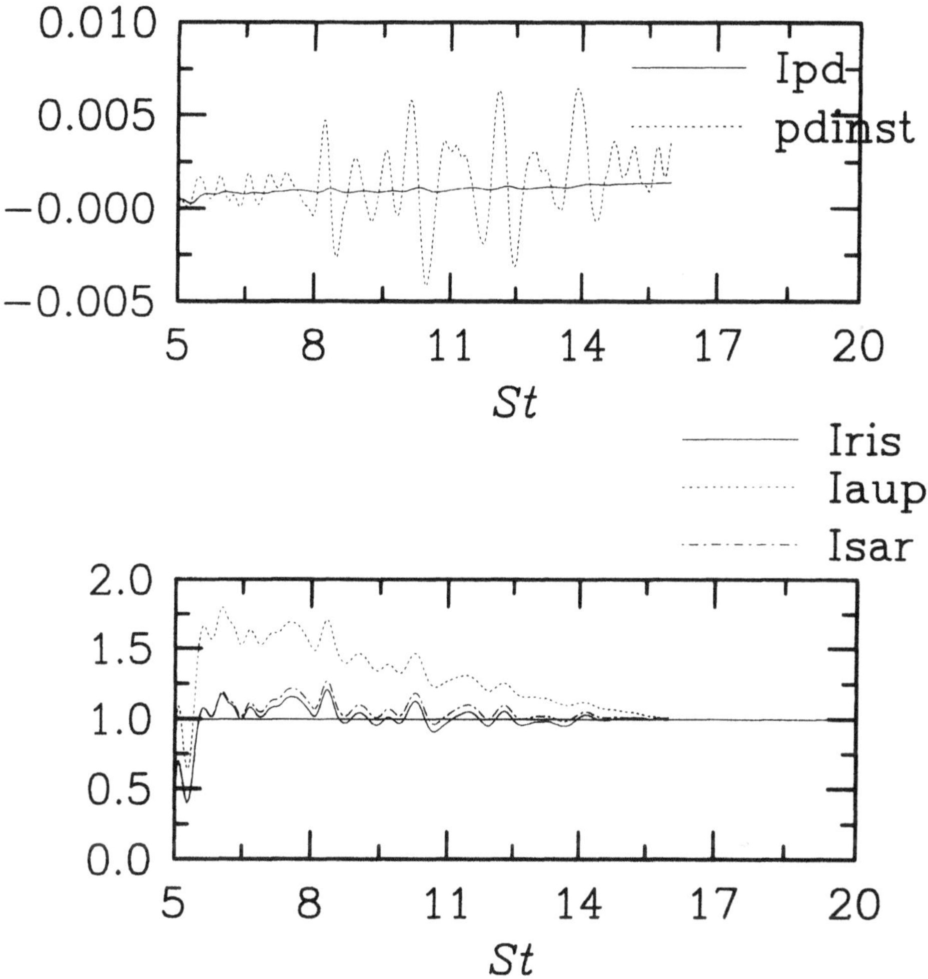

Figure 6. Comparisons of the different models.

TRANSPORT COEFFICIENTS IN ROTATING WEAKLY COMPRESSIBLE TURBULENCE

ROBERT RUBINSTEIN
Institute for Computer Applications in Science and Engineering
NASA Langley Research Center
Hampton, Virginia

YE ZHOU
IBM Research Division
T.J. Watson Research Center
Yorktown Heights, New York

AND

GORDON ERLEBACHER
Program in Computational Science and Engineering
Florida State University
Tallahassee, Florida

Abstract. Analytical studies of compressible turbulence have found that compressible velocity fluctuations create both effective fluid transport properties and an effective equation of state. This paper investigates the additional effects of rotation on compressible turbulence. It is shown that rotation modifies the transport properties of compressible turbulence by replacing the turbulence time scale by a rotational time scale, much as rotation modifies the transport properties of incompressible turbulence. But thermal equilibrium properties are modified in a more complex manner. Two regimes are possible: one dominated by incompressible fluctuations, in which the sound speed is modified as it is in incompressible turbulence, and a rotation dominated regime in which the sound speed enhancement is rotation dependent. The dimensionless parameter which discriminates between regimes is identified. In general, rotation is found to suppress the effects of compressibility.

M. D. Salas et al. (eds.), Modeling Complex Turbulent Flows, 329–347.

1. Introduction

Rotating weakly compressible turbulence occurs in astrophysical flows and in engineering flows like swirling jets. The development of transport models and sub-grid scale models for such flows is difficult because the coupling between compressibility and rotation introduces at least two natural dimensionless parameters: the turbulent Mach number $M_t = K^{1/2}/c$ and the rotation parameter $\zeta = \Omega K/\varepsilon$. It is not easy to assess the combined effects of these parameters heuristically. Recourse to DNS will require an unrealistically large number of simulations to cover a useful range of parameters. LES will not be practical without sub-grid scale models which are sensitive to both rotation and compressibility: the development of such models is a goal of this paper.

The present work addresses these problems using the rational statistical turbulence closure provided by Yoshizawa's (1986,1995) two-scale direct interaction approximation (TSDIA), which is a natural and analytically tractable extension of Kraichnan's (1959) direct interaction approximation (DIA) to inhomogeneous turbulence. It is based on previous work on the transport theory of weakly compressible turbulence (Rubinstein and Erlebacher, 1997, RE in what follows) and on the analytical theory of rotating turbulence initiated by Zhou (1995).

The problem of weakly compressible turbulence, in which the equations for compressible quantities are linearized about a basic state of incompressible turbulence was considered in RE. TSDIA was applied to describe weak compressibility effects on the mean flow equations by a set of transport coefficients: this description generalizes the eddy viscosity description of incompressible turbulence. Following Yoshizawa, the transport coefficients are expressed in terms of the fundamental statistical descriptors of weakly compressible turbulence: the response and correlation functions of incompressible turbulence (Kraichnan, 1959) and the additional statistical descriptors of the compressible part of the motion. By assuming that the basic state of incompressible turbulence is in a Kolmogorov steady state, it is possible to derive both formulas for the transport coefficients in terms of single point descriptors of turbulence and, following Yakhot and Orszag (1986), sub-grid scale models for LES of weakly compressible turbulence.

The analysis of rotation effects will be based on Zhou's (1995) scaling theory of rotating turbulence, which demonstrates the existence of a steady state with constant inertial range energy flux in rotating turbulence, but with non-Kolmogorov scaling exponents. In the limit of strong rotation, this theory is closely related to the theory of *weak turbulence* (Zakharov *et al.*, 1992) in which nonlinear temporal decorrelation of Fourier modes is dominated by linear dispersive decorrelation. The extension of the theory of RE to rotating turbulence will be accomplished by perturbing about this state of rotating incompressible turbulence instead of about the Kolmogorov state.

Like previous investigations of compressible turbulence (Staroselsky *et al.*, 1990; Chandrasekhar, 1951), RE found two kinds of compressibility effects: modified transport coefficients and modified equilibrium properties. Turbulence creates diffusivities for entropy and pressure and enhances the molecular shear and bulk viscosities. But turbulence also modifies the specific heat ratio and creates an effective turbulent pressure. These combined effects lead to an enhanced propagation speed for sound waves in the mean flow; they can be considered as a modified equation of state for compressible turbulence. Rotation alters the transport properties much as it does in incompressible turbulence, by replacing the turbulent time scale by a rotational time scale. It will be shown that rotating compressible turbulence can exhibit two different effective equations of state, depending on the size of the parameter $\zeta^{1/2}M_t$. In general, rotation tends to suppress or counteract the effects of compressibility.

2. Review of the TSDIA Theory of Weakly Compressible Turbulence

The basic result of RE is the model for weakly compressible turbulence

$$\frac{\partial S}{\partial t} + \mathbf{U}\cdot\nabla S = \nabla\cdot(\nu^{ss}\nabla S) \tag{1}$$

$$\begin{aligned}\frac{\partial \mathbf{U}}{\partial t} &+ \mathbf{U}\cdot\nabla\mathbf{U} + \left(\frac{1}{R} - \frac{1}{\hat{R}}\right)\nabla P = \nabla\cdot\{-\frac{2}{3}K\mathbf{I} + \nu[\nabla\mathbf{U} + (\nabla\mathbf{U})^T]^S\} \\ &+ \nabla(\nu^{uu}\nabla\cdot\mathbf{U})\end{aligned} \tag{2}$$

$$\frac{\partial P}{\partial t} + \mathbf{U}\cdot\nabla P + (\gamma - \hat{\gamma})P\nabla\cdot\mathbf{U} = \nabla\cdot(\nu^{pp}\nabla P) \tag{3}$$

where the turbulence dependent properties are defined by the integrals

$$\nu = \frac{4}{15}\int_0^\infty 4\pi k^2 dk \int_0^\infty d\tau\ G(k,\tau)Q(k,\tau) \tag{4}$$

$$\nu^{ss} = \frac{2}{3}\int_0^\infty 4\pi k^2 dk \int_0^\infty d\tau\ G^{ss}(k,\tau)Q(k,\tau) \tag{5}$$

$$\begin{aligned}\nu^{pp} &= \frac{2}{3}\int_0^\infty 4\pi k^2 dk \int_0^\infty d\tau G^{pp}(k,\tau)Q(k,\tau) \\ &- \frac{1}{c^2}\int_0^\infty 4\pi k^2\ dk \int_0^\infty d\tau\{\frac{2}{3}G^{ww}(k,\tau) + \frac{1}{3}G^{\phi\phi}(k,\tau)\}Q^p(k,\tau)\end{aligned} \tag{6}$$

$$\hat{\gamma} = (\gamma - 1)\Pi/P \tag{7}$$

$$\Pi = \gamma\int_0^\infty 4\pi k^2 dk \int_0^\infty d\tau\ kG^{\phi p}(k,\tau)Q^p(k,\tau) \tag{8}$$

$$
\begin{aligned}
\nu^{uu} &= \frac{1}{3}\int_0^\infty 4\pi k^2 dk \int_0^\infty d\tau \frac{2}{3} G^{ww}(k,\tau)Q(k,\tau) \\
&- \frac{\gamma}{R^2c^2}\int_0^\infty 4\pi k^2 dk \int_0^\infty d\tau\ G^{pp}(k,\tau)Q^p(k,\tau) \qquad (9)\\
\frac{1}{\hat{R}} &= \int_0^\infty 4\pi k^2 dk \int_0^\infty d\tau\ kG^{\phi p}(k,\tau)Q(k,\tau) \qquad (10)
\end{aligned}
$$

The quantities ν, ν^{ss}, ν^{pp} and ν^{uu} are effective transport properties, whereas Π and $\hat{R}$ define the modified equation of state of compressible turbulence. In Eqs. (4)-(10), G and Q are the response and correlation functions, the fundamental statistical descriptors of isotropic turbulence in DIA (Kraichnan, 1959) and Q^p is the two-time pressure correlation function. The compressible response functions $G^{\phi\phi}, G^{\phi p}, G^{pp}, G^{p\phi}, G^{ww}$ are defined in terms of the normalized Helmholtz decomposition

$$u_i(\mathbf{k},t) = w_i(\mathbf{k},t) + ik_ik^{-1}\phi(\mathbf{k},t) \qquad (11)$$

They satisfy the DIA response equations, written in terms of time Fourier transforms,

$$i\omega G^{ww}(\mathbf{k},\omega) + \eta^{ww}G^{ww}(\mathbf{k},\omega) = 1 \qquad (12)$$

and

$$\left\{\begin{bmatrix} -i\omega & k/\rho_0 \\ -\rho_0 c^2 k & -i\omega \end{bmatrix} + \begin{bmatrix} \eta^{\phi\phi} & \eta^{\phi p} \\ \eta^{p\phi} & \eta^{pp} \end{bmatrix}\right\}\begin{bmatrix} G^{\phi\phi} & G^{\phi p} \\ G^{p\phi} & G^{pp} \end{bmatrix} = \begin{bmatrix} 1 & 0 \\ 0 & 1 \end{bmatrix} \qquad (13)$$

Following RE, the response equations will be solved in a Markovianized limit in which the damping functions η in Eqs. (12), (13) are defined by

$$
\begin{aligned}
\eta^{ww}(\mathbf{k}) &= -\frac{1}{2}P_{ij}(\mathbf{k})\int_0^\infty d\tau \int_{\mathbf{k}=\mathbf{p}+\mathbf{q}} d\mathbf{p}d\mathbf{q}\Gamma^1_{imn}(\mathbf{k},\mathbf{p},\mathbf{q})G^{ww}(\mathbf{q},\tau)\times \\
&\quad P_{np}(\mathbf{q})\Gamma^1_{prj}(\mathbf{q},-\mathbf{p},\mathbf{k})Q_{mr}(\mathbf{p},\tau) \\
&- \frac{1}{2}P_{ij}(\mathbf{k})\int_0^\infty d\tau \int_{\mathbf{k}=\mathbf{p}+\mathbf{q}} d\mathbf{p}d\mathbf{q}\Gamma^1_{imn}(\mathbf{k},\mathbf{p},\mathbf{q})G^{\phi\phi}(\mathbf{q},\tau)\times \\
&\quad P^*_{np}(\mathbf{q})\Gamma^1_{prj}(\mathbf{q},-\mathbf{p},\mathbf{k})Q_{mr}(\mathbf{p},\tau) \qquad (14)\\
\eta^{\phi\phi}(\mathbf{k}) &= -P^*_{ij}(\mathbf{k})\int_0^\infty d\tau \int_{\mathbf{k}=\mathbf{p}+\mathbf{q}} d\mathbf{p}d\mathbf{q}\Gamma^1_{imn}(\mathbf{k},\mathbf{p},\mathbf{q})G^{ww}(\mathbf{q},\tau)\times \\
&\quad P_{np}(\mathbf{q})\Gamma^1_{prj}(\mathbf{q},-\mathbf{p},\mathbf{k})Q_{mr}(\mathbf{p},\tau) \\
&- P^*_{ij}(\mathbf{k})\int_0^\infty d\tau \int_{\mathbf{k}=\mathbf{p}+\mathbf{q}} d\mathbf{p}d\mathbf{q}\Gamma^1_{imn}(\mathbf{k},\mathbf{p},\mathbf{q})G^{\phi\phi}(\mathbf{q},\tau)\times \\
&\quad P^*_{np}(\mathbf{q})\Gamma^1_{prj}(\mathbf{q},-\mathbf{p},\mathbf{k})Q_{mr}(\mathbf{p},\tau) \qquad (15)\\
\eta^{\phi p}(\mathbf{k}) &= ik_ik^{-1}\int_0^\infty d\tau \int_{\mathbf{k}=\mathbf{p}+\mathbf{q}} d\mathbf{p}d\mathbf{q}\Gamma^1_{imn}(\mathbf{k},\mathbf{p},\mathbf{q})\times
\end{aligned}
$$

$$
\begin{aligned}
& G_n^{up}(\mathbf{q},\tau)\Gamma_r^2(\mathbf{q},-\mathbf{p},\mathbf{k})Q_{mr}(\mathbf{p},\tau) && (16)\\
\eta^{p\phi}(\mathbf{k}) &= -ik_jk^{-1}\int_0^\infty d\tau\int_{\mathbf{k}=\mathbf{p}+\mathbf{q}} d\mathbf{p}d\mathbf{q}\Gamma_m^2(\mathbf{k},\mathbf{p},\mathbf{q})\times \\
& G_n^{pu}(\mathbf{q},\tau)\Gamma_{nrj}^1(\mathbf{q},-\mathbf{p},\mathbf{k})Q_{rm}(\mathbf{p},\tau) && (17)\\
\eta^{pp}(\mathbf{k}) &= -\int_0^\infty d\tau\int_{\mathbf{k}=\mathbf{p}+\mathbf{q}} d\mathbf{p}d\mathbf{q}\Gamma_m^2(\mathbf{k},\mathbf{p},\mathbf{q})G^{pp}(\mathbf{q},\tau)\times \\
& \Gamma_n^2(\mathbf{q},-\mathbf{p},\mathbf{k})Q_{mn}(\mathbf{p},\tau) && (18)
\end{aligned}
$$

where the quantities Γ, which characterize the interactions between the fields, are given by

$$
\begin{aligned}
\Gamma_{imn}^1(\mathbf{k},\mathbf{p},\mathbf{q}) &= i(q_m\delta_{in}+p_n\delta_{im})\\
\Gamma_n^2(\mathbf{k},\mathbf{p},\mathbf{q}) &= ik_n
\end{aligned}
$$

These equations are solved approximately in RE by a series expansion in turbulent Mach number. Weak compressibility assumes the existence of a basic state of incompressible turbulence. If this turbulence is in a Kolmogorov steady state, in which isotropy implies that

$$
\begin{aligned}
G_{ij}(\mathbf{k},\tau) &= G(k,\tau)P_{ij}(\mathbf{k}) && (19)\\
Q_{ij}(\mathbf{k},\tau) &= Q(k,\tau)P_{ij}(\mathbf{k}) && (20)
\end{aligned}
$$

the approximation leads, at lowest nontrivial order, to the conclusion that G^{ww} is the response function of a passive vector field, and to the expressions

$$
\begin{aligned}
G^{\phi p}(k,\tau) &= -\frac{k}{\rho_0\mathcal{S}}\exp(-\eta^{\phi\phi}\tau/2)\sin(\mathcal{S}\tau)H(\tau) && (21)\\
G^{p\phi}(k,\tau) &= \rho_0\frac{c^2k}{\mathcal{S}}\exp(-\eta^{\phi\phi}\tau/2)\sin(\mathcal{S}\tau)H(\tau)\\
G^{pp}(k,\tau) &= \exp(-\eta^{\phi\phi}\tau/2)\cos(\mathcal{S}\tau)H(\tau)\\
&+ \frac{\eta^{\phi\phi}}{2\mathcal{S}}\exp(-\eta^{\phi\phi}\tau/2)\sin(\mathcal{S}\tau)H(\tau)\\
G^{\phi\phi}(k,\tau) &= \exp(-\eta^{\phi\phi}\tau/2)\cos(\mathcal{S}\tau)H(\tau)\\
&- \frac{\eta^{\phi\phi}(k)}{2\mathcal{S}}\exp(-\eta^{\phi\phi}\tau/2)\sin(\mathcal{S}\tau)H(\tau) && (22)
\end{aligned}
$$

where

$$
\mathcal{S}=\{c^2k^2-\frac{1}{4}\eta^{\phi\phi}(k)^2\}^{1/2} \qquad (23)
$$

In non-rotating turbulence, the damping function $\eta^{\phi\phi}$ is determined by the Reynolds analogy

TABLE 1. **Results of RE**

	Single-point model	General subgrid model	Smagorinsky model
ν^{pp}	$M_t^2 K^2/\varepsilon$	$\varepsilon\Delta^2/c^2$	$(S^<)^3\Delta^4/c^2$
ν^{uu}	K^2/ε	$\varepsilon^{1/3}\Delta^{4/3}$	$(S^<)\Delta^2$
Π	KM_t^2	$\varepsilon^{4/3}\Delta^{4/3}/c^2$	$(S^<)^4\Delta^4/c^2$
$1/\hat{R}$	M_t^2	$\varepsilon^{2/3}\Delta^{2/3}/c^2$	$(S^<)^2\Delta^2/c^2$

$$\eta^{\phi\phi}(k) = \alpha_\phi \eta(k) \tag{24}$$

where α_ϕ is an inverse Prandtl number, and the incompressible damping function $\eta(k)$ takes the Kolmogorov scaling form

$$\eta(k) = C_D \varepsilon^{1/3} k^{2/3} \tag{25}$$

Analogs of these formulas for rotating turbulence will be developed subsequently. Once such formulas are known, the integrals for transport coefficients Eqs. (4)-(10) can be evaluated explicitly. For turbulence transport models, the integration will extend over the entire inertial range $k \geq k_0$ where k_0 is the inverse integral scale. Subgrid models are derived instead (Yakhot and Orszag, 1986) by integrating only over wavevectors which satisfy the condition $k \geq 2\pi/\Delta$, where Δ is the filter size; as usual, this derivation assumes that these scales are in the inertial range. The quantity ε can be closed by Smagorinsky's hypothesis of local energy equilibrium. An interesting alternative, related to postulating the *sweeping hypothesis* for turbulent time correlations, has been suggested recently by Yoshizawa *et al.* (1996).

The results of RE are summarized in Table 1. The effective properties are given without the appropriate constants, all of which could in principle be evaluated theoretically. The column labeled "general subgrid model" contains the result of integrating the integral expression over the subgrid scales; the corresponding "Smagorinsky model" is obtained by equating the dissipation rate to the resolved production. Throughout the table, $S^<$ is defined by $(S^<)^2 = S^<_{ij} S^<_{ij}$, where $S^<_{ij}$ is the resolved fluctuating strain rate.

3. Review of the Statistical Theory of Rotating Turbulence

Kolmogorov scaling applies to a turbulent steady state with constant energy flux. External effects like rotation will modify this steady state. Rotation is an especially simple external effect, since energy remains an inviscid invariant under rotation, and a steady state with constant energy flux remains possible. But Kolmogorov scaling may no longer apply to this steady state and the occurrence of a distinguished time scale precludes the deduction of the applicable scaling law by dimensional analysis alone. To deduce the scaling, we will appeal to closure in the form of the direct interaction approximation.

For rotating turbulence, the DIA equations of motion take the form

$$\begin{aligned} &\dot{G}_{ij}(\mathbf{k},t,s) + 2P_{ip}(\mathbf{k})\Omega_{pq}G_{qj}(\mathbf{k},t,s) \\ + &\int_s^t dr\ \eta_{ip}(\mathbf{k},t,r)G_{pj}(\mathbf{k},r,s) = 0 \end{aligned} \tag{26}$$

$$\begin{aligned} &\dot{Q}_{ij}(\mathbf{k},t,s) + 2P_{ip}(\mathbf{k})\Omega_{pq}Q_{qj}(\mathbf{k},t,s) \\ + &\int_s^t dr\ \eta_{ip}(\mathbf{k},t,r)Q_{pj}(\mathbf{k},r,s) \\ = &\int_0^t dr\ G_{ip}(\mathbf{k},t,r)F_{pj}(\mathbf{k},s,r) \end{aligned} \tag{27}$$

where the eddy damping η and forcing F are defined by

$$\begin{aligned} \eta_{ir}(\mathbf{k},t,s) &= \int_{\mathbf{k}=\mathbf{p}+\mathbf{q}} d\mathbf{p}d\mathbf{q}P_{imn}(\mathbf{k})P_{\mu rs}(\mathbf{p}) \times \\ &\quad G_{m\mu}(\mathbf{p},t,s)Q_{ns}(\mathbf{q},t,s) \end{aligned} \tag{28}$$

$$\begin{aligned} F_{ij}(\mathbf{k},t,s) &= \int_{\mathbf{k}=\mathbf{p}+\mathbf{q}} d\mathbf{p}d\mathbf{q}P_{imn}(\mathbf{k})P_{jrs}(\mathbf{k}) \times \\ &\quad Q_{ns}(\mathbf{p},t,s)Q_{mr}(\mathbf{q},t,s) \end{aligned} \tag{29}$$

In Eqs. (28) and (29),

$$\begin{aligned} P_{imn}(\mathbf{k}) &= k_m P_{in}(\mathbf{k}) + k_n P_{im}(\mathbf{k}) \\ P_{im}(\mathbf{k}) &= \delta_{im} - k^{-2}k_i k_m \end{aligned}$$

and Ω_{pq} is the antisymmetric rotation matrix.

The solution of these equations in complete generality is not known. A natural perturbation theory treats the rotation terms as small, and perturbs about an isotropic turbulent state. This approach is adopted by Shimomura and Yoshizawa (1986), who derive a TSDIA theory in which inhomogeneity and rotation are both described by small parameters.

A complementary limit is also of interest. Namely, in the response equation, balance the time derivative by the rotation term, and treat the eddy damping as small. This linear theory of the response equation treats strongly rotating turbulence as a case of *weak turbulence* (Zakharov *et al.*, 1992) in which nonlinear decorrelation of Fourier modes is dominated by linear dispersive decorrelation (Waleffe, 1993). The result is conveniently expressed in terms of the *Craya-Herring* basis

$$\begin{aligned}
\mathbf{e}^{(1)}(\mathbf{k}) &= \mathbf{k}\times\boldsymbol{\Omega}/\mid \mathbf{k}\times\boldsymbol{\Omega} \mid \\
\mathbf{e}^{(2)}(\mathbf{k}) &= \mathbf{k}\times(\mathbf{k}\times\boldsymbol{\Omega})/\mid \mathbf{k}\times(\mathbf{k}\times\boldsymbol{\Omega}) \mid
\end{aligned}$$

or the equivalent helical mode basis used, for example, by Waleffe (1993) and Cambon and Jacquin (1989), and the corresponding tensors

$$\begin{aligned}
\xi^0_{ij} &= e^{(1)}_i e^{(2)}_j - e^{(1)}_j e^{(2)}_i \\
\xi^1_{ij} &= e^{(1)}_i e^{(2)}_j + e^{(1)}_j e^{(2)}_i \\
\xi^2_{ij} &= e^{(1)}_i e^{(1)}_j - e^{(2)}_i e^{(2)}_j \\
\xi^3_{ij} &= e^{(1)}_i e^{(1)}_j + e^{(2)}_i e^{(2)}_j
\end{aligned} \tag{30}$$

Note that $\xi^3_{ij} = P_{ij}(\mathbf{k})$.

In this approximation, the leading order solution of Eq. (26) is

$$G_{ij}(\mathbf{k},\tau) = \{\cos(\Theta\tau)P_{ij}(\mathbf{k}) + \sin(\Theta\tau)\xi^0_{ij}(\mathbf{k})\}H(\tau) \tag{31}$$

where

$$\Theta(\mathbf{k}) = \mathbf{k}\cdot\boldsymbol{\Omega}/k$$

$\tau = t - s$ is time difference, and H is the unit step function.

Let us adopt the *fluctuation-dissipation* hypothesis relating the two-time correlation function to the response function and single-time correlation function

$$Q_{ij}(\mathbf{k},\tau) = G_{im}(\mathbf{k},\tau)Q_{mj}(\mathbf{k}) + G_{jm}(\mathbf{k},-\tau)Q_{mi}(\mathbf{k}) \tag{32}$$

Conditions under which this approximation is reasonable are discussed by Woodruff (1992, 1994). Substituting Eqs. (31) and (32) in Eq. (27) suggests that the single-time correlation function should take the general form containing all of the ξ tensors of Eq. (30)

$$Q_{ij}(\mathbf{k}) = \sum_{0\le p\le 3} Q^p(\mathbf{k})\xi^p_{ij}(\mathbf{k}) \tag{33}$$

which is equivalent to the form of the correlation function noted by Cambon and Jacquin (1989). But substituting Eq. (33) in Eq. (32), we find instead that the symmetry

$$Q_{ij}(\mathbf{k},\tau) = Q_{ji}(\mathbf{k},-\tau)$$

leads at lowest order in the weak turbulence approximation to the simpler expression

$$Q_{ij}(\mathbf{k},\tau) = \sin(\Theta\tau)Q(\mathbf{k})\xi^0_{ij}(\mathbf{k}) + \cos(\Theta\tau)Q(\mathbf{k})\xi^3_{ij}(\mathbf{k}) \tag{34}$$

This simplification is a consequence of the approximate treatment of time correlations; refinement of the approximation will recover the full expression Eq. (33).

The DIA inertial range energy balance (Kraichnan, 1971), which states that a steady state with constant energy flux exists, is

$$\begin{aligned}\varepsilon &= [I^+ - I^-]P_{imn}(\mathbf{k})\int_0^\infty d\tau\ 2P_{\mu rs}(\mathbf{p})G_{m\mu}(\mathbf{p},\tau)Q_{ns}(\mathbf{q},\tau)Q_{ir}(\mathbf{k},\tau)\\ &\quad -P_{jrs}(\mathbf{k})G_{ij}(\mathbf{k},\tau)Q_{ns}(\mathbf{p},\tau)Q_{mr}(\mathbf{q},\tau)\end{aligned} \tag{35}$$

where the integration operators in Eq. (35) are defined by

$$\begin{aligned} I^+(k_0) &= \int_{k\geq k_0} d\mathbf{k}\int_{p,q\leq k_0}\delta(\mathbf{k}-\mathbf{p}-\mathbf{q})d\mathbf{p}d\mathbf{q}\\ I^-(k_0) &= \int_{k\leq k_0} d\mathbf{k}\int_{p,q\geq k_0}\delta(\mathbf{k}-\mathbf{p}-\mathbf{q})d\mathbf{p}d\mathbf{q}\end{aligned}$$

Substituting Eqs. (31) and (34) in Eq. (35) shows that the time integrals in Eq. (35) will contain the factors

$$T(\mathbf{k},\mathbf{p},\mathbf{q}) = \Omega^{-1}\delta(\pm\mathbf{p}\cdot\boldsymbol{\Omega}/p\Omega \pm \mathbf{q}\cdot\boldsymbol{\Omega}/q\Omega \pm \mathbf{k}\cdot\boldsymbol{\Omega}/k\Omega) \tag{36}$$

where

$$\Omega = (\Omega_{pq}\Omega_{pq})^{1/2}$$

Thus, wave-vector integrations take place over resonant triads only (Waleffe, 1993) and these integrals scale as Ω^{-1} (Zhou, 1995).

Introduce the *ansatz*

$$Q(\mathbf{k}) = k^{-\alpha-2}f(\boldsymbol{\Omega}\cdot\mathbf{k}/k\Omega) \tag{37}$$

In view of Eq. (36), the energy balance Eq. (35) requires the scaling $\alpha = 2$ and the proportionality $f \sim \sqrt{\varepsilon\Omega}$. Define the shell-averaged energy spectrum

$$E(k) = \oint d\mathbf{k}\ Q(\mathbf{k}) \tag{38}$$

then (Zhou, 1995)

$$E(k) = C_K^\Omega \sqrt{\varepsilon\Omega} k^{-2} \tag{39}$$

It is easily verified that the flux integral Eq. (35) converges for an infinite k^{-2} inertial range. Thus, the scaling law of Eq. (39) is the scaling of a solution of the DIA equations of motion in the weak turbulence approximation.

The anisotropy of rotating turbulence implies that the eddy viscosity is a tensor defined so that

$$< u_i u_j > = \; -\nu_{ijrs} \frac{\partial U_r}{\partial X_s}$$

where TSDIA gives

$$\nu_{ijrs} = \frac{1}{4} \int d\mathbf{k} \int_0^\infty d\tau \; G_{ir}(\mathbf{k}, \tau) Q_{js}(\mathbf{k}, \tau) + (ij)(rs) \tag{40}$$

and the parentheses indicate index symmetrization. Using the general form of the single-time correlation function Eq. (33) and performing the time integration,

$$\nu_{ijrs} = \int d\mathbf{k} \; \frac{1}{\Omega} \delta(\mathbf{k} \cdot \mathbf{\Omega}/k\Omega) \sum_{\lambda,\mu} \xi_{ir}^{\lambda}(\mathbf{k}) \xi_{js}^{\mu}(\mathbf{k}) Q^{\lambda}(\mathbf{k}) \tag{41}$$

The delta function dependence of the integrand of Eq. (41) implies that this integrand is nonvanishing only for vectors perpendicular to the rotation axis. This fact originates in the resonance condition noted earlier: since TSDIA treats the effect of small turbulent scales on large scales characterizing the mean flow, it evaluates turbulence transport properties in the *distant interaction approximation* (Kraichnan, 1987), in which interactions among modes with wavevectors $\mathbf{k}, \mathbf{p}, \mathbf{q}$ are all treated as distant interactions for which $k/p, k/q \to 0$. In this limit, only planar triads such that $\mathbf{k} \cdot \mathbf{\Omega} = \mathbf{p} \cdot \mathbf{\Omega} = \mathbf{q} \cdot \mathbf{\Omega} = 0$ can satisfy the resonance condition. This states the "two-dimensionalization" of energy transfer in strongly rotating turbulence, an observation which is consistent with the more comprehensive theory of Mahalov and Zhou (1996).

The isotropic contribution to the eddy viscosity is characterized by the scalar

$$\nu = C_\nu^\Omega \frac{K}{\Omega} \tag{42}$$

For a subgrid-scale model, instead

$$\nu(\Delta) = C_S^\Omega \sqrt{\varepsilon/\Omega} \Delta \tag{43}$$

Closing the dissipation rate ε by the resolved production

$$\varepsilon = \frac{1}{2}\nu(\Delta)S_{ij}^{<}S_{ij}^{<} = \nu(\Delta)(S^{<})^2 \tag{44}$$

where $S_{ij}^{<}$ is the resolved fluctuating strain rate, leads to the *Smagorinsky model for strongly rotating turbulence*

$$\nu(\Delta) = C_{S2}^{\Omega}(S^{<})^2\Delta^2/\Omega \tag{45}$$

Thus, the subgrid scale viscosity is suppressed by rotation. The eddy viscosity has the same Δ^2 dependence on grid size as the standard Smagorinsky model for non-rotating turbulence.

It is important to distinguish clearly between this weak turbulence approximation and a fully linear theory like the rapid distortion theory. The weak turbulence approximation ignores nonlinearity only in the response equation but retains the nonlinearity of the energy flux condition. The weak turbulence approximation is rational because the corrections due to nonlinearity of the response equation can be systematically evaluated. Namely, substituting Eqs. (31) and (39) in Eq. (28) for the eddy damping factor and taking the distant interaction limit,

$$\eta \sim k^2 \int_k^\infty dp\, \frac{1}{\Omega}\sqrt{\varepsilon\Omega}p^{-2} \sim k\sqrt{\varepsilon/\Omega} \tag{46}$$

The corrections to the time scale and energy spectrum therefore have the form

$$\begin{aligned} T &\sim \frac{1}{\Omega}\{1 + O(\Omega^{-3/2})\}^{-1} \\ E &\sim \sqrt{\varepsilon\Omega}k^{-2}\{1 + O((k^2\varepsilon/\Omega^3)^{1/2}\} \end{aligned} \tag{47}$$

The low rotation rate expansion of Shimomura and Yoshizawa (1986) gives the complementary expansions in positive powers of Ω,

$$\begin{aligned} T &\sim \varepsilon^{-1/3}k^{-2/3}\{1 + O(\Omega)\} \\ E &\sim \varepsilon^{2/3}k^{-5/3}\{1 + O(\Omega/\varepsilon^{1/3}k^{2/3}\} \end{aligned} \tag{48}$$

4. The Effect of Rotation on Weakly Compressible Turbulence

In the presence of rotation, the centrifugal and Coriolis terms must be added to the momentum equations, but the entropy and pressure equations remain unchanged. The modified momentum equations have the form

$$\dot{u}_i + 2\Omega_{ip}u_p - (1 + \frac{\rho}{R})\Omega^2 x_i = \cdots \tag{49}$$

where terms independent of rotation have not been explicitly written. The mean contribution to the centrifugal term should be added to the mean momentum equation. The fluctuating component has both solenoidal and potential parts, which will appear in the equations for w_i and ϕ.

The Coriolis term must be treated by the Helmholtz decomposition Eq. (12) followed by a second decomposition into solenoidal and potential parts,

$$\begin{aligned} \Omega_{ip}u_p &= [P_{ij}(\mathbf{k}) + P^*_{ij}(\mathbf{k})]\Omega_{jp}[\mathbf{u}^\infty_p + w_p + ik^{-1}k_p\phi] \\ &= P_{ij}(\mathbf{k})\Omega_{jp}\mathbf{u}^\infty_p + P_{ip}\Omega_{pj}w_p + ik^{-1}\Omega_{ip}k_p\phi \\ &+ P^*_{ip}\Omega_{jp}\mathbf{u}^\infty_p + P^*_{ip}\Omega_{pj}w_p \end{aligned} \tag{50}$$

The first term in Eq. (50) is the usual Coriolis term which causes the modified statistically steady state described in Sect. 3. The second term is the analogous Coriolis term for the compressible field w_i. The third term is transverse; therefore, it couples the field ϕ to the w_i equation. This coupling does not occur in non-rotating turbulence. The fourth term shows that the incompressible field is a source for the compressible potential field ϕ. Finally, the fifth term couples the field w_i to the ϕ equation. Symmetry breaking by rotation makes this coupling possible. Note the important fact that ϕ cannot be coupled linearly to itself through rotation.

In the analysis of non-rotating weakly compressible turbulence, it was not possible to treat all couplings exactly; likewise, the present analysis will treat all rotation-dependent couplings perturbatively, except for the rotational self-couplings of the solenoidal fields w and $\mathbf{u}^\infty$. This defines the *diagonal approximation* for rotational effects

$$\begin{aligned} \dot{\mathbf{u}}^\infty_i + 2P_{ij}\Omega_{jp}\mathbf{u}^\infty_p &= \cdots \\ \dot{w}_i + 2P_{ij}\Omega_{jp}w_p &= -ik^{-1}\Omega_{ip}k_p\phi + \cdots \\ \dot{\phi} &= P^*_{ip}\Omega_{jp}(\mathbf{u}^\infty_p + w_p)\cdots \end{aligned} \tag{51}$$

where dots indicate centrifugal and rotation-independent terms. In this approximation, it is appropriate to replace the Kolmogorov steady state for the basic incompressible field $\mathbf{u}^\infty$ by the k^{-2} steady state for strong rotation described in Sect. 3.

5. Evaluation of the Transport Coefficients

5.1. THE ENTROPY DIFFUSIVITY

Since the entropy equation is not altered by rotation, G^{ss} remains the response function of a passive scalar. The usual Reynolds analogy, which postulates the proportionality of the time correlations of the velocity and passive scalar fields leads, following the derivation of Eq. (42) to

$$\nu^{ss} = C^{\Omega}_{ss}\frac{K}{\Omega} \tag{52}$$

in the strong rotation limit.

5.2. THE QUANTITIES Π AND $\hat{R}$

According to Eqs. (8) and (10), the evaluation of these quantities requires the response function $G^{\phi p}$ and the two-time pressure correlation in rotating incompressible turbulence. The perturbation theory of RE begins by observing that G^{ww} is the response function of a passive vector field. Whereas this could be asserted exactly in non-rotating turbulence, the couplings which occur in Eq. (51) restrict this assertion to the lowest order in the diagonal approximation for rotation effects.

Thus, we assume that to lowest order G^{ww} takes the same form as the response function in Eq. (31)

$$G^{ww}_{ij}(\mathbf{k},\tau) = \{\cos(\Theta\tau)P_{ij}(\mathbf{k}) + \sin(\Theta\tau)\xi^{0}_{ij}(\mathbf{k})\}H(\tau) \tag{53}$$

As in RE, the damping functions except for $\eta^{\phi\phi}$ all vanish to this order. We must first check that the integral in Eq. (15) which defines this function converges for the k^{-2} spectral scaling. As usual, this convergence check requires the evaluation of the geometric factors in this integral in the limits $p, q \to \infty$. This factor has the form

$$\begin{aligned}
& (q_m\delta_{in} + k_m\delta_{in})(k_r\delta_{jp} - p_j\delta_{pr})\xi^{\lambda}_{mr}(\mathbf{p})\xi^{\mu}_{np}(\mathbf{q}) \\
= \;& k_m k_r \xi^{\lambda}_{mr}(\mathbf{p})\xi^{\mu}_{ij}(\mathbf{q}) - k_m p_j \xi^{\lambda}_{mp}(\mathbf{p})\xi^{\mu}_{ip}(\mathbf{q}) \\
+ \;& k_n k_r \xi^{\lambda}_{ir}(\mathbf{p})\xi^{\mu}_{nj}(\mathbf{q}) - k_n p_j \xi^{\lambda}_{ip}(\mathbf{p})\xi^{\mu}_{np}(\mathbf{q})
\end{aligned} \tag{54}$$

The second and fourth terms, which power counting would suggest cause logarithmic divergence in Eq. (15), instead vanish by symmetry. The remaining terms are of order p^{-2} and therefore converge for large p. The infrared divergence when $p \to 0$ is explained by the sweeping effect on Eulerian time correlations to which Kraichnan (1964) first called attention. Rather than evaluate the transport coefficients in a Lagrangian statistical theory (Kraichnan, 1965; Kaneda, 1981) in which this divergence does not occur, we shall simply regularize the Eulerian theory by integrating over the range $p, q \geq k$ only (Kraichnan, 1963). This regularization leads finally to the result

$$\eta^{\phi\phi}(\mathbf{k},\tau) = C^{\Omega}_{\phi\phi}\sqrt{\varepsilon/\Omega}k \tag{55}$$

which can be compared to the correction term computed in Eq. (46).

The evaluation of the two-time pressure correlation function is actually simpler in the weak turbulence approximation for rotating turbulence that it is for non-rotating turbulence because the resonance condition and even time parity of time correlation functions force the time dependence $\cos(\Theta\tau)$. As in the non-rotating case, convergence of the single-time correlation under the quasi-normal approximation appropriate to the direct interaction approximation is easily established, and we conclude that

$$Q^p(k,\tau) = C_{B,\Omega}\cos(\Theta\tau)\varepsilon\Omega k^{-3}/4\pi k^2 \tag{56}$$

It is now straightforward to verify the calculation

$$\begin{aligned}\Pi/\gamma &= \frac{1}{2}\int d\mathbf{k}\int_0^\infty d\tau\, \exp(-\sqrt{\theta}k\tau)\times \\ &\quad [\sin(Vk+\Omega)\tau + \sin(Vk-\Omega)\tau]Q^p(k) \\ &= \int_0^\infty 4\pi dk\, 2\frac{\varepsilon\Omega}{V}k^{-2}\frac{Vk^3(\theta)+V^2)-Vk\Omega^2}{[\theta+V^2]^2k^4+2(\theta-V^2)\Omega^2k^2+\Omega^4}\end{aligned} \tag{57}$$

where

$$V = \sqrt{c^2-\theta}$$

and

$$\theta = C^{\Omega}_{\phi\phi}\frac{\varepsilon}{\Omega}$$

The integral in Eq. (57) could be evaluated in closed form. However, its limits for large and small wavenumber k are easily evaluated directly. Assume that the k^{-2} spectrum will be cut off at wavenumber k_0, which is either the inverse integral scale, or inverse filter size. If the cutoff scale k_0 is large, then approximately

$$\Pi/\gamma = 4\pi\int_{k_0}^\infty C(Vk)^{-1}\frac{1}{c}\varepsilon\Omega k^{-2} = C'\frac{\varepsilon\Omega}{c^2}k_0^{-2} \tag{58}$$

Setting k_0 to the inverse integral scale leads to the single-point model

$$\Pi/\gamma = C^{\Omega}_{\Pi>}\frac{K^2}{c^2} \tag{59}$$

These results coincide with those of RE except that the energy spectrum and total kinetic energy are determined from the k^{-2} spectrum instead of from the Kolmogorov spectrum. However, the large scales contribute a different limit, namely,

$$\Pi/\gamma = 4\pi\int_{k_0}^{k^*}\frac{Vk}{c\Omega^2}\varepsilon\Omega K^{-2} = C^{\Omega}_{\Pi<}\frac{\varepsilon}{\Omega}\log(k^*/k_0) \tag{60}$$

The scale k^* is the scale at which the high wavenumber limit Eq. (58) applies; its value is calculated below. Note the logarithmic contribution.

The approximation Eq. (58) applies to small scales which are relatively insensitive to rotation. Whereas the theory of weak compressibility requires the condition $K << c^2$, the limit Eq. (58) requires the much stronger condition

$$K << c\sqrt{\varepsilon/\Omega} \tag{61}$$

At very high rotation rates, a range of large scales such that $\Omega >> ck$ can exist. Such scales are described instead by the limit Eq. (60). The transition between spectral regions is determined by the parameter

$$k^* = \frac{\Omega}{c} \tag{62}$$

Since Eq. (60) describes a condition in which compressibility effects have been reduced, we conclude that rotation tends to counteract the effects of compressibility. The single-point parameter which discriminates between regimes is the ratio

$$\frac{K}{c\sqrt{\varepsilon/\Omega}} = \frac{K^{1/2}}{c}\sqrt{\Omega K/\varepsilon} = \zeta^{1/2}M_t \tag{63}$$

Since rotation reduces the energy transfer ε, large values of the parameter ζ are possible in rotating compressible turbulence.

The evaluation of the coefficient $\hat{R}$ proceeds similarly. However, it should be noted that the anisotropy of rotating turbulence actually makes this quantity a tensor, so that the mean equations of motion contain the pressure term

$$[\frac{1}{R}\delta_{ij} - \frac{1}{\hat{R}}\alpha_{ij}]\frac{\partial P}{\partial X_j}$$

instead of the isotropic pressure term of Eq. (3). The possibility that sound waves in the mean velocity field become dispersive as a result of the direction dependence of the effective sound speed deserves further investigation. For the present, the sound speed will be treated as a scalar. By comparing Eqs. (8) and (10), it is evident that $\hat{R}$ is found from the previous calculation by multiplying the integrand in Eq. (8) by $k\sqrt{\varepsilon/\Omega}$. The small scale limit becomes

$$\frac{1}{\hat{R}} = C^{\Omega}_{R>}\frac{\varepsilon\Omega k_0^{-2}}{c^2}\frac{k_0}{\sqrt{\varepsilon\Omega}} = C^{\Omega}_{R>}\frac{\sqrt{\varepsilon\Omega}k_0^{-1}}{c^2} \tag{64}$$

and the large scale limit becomes

$$\frac{1}{\hat{R}} = C^{\Omega}_{R<}\frac{\varepsilon}{\Omega}\frac{k_0}{\sqrt{\varepsilon\Omega}} = C^{\Omega}_{R<}\sqrt{\varepsilon/\Omega^3}k_0 \tag{65}$$

The single-point quantities corresponding to Eqs. (64) and (65) are simply

$$\frac{1}{\hat{R}} = \begin{cases} M_t^2 & \text{for } k_0 \to \infty \\ 1/\zeta & \text{for } k_0 \to 0 \end{cases} \tag{66}$$

Eqs. (65) and (66) show that the compressibility correction $1/\hat{R}$ actually vanishes for very large rotation rates, again suggesting that rotation tends to counteract compressibility effects.

5.3. THE PRESSURE AND BULK VISCOSITIES

Like $\hat{R}$, these quantities are tensors in rotating turbulence, but only scalar values will be computed here. Eq. (6) shows that the pressure diffusivity ν^{pp} is the sum of two terms. For the first term, in the low turbulent Mach number limit,

$$\begin{aligned} \nu_1^{pp} &= \int_{k_0}^{\infty} dk \, \frac{\eta^3/4 + 2\eta S^2}{(\eta/2)^4 + 2(\eta/2)^2(S^2 + \Omega^2) + (S^2 - \Omega^2)^2} \sqrt{\varepsilon\Omega} k^{-2} \\ &\sim \int_{k_0}^{\infty} dk \, \sqrt{\varepsilon/\Omega}(c^2 - 3\theta/4)^{-1} k^{-1} \sqrt{\varepsilon\Omega} k^{-2} \\ &= C_1 \frac{K^2}{\Omega(c^2 - 3\theta/4)} \end{aligned} \tag{67}$$

where η is the incompressible damping factor of Eq. (25) This expression applies to both large and small values of the cutoff scale k_0. The second term is evaluated like the entropy diffusivity as

$$\nu_2^{pp} = \frac{\varepsilon\Omega}{c^2} k_0^{-2} = C_2 \frac{K^2}{\Omega c^2} \tag{68}$$

In summing these contributions, care must be taken that cancellations between these terms do not change the limits of the transport coefficients. In the absence of such cancellations,

$$\nu^{pp} = C_{pp}^{\Omega} \frac{K}{c^2} \frac{K}{\Omega} \tag{69}$$

This result can be compared with the non-rotating result in RE. Rotation has the same effect that it has on the incompressible eddy viscosity: the turbulence time scale K/ε is replaced by $1/\Omega$.

A similar calculation leads to an explicit expression for the bulk viscosity,

$$\nu^{uu} = C_{\Omega}^{uu} \frac{K}{\Omega} \tag{70}$$

TABLE 2. **Main Results**

	Single-point model	General subgrid model	Smagorinsky model
ν^{pp}	$M_t^2 K/\Omega$	$\varepsilon\Delta^2/c^2$	$(S^<)^3\Delta^4/c^2$
ν^{uu}	K/Ω	$\sqrt{\varepsilon/\Omega}\Delta$	$(S^<)^2\Delta^2/\Omega$
Π (large k)	KM_t^2	$\varepsilon\Omega\Delta^2/c^2$	$\Omega(S^<)^3\Delta^4/c^2$
Π (small k)	ε/Ω	$\varepsilon\Delta^0/\Omega$	$(S^<)^3\Delta^2/\Omega$
$1/\hat{R}$ (large k)	M_t^2	$\sqrt{\varepsilon\Omega}\Delta/c^2$	$(S^<)^{3/2}\Omega^{1/2}\Delta^2/c^2$
$1/\hat{R}$ (small k)	$\varepsilon/K\Omega$	$\sqrt{\varepsilon/\Omega^3}\Delta^{-1}$	$(S^<)^{3/2}\Omega^{-3/2}\Delta^0$

Note that in RE, this quantity was of order M_t^0. Again, the consequence of rotation is the simple replacement of the incompressible turbulence time scale by $1/\Omega$.

The results of this section are summarized in Table 2. The format follows Table 1, except that both large and small scale limits are given for the equilibrium properties Π and $\hat{R}$.

6. The Equation of State of Rotating Weakly Compressible Turbulence

Beginning with the first mixing length models, turbulence modeling has always postulated that turbulent fluctuations modify fluid transport properties by replacing thermal fluctuations with scale dependent macroscopic turbulent fluctuations. Analyses of compressible turbulence (Chandrasekhar, 1952, Staroselsky *et al.*, 1990) have noted analogous turbulence effects on thermal equilibrium properties: velocity fluctuations lead to a modified equation of state of compressible turbulence. Thus, Eqs. (6), (8) and (10) show that fluctuations modify γ and lead to changes of the effective pressure and density represented by the terms Π and $\hat{R}$.

In rotating compressible turbulence, the picture is somewhat more complex. The transport coefficients ν are all modified like the incompressible eddy viscosity: the turbulence time scale K/ε is replaced by the rotation time scale $1/\Omega$. But thermal equilibrium properties are modified so that two different regimes are possible: a regime dominated by incompressible fluctuations, in which the sound speed is modified as it is in non-rotating turbulence, and a rotation dominated regime in which the sound speed enhancement is proportional to $\sqrt{\varepsilon/\Omega}$ instead. As noted earlier, this limit indicates a reduction of compressibility effects on large scales by rotation.

7. Conclusions

Yoshizawa's TSDIA formalism provides a systematic approach to the complex coupled field problem of rotating weakly compressible turbulence. In this problem, the appearance of two natural dimensionless parameters would frustrate attempts to derive appropriate models heuristically. In particular, although the effects of rotation on turbulent transport coefficients could perhaps be guessed, the modified equation of state in rotating weakly compressible turbulence is less intuitively obvious.

The main defect of this calculation is the assumption of a diagonal approximation for rotation effects, although like RE, we do not make a diagonal approximation for compressibility effects. We should attempt either to verify that the non-diagonal terms are small, or look for a more sophisticated approximation.

A novel feature of the present analysis is the use of a non-Kolmogorov steady state as the reference state in a TSDIA calculation. The introduction of such steady states expands the power and utility of TSDIA to a wider range of problems.

References

Cambon, C. and Jacquin, L., 1989. "Spectral approach to non-isotropic turbulence subjected to rotation," *J. Fluid Mech.* **202**, p. 295.

Chandrasekhar, S., 1951. "The fluctuation of density in isotropic turbulence," *Proc. Roy. Soc. London A* **210**, p. 18.

Kaneda, Y., 1981. "Renormalized expansions in the theory of turbulence with the use of the Lagrangian position function," *J. Fluid Mech.* **107**, p. 131.

Kraichnan, 1959. "The structure of turbulence at very high Reynolds number," *J. Fluid Mech.* **5**, p. 497.

Kraichnan, 1964. "Kolmogorov's hypotheses and Eulerian turbulence theory," *Phys. Fluids* **7**, p. 1723.

Kraichnan, R.H., 1965. "Lagrangian-history closure approximation for turbulence," *Phys. Fluids* **8**, p. 575.

Kraichnan, R.H., 1971. "Inertial range transfer in two and three dimensional turbulence," *J. Fluid Mech.* **47**, p. 525.

Kraichnan, R.H., 1987. "An interpretation of the Yakhot-Orszag turbulence theory," *Phys. Fluids* **30**, p. 2400.

Mahalov, A. and Zhou, Y., 1996. "Analytical and phenomenological studies of rotating turbulence," *Phys. Fluids* **8**, p. 2138.

Rubinstein, R. and Erlebacher, G., 1997. "Transport coefficients in weakly compressible turbulence," *Phys. Fluids* **9**.

Shimomura, Y. and Yoshizawa, A., 1986. "Statistical analysis of anisotropic turbulent viscosity in a rotating system," *J. Phys. Soc. Japan* **55**, p. 1904.

Staroselski, I., Yakhot, V., Orszag, S.A., and Kida, S., 1990. "Long time large scale properties of a randomly stirred compressible fluid," *Phys. Rev. Lett.* **65**, p. 171.

Waleffe, F., 1993. "Inertial transfers in the helical decomposition," *Phys. Fluids* **5**, p. 677.

Woodruff, S.L., 1992. "Dyson equation analysis of inertial-range turbulence," *Phys. Fluids A* **5**, p. 1077.

Woodruff, S.L., 1994. "A similarity solution for the direct interaction approximation and its relationship to renormalization group analyses of turbulence," *Phys. Fluids* **6**, p. 3051.

Yakhot, V. and Orszag, S.A., 1986. "Renormalization group analysis of turbulence," *J. Sci. Comput.* **1**, p. 3.

Yoshizawa, A., 1984. "Statistical analysis of the deviation of the Reynolds stress from its eddy viscosity representation," *Phys. Fluids* **27**, p. 3177.

Yoshizawa, A., 1995. "Simplified statistical approach to complex turbulent flows and ensemble-mean compressible turbulence modeling," *Phys. Fluids* **7**, p. 3105.

Yoshizawa, A., Tsubokura, M., Kobayashi, T., and Taniguchi, M., 1996. "Modeling of the dynamic subgrid scale viscosity in large eddy simulation," *Phys. Fluids* **8**, p. 2254.

Zakharov, V.E., L'vov, V.S., and Falkovich, G., 1992. Kolmogorov Spectra of Turbulence I, Springer.

Zhou, Y., 1995. "A phenomenological treatment of rotating turbulence," *Phys. Fluids* **7**, p. 2092.

DEVELOPMENT OF A TURBULENCE MODEL FOR FLOWS WITH ROTATION AND CURVATURE

S. THANGAM AND X. WANG
Stevens Institute of Technology
Castle Point on the Hudson, Hoboken, New Jersey

AND

Y. ZHOU
IBM Research Division, T.J. Watson Research Center
Yorktown Heights, New York and
Institute for Computer Applications in Science and Engineering
NASA Langley Research Center, Hampton, Virginia

Abstract. A generalized eddy viscosity model is formulated by using the rotation modified energy spectrum. Rotation and mean shear effects are directly included in the eddy viscosity without the use of the local equilibrium assumption. Since the model is of a general form, additional modifications to the eddy viscosity for non-equilibrium effects are not needed. The formulation also includes modeling vortex stretching and viscous destruction terms of the dissipation rate equation based on the limit of rotating isotropic turbulence at high Reynolds numbers. Since the rotation modified energy spectrum includes the contribution of rotation effects on the dissipation and structure of turbulence, the model coefficients in the dissipation rate equation are also modified. The model is shown to reproduce the rotation effects in isotropic turbulence. The implications of the model and its general applicability for turbulent flows are also addressed.

1. Introduction

Traditional two-equation models of eddy viscosity type (e.g. $k - \epsilon$ model) cannot account for the effects of anisotropy of turbulent stresses, system rotation or of streamline curvature. On the other hand, the simplicity of two-equation models makes them attractive from the point of implemen-

M. D. Salas et al. (eds.), Modeling Complex Turbulent Flows, 349–360.

tation. This has also motivated considerable work in the development of "anisotropic eddy viscosity" models that can capture the effects of anisotropy in turbulent stresses, streamline curvature and rotation. Examples of such two-equation models include those that rely on the analogy with non-Newtonian flows (for a review, see, Speziale, 1991), the two-scale DIA model of Yoshizawa (1984), the small parameter renormalization group (ϵ-RNG) models of Rubinstein and Barton (1990) and Yakhot *et al.* (1992) and the recursion Renormalization Group theory based models of Zhou *et al.* (1994). These models are capable of predicting the anisotropy of normal stresses in shear flows; but they are not effective in the prediction of rotating flows since they do not contain any mechanism for the inclusion of streamline curvature and rotation. On the other hand, explicit algebraic stress models (ASM) which are obtained for a hierarchy of second-order closures by using the standard local equilibrium hypothesis, have had some success in capturing rotation effects (see, for example, Gatski and Speziale, 1993; which is developed using the procedure suggested by Pope, 1975). But the primary difficulty with such explicit models are due to the use of local equilibrium hypothesis.

Another issue at hand involves the dissipation (ϵ) equation. In its standard form, the ϵ equation does not depend explicitly on rotation. However, it is well established that a consequence of the effect of rotation on isotropic turbulence is the inhibition of the energy transfer from large to small scales, which leads to a reduction of the dissipation rate. Earlier attempts by Hanjalić *et al.* (1980), Bardina *et al.* (1985) were based on phenomenological considerations. For the decay of rotating isotropic turbulence, the model by Hanjalić *et al.* (1980) does not yield any improvement in predictions, while the model of Bardina *et al.* (1988) presents difficulty at high rotation rates. In this regard, recent work by Girimaji (1995, 1996) which utilize a weakly non-local hypothesis for the development of explicit algebraic stress models that are suitable for curved flows and that by Speziale (1997) which involves the development of an explicit algebraic stress model based on phenomenological considerations show considerable promise.

In the present paper, an alternative method that avoids the use of the local equilibrium assumption and which at the same time explicitly accounts for the effects of rotation rate is employed to derive the eddy viscosity in rotating flows. The model is shown to accurately predict the decay of rotating isotropic turbulent flows.

2. Model Development

2.1. ROTATION CORRECTED ENERGY SPECTRUM

In the inertial range of isotropic turbulence at high Reynolds numbers, assumptions of local cascade and dimensional analysis yield (Kraichnan, 1965; Zhou, 1995)

$$\epsilon = Bt_3[E(k)]^2k^4, \tag{1}$$

where t_3 is the correlation time scale of nonlinear triadic interaction, $E(k)$ is the three dimensional energy spectrum, and B is a constant.

For nonrotating decay of isotropic turbulence at high Reynolds numbers, the only time scale available is the turbulent nonlinear time scale t_{nl} (which corresponds to the eddy turnover time, $[E(k)k^3]^{-1/2}$). Thus, when $t_3 = t_{nl}$, (1) will yield the Kolmogorov spectrum

$$E(k) = C_k\epsilon^{2/3}k^{-5/3}. \tag{2}$$

On the other hand, rotating isotropic turbulence is characterized by two disparate timescales — a turbulent nonlinear time scale and a time scale associated with rotation rate Ω. At high rotation rates, t_3 is directly proportional to the short timescale $1/\Omega$ (Squires *et al.*, 1994). Therefore, from (1) we obtain the energy spectrum for strong rotating turbulence

$$E(k) = C_\Omega(\Omega\epsilon)^{1/2}k^{-2}. \tag{3}$$

From (1),(2) and (3), we find that the constants B and C_Ω can be related to C_k:

$$B = C_k{}^{-3/2}, C_\Omega = B^{-1/2} = C_k{}^{3/4}.$$

For moderate rotation rates, t_3 depends on t_{nl} and $1/\Omega$, but the dependence of these two scales cannot be uniquely determined. A simple choice, which satisfies the limiting cases: $t_3 \rightarrow t_{nl}$ without rotation and $t_3 \rightarrow t_\Omega$ for high rotation rates, and one which ensures t_3 to be independent of the direction of rotation as $\Omega \rightarrow 0$ ($\frac{dt_3}{d\Omega}|_{\Omega=0} \rightarrow 0$) is given by:

$$\frac{1}{t_3^2} = \frac{1}{t_{nl}^2} + \frac{1}{t_\Omega^2}, \tag{4}$$

Using equations (1)-(4), we obtain the generalized energy spectrum:

$$\epsilon = \frac{BE^{3/2}k^{5/2}}{\sqrt{1+\Omega^2/Ek^3}}. \tag{5}$$

It should be noted here that the cutoff wave number for rotation effects, $k_\Omega = C_a(\Omega^3/\epsilon)^{1/2}$ delimits the region of spectrum where the rotation effects are important. In the region where $k \ll k_\Omega$, equation (5) leads to the strong rotation energy spectrum $E(k) = C_\Omega(\Omega\epsilon)^{1/2}k^{-2}$, while for $k \gg k_\Omega$, the Kolmogorov inertial subrange $E(k) = C_k\epsilon^{2/3}k^{-5/3}$ is recovered. A schematic representation of this simplified spectrum (shown in Figure 1) is used in the present work for the development of the ϵ equation.

2.2. EDDY VISCOSITY

If one examines the energy transfer in nearly isotropic turbulence, it is seen that spectral energy flux $T(k)$ across the wave number k in the inertial subrange is affected mainly by local interaction (Zeman, 1994). Therefore, $T(k) \propto -S_{ij}(k')\langle u_i u_j(k'')\rangle \propto \nu_T(k'')S^2(k')$ where $S(k')$ is the turbulent strain rate for eddies of size $1/k'$, approximately ($k/3 < k' < k$ and $k'' = 3k'$). The quantities at k' and k'' are similarly related and $S(k') \propto [E(k)k^3]^{1/2} \propto t_{nl}{}^{-1}(k'')$. In the inertial subrange the spectral energy flux $T(k)$ across each wave number is constant and equal to the dissipation ϵ, therefore,

$$T(k) = \epsilon \propto \nu_T(k)E(k)k^3. \tag{6}$$

By utilizing (5) and the generalized energy spectrum, $\epsilon = \frac{BE^{3/2}k^{5/2}}{\sqrt{1+\Omega^2/Ek^3}}$, we obtain spectral eddy viscosity at a given k,

$$\nu_T(k) = \frac{(E/k)^{1/2}}{C_5\sqrt{1 + C_4/y(k)}}, \tag{7}$$

where C_4, C_5 are constants, $y = Ek^3/\Omega^2$. The parameter Ω^2/Ek^3 is the measure of damping effect of rotation on the local eddy viscosity.

The above expression for eddy viscosity is now applied for the energy-containing range of the spectrum. In this range the characteristic velocity, length and time scales of eddies at wave number k are $U_k = [E(k)k]^{1/2}$, $l_k = 1/k$, and $t_k = [E(k)k^3]^{-1/2}$, respectively. So that in (7), $(E/k)^{1/2} = \frac{Ek}{\sqrt{Ek^3}} \propto \frac{U_k^2}{t_k}, \frac{\Omega^2}{Ek^3} \propto \frac{\Omega^2}{t_k^2}$. Making the substitution, $t_k \rightarrow t_{nl}, U_k^2 \rightarrow K$, we have $(E/k)^{1/2} \propto \frac{K}{t_{nl}}, \frac{\Omega^2}{Ek^3} \propto \frac{\Omega^2}{t_{nl}{}^2}$. If we use K/ϵ as turbulent time scale of energy-containing eddies, we obtain

$$\nu_T(x) = \frac{C_6\frac{K^2}{\epsilon}}{\sqrt{1 + C_7(\frac{K\Omega}{\epsilon})^2}}. \tag{8}$$

The approach of Zhou (1995) involves the estimation of turbulent kinetic energy K past the cutoff wave number k_o by

$$K = \int_{k_o}^{\infty} E(k)dk = \frac{3}{2}C_k\frac{\epsilon^{2/3}}{k_o^{2/3}}, \tag{9}$$

Here, the Kolmogorov inertial range spectrum was used for $E(k)$ and this approach yields $C_7 = \frac{4}{9}C_k{}^{-3}$. In the present work, the low band value of C_k of 1.3 (the typical range of Kolmogorov constant C_k is 1.3 ~ 2.3) is used to estimate C_7.

To obtain eddy viscosity for general rotating flows where shear is present, the rotation rate may be replaced by $(W_{ij}W_{ij})^{1/2}$, where W_{ij} is the intrinsic mean vorticity, defined by

$$W_{ij} = \frac{1}{2}(\frac{\partial U_i}{\partial x_j} - \frac{\partial U_j}{\partial x_i}) + \alpha e_{mji}\Omega_m. \tag{10}$$

The constant $\alpha = 2 \sim 2.25$ is chosen to be consistent with linear stability analysis and LES (Speziale, 1991). Considering a homogeneous rotating shear flow with $S_{12} = R_{12} = \frac{1}{2}U_{1,2} = \frac{1}{2}S$, the linear stability analysis and LES results indicate the maximum turbulence amplification at $(\Omega/S)_{max} = 0.25$. The turbulent constitutive relation in the presence of rotation becomes

$$\nu_T = \frac{C_6\frac{K^2}{\epsilon}}{\sqrt{1+0.2W_{ij}W_{ij}}}, \tag{11}$$

where $W_{ij} = [\frac{1}{2}(\frac{\partial U_i}{\partial x_j} - \frac{\partial U_j}{\partial x_i}) + 2.25e_{mji}\Omega_m]\frac{K}{\epsilon}$.

The modeling constant C_6 deserves more attention. The strategy is to ensure that the fixed point of modeling equations match the experimentally determined fixed points for homogeneous flows. In a homogeneous shear flow, $S = \partial u/\partial y = \text{const.}$, the dimensionless shear parameter $\frac{SK}{\epsilon}$ achieves equilibrium value ($\frac{SK}{\epsilon}|_\infty = 5.6$) which is independent of the initial conditions (Speziale, 1991). Using the transport equations for K and ϵ in homogeneous shear flows, $\nabla^2 K = /\nabla^2\epsilon = 0$, $\tau_{ij} = -2\nu_T S_{ij}$ one can obtain

$$\frac{d\eta}{dt} = -\frac{C_6}{\sqrt{1+\beta\eta^2}}(C_{\epsilon 1} - 1)\eta^2 + (C_{\epsilon 2} - 1), \tag{12}$$

where $\eta = \frac{SK}{\epsilon}, S = (2S_{ij}S_{ij})^{1/2}$. The constant C_6 can be determined by solving $\frac{d\eta}{dt} = 0$ to match the fixed point $\eta_0 = \frac{SK}{\epsilon}|_\infty = 5.6$. Then C_6 and modeling coefficients $C_{\epsilon 1}$, $C_{\epsilon 2}$ are related by

$$\frac{C_6(C_{\epsilon 1} - 1)}{C_{\epsilon 2} - 1} = \frac{\eta_0^2}{\sqrt{1+0.1\eta_0^2}}. \tag{13}$$

Thus, for standard $K-\epsilon$ equation, with model coefficients, $C_{\epsilon 1} = 1.44, C_{\epsilon 2} = 1.83, C_\mu = 0.09, \sigma_K = 1, \sigma_\epsilon = 1.3$, the coefficient C_6 is 0.13. For the RNG based two-equation models (Zhou *et al.*, 1995; Thangam *et al.*, 1998) the coefficients are: $C^*_{\epsilon 1} = 1.42 - \frac{\eta(1-\eta/\eta_0)}{1+\beta\eta^2}(1 + 0.07\frac{P}{\epsilon}), C_{\epsilon 2} = 1.68, C_\mu = 0.085, \alpha_K = \alpha_\epsilon = 1.39$, which yields $C_6 = 0.11$.

2.3. THE DISSIPATION EQUATION

For turbulent flows at high Reynolds numbers (defined by using the turbulence Reynolds number, $R_T = K^2/\nu\epsilon$), the production and destruction of dissipation balance each other. Thus, in the transport equation for dissipation, the production and destruction terms are of $O(R_T{}^{1/2})$ providing a net $O(1)$ effect independent of the Reynolds number. In isotropic turbulence, the equations describing the evolution of turbulence reduce to:

$$\dot{K} = -\epsilon, \dot{\epsilon} = P_{\epsilon 4} - Y = -W. \tag{14}$$

When there is no rotation, the net balance between production and destruction, W in the budget of ϵ is a function of ϵ, K, and ν and becomes independent of ν at high Reynolds numbers. From dimensional analysis, the closure of W based on phenomenological consideration yields:

$$\dot{\epsilon} = -C_{\epsilon 2}\frac{\epsilon^2}{K}.$$

The solution of the above for isotropic turbulence results in a power-law decay,

$$q^2 \sim t^{-n}, \epsilon \sim nt^{-(n+1)},$$

the constant $C_{\epsilon 2}$ can be related to the exponent of decay law,

$$C_{\epsilon 2} = \frac{n+1}{n}.$$

The energy spectrum shown in Figure 2 corresponds to that of non-rotating isotropic turbulence. It is characterized by the Kolmogorov spectrum in the inertial subrange and Ak^s behavior in the low-wave number region. The corresponding value of the index $n = 6/5$ and the resulting value of $C_{\epsilon 2} = 11/6$ has been shown to be appropriate for a variety of engineering applications dominated by strong shear layers (Reynolds, 1976).

In the present work, we propose that from dimensional arguments, one simple form of W to account for rotation is:

$$W = C^*_{\epsilon 2}(\frac{K\Omega}{\epsilon})\frac{K^2}{\epsilon}, \tag{15}$$

where $C^*_{\epsilon 2}$ is the function of the new dimensionless group $\frac{K\Omega}{\epsilon}$, and approaches $C_{\epsilon 2} = 11/6$ at zero rotation.

The simplified rotation modified spectrum (Figure 1) will be used next to determine $C^*_{\epsilon 2}$ following the method proposed by Reynolds (1976) for the nonrotating case. The energy contained in our model spectrum is the sum of the contributions from the three segments of the spectrum. Let $k_0 \ll k_m < k_\Omega \ll k_d$); the corresponding description of the turbulent kinetic energy K is given by:

$$K = \int_0^\infty E(k)dk \tag{16}$$

$$= \int_0^{k_m} Ak^s dk + \int_{k_m}^{k_\Omega} C_\Omega(\Omega\epsilon)^{1/2}k^{-2}dk + \int_{k_\Omega}^{k_d} \alpha\epsilon^{2/3}k^{-5/3}dk \tag{17}$$

$$= \frac{A}{s+1}k_m^{s+1} - C_\Omega(\Omega\epsilon)^{1/2}(k_\Omega^{-1} - k_m^{-1}) + \frac{3}{2}\alpha\epsilon^{2/3}k_\Omega^{-2/3} \tag{18}$$

Here, k_m and k_Ω can be obtained by matching the three segments of the spectrum

$$k_m = [\frac{C_\Omega}{A}(\Omega\epsilon)^{1/2}]^{1/(s+2)},$$

$$k_\Omega = (\frac{C_\Omega}{C_k})^3(\frac{\Omega^3}{\epsilon})^{1/2},$$

thus

$$K = C_1(\Omega\epsilon)^{3/8} + C_2\epsilon\Omega^{-1}, \tag{19}$$

where constants

$$C_1 = \frac{A}{3}(C_\Omega/A)^{3/4} + C_\Omega(C_\Omega/A)^{-1/4},$$

$$C_2 = \frac{1}{2}C_k^{3/2}.$$

Substituting (19) in equations (14) and (15), and requiring that for zero rotation, as $\Omega \to 0, C^*_{\epsilon 2} \to 11/6$, and $C^*_{\epsilon 2} \to \frac{\frac{8}{3}\frac{K\Omega}{\epsilon}}{\frac{5}{3}C_2 + \frac{K\Omega}{\epsilon}}$ for strong rotation, we obtain:

$$C^*_{\epsilon 2} = \frac{C_3 C_{\epsilon 2} + \frac{8}{3}\frac{K\Omega}{\epsilon}}{C_3 + \frac{K\Omega}{\epsilon}} = \frac{C_3 C_{\epsilon 2} + \frac{8}{3}Ro^{-1}}{C_3 + Ro^{-1}}, \tag{20}$$

where $Ro = \frac{\epsilon}{K\Omega}$ is the Rossby number; $C_3 = \frac{5}{3}C_2 = \frac{5}{6}C_k^{3/2}$, C_k is the Kolmogorov constant. In the next section, we consider the application of this model and discuss its implications.

3. Results and Discussion

Computations were performed for isotropic rotating turbulent flow using the proposed two-equation model as well as the standard two-equation model. The evolution of the turbulent kinetic energy for the three different rotation rates are shown in Figure 3 along with experimental and DNS data. The first two moderate rotation rates correspond to the experiments of Wigeland and Nagib (1978), and it is clear that the standard model is unable to capture the effects of rotation on isotropic turbulence. On the other hand, the proposed model based on the the rotation corrected energy spectrum is seen to accurately predict the turbulence decay for moderate rotations. Furthermore, the rapidly rotating isotropic turbulence is also well-predicted by the proposed model for which the computational results are compared with the direct numerical simulation of Speziale, Mansour and Rogallo (1987). Here again, the standard model with its uncorrected energy spectrum is seen to yield results that are not consistent with the observed data.

For the case of isotropic turbulence at high Reynolds numbers, the dynamics of turbulence dissipation, is to a large measure controlled by turbulence production due to vortex-self stretching $P_{\epsilon 4}$ and viscous destruction Y. The success of modeling the dissipation equation depends on describing properly the difference between these two terms, which represents the major source of dissipation rate ϵ. In general, isotropic grid turbulence is often considered in model development because it is probably the simplest of the turbulent flows and yet it contains these two major features of turbulence.

It is obvious that the failing of isotropic eddy viscosity $(C_\mu K^2/\epsilon)$ in rotating turbulence is due to its inability to account for the effect of multiple scales. The standard $K-\epsilon$ model is obtained by constructing model transport equations for K and ϵ, and using the relation $K = \frac{3}{2}U^2$, $\epsilon = U^3/l$:

$$\nu_T \propto K^{1/2} l \propto K^{1/2} \frac{U^3}{\epsilon} \propto C_\mu K^2/\epsilon$$

The statement $\epsilon = \alpha(U^3(k)/l(k))$ can be recognized from the Kolmogorov energy spectrum $E(k) = \alpha\epsilon^{2/3}k^{-5/3}$ by defining the velocity and length scales for a given wave number k, $U(k) = [kE(k)]^{1/2}$, $l(k) = 2\pi/k$. Extending $\epsilon = \alpha(U^3(k)/l(k))$ to energy containing range, we have the Taylor approximation $\epsilon = U^3/l$. These assumptions are implicit in the development of standard $K-\epsilon$ model.

The concept of modeling complex turbulent flows with the use of three time scales has been introduced to analyze energy transfer in rotating isotropic turbulence. t_3 is the correlation time scale of nonlinear triadic interaction (defined by $\epsilon = Bt_3E^2k^4$); t_{nl} is turbulent nonlinear time scale (eddy turnover time or characteristic time scale at given wave number k, $[E(k)k^3]^{-1/2}$); t_s, the spectral transfer time, is defined by $\epsilon = U_k^2/t_s(k)$, where $U_k = [kE(k)]^{1/2}$ is the characteristic velocity of eddies at wave number k. These three time scales can be related by

$$t_s(k) = U_k^2/\epsilon = \frac{Ek}{Bt_3E^2k^4} = \frac{t_{nl}^2(k)}{Bt_3(k)}$$

With zero rotation, $t_3 = t_{nl}$, hence $t_s = B^{-1}t_{nl}$, consequently $\epsilon = \alpha(U^3(k)/l(k))$ and $E(k) = C_k\epsilon^{2/3}k^{-5/3}$. In the presence of rotation, $t_3 = f(t_\Omega, t_{nl}), t_s \neq B^{-1}t_{nl}$, then $\epsilon = U_k^2/t_s \neq \alpha(U^3(k)/l(k))$, the Taylor approximation as well as the Kolmogorov spectrum break down. By introducing the model, $1/t_3^2 = 1/t_{nl}^2 + 1/t_\Omega^2$, we reconstruct a generalized energy spectrum and reformulate eddy viscosity to account for the interaction between eddies of different scales. In order to extend the rotation-modified eddy viscosity to include the turbulence in shear flows, the dimensionless vorticity parameter is introduced in the description of eddy viscosity. Since such a model is of a general form, it has been known to capture key elements to the RDT solution for homogeneous shear flows (Speziale, 1997). It should be also noted in this context that no special modifications to the eddy viscosity for large non-equilibrium effects are needed in the proposed model.

4. Conclusions

The effects of rotation on isotropic turbulence observed by experiments cannot be reproduced by most of the existing models and this was the primary motivation for the present work. In the model proposed, rotation effects are characterized by a suitably modified energy spectrum and the resulting model coefficient $C_{\epsilon 2}$ includes the effect of rotation on the dissipation and the structure of turbulence. It should be emphasized that there are no additional coefficients or correction terms in the proposed model for the dissipation equation. Validation of the new model for rotating isotropic turbulence against physical experiments has shown that with the proposed model coefficient $C_{\epsilon 2}^*$, the observed rotation effects can be accurately reproduced.

References

Bardina, J., Ferziger, J.H. and Rogallo, R.S., 1985. "Effects of rotation on isotropic turbulence: Computation and modelling," *Journal of Fluid Mechanics* **154**, p. 32.

Cambon, C., Jacquin, L., and Lubrang, J.-L., 1992. "Towards a new Reynolds stress model for rotating turbulent flows," *Physics of Fluids A* **4**, p. 812.

Gatski, T.B. and Speziale, C.G., 1993. "On explicit algebraic stress models for complex turbulent flows," *Journal of Fluid Mechanics* **254**, p. 59.

Girimaji, S.S., 1995. "Fully explicit and self-consistent algebraic Reynolds stress model," ICASE Report 95-82.

Girimaji, S.S., 1996. "Improved algebraic Reynolds stress models for engineering flows," *Engineering Turbulence Modeling and Experiments*, eds. W. Rodi and G. Bergeles, Elsevier Science B.V., p. 121.
Hanjalić, K., Launder, B.E., and Schiestel, R., 1980. "Multiple time-scale concepts in turbulent transport modeling," *Turbulent Shear Flows* **2**, p. 36.
Hanjalić, K., 1994. "Advanced turbulence closure models: a view of current status and future prospects," *International Journal of Heat and Fluid Flow* **15**, p. 178.
Kraichnan, R.H., 1965. "Inertial-range spectrum of hydromagnetic turbulence," *Physics of Fluids* **7**, p. 1385.
Mansour, N.N., Cambon, C., and Speziale, C.G., 1991. "Theoretical and computational study of rotating isotropic turbulence," *Studies in Turbulence*, eds. Gatski, T.B., Sarkar, S., and Speziale, C.G., Springer-Verlag, Heidelberg.
Pope, S.B., 1975. "A more general eddy viscosity hypothesis," *Journal of Fluid Mechanics* **72**, p. 331.
Reynolds, W.C., 1976. "Computation of turbulent flows," *Annual Review of Fluid Mechanics* **8**, p. 183.
Reynolds, W.C., 1989. "Effects of rotation on homogeneous turbulence," *10th Australian Fluid Mechanics Conference*, Melbourne, Australia.
Speziale, C.G., Mansour, N.N., and Ragallo, R.S., 1987. "Decay of turbulence in a rapidly rotating frame," *Proceedings of 1987 CTR Summer Program* **CTR-S87**, p. 205.
Speziale, C.G., 1991. "Analytical methods for the development of Reynolds stress closures in turbulence," *Annual Review of Fluid Mechanics* **23**, p. 107.
Speziale, C.G., 1997. "Modeling non-equilibrium turbulent flows," *ICASE/LaRC/AFOSR Symposium on Modeling Complex Turbulent Flows*, Hampton, Virginia.
Squires, K.D., Chasnov, J.R., Mansour, N.N., and Cambon, C., 1994. "The asymptotic state of rotating homogeneous turbulence at high Reynolds numbers," *74th Fluid Dynamics Symposium on the application of Direct and Large Eddy Simulation to Transition and Turbulence*, Crete, Greece.
Thangam, S., Zhou, Y., Zigh, A., and Wang, X., 1998. "Modeling of complex turbulent flows in the presence of rotation and curvature," ICASE Report (in preparation).
Wigland, R.A. and Nagib, H.M., 1978. "Grid generated turbulence with and without rotation about the streamwise direction," *IIT Fluids & Heat Transfer Report R78-1*.
Yakhot, Y., Orszag, S.A., Thangam, S., Gatski, T.B., and Speziale, C.G., 1992. "Development of turbulence models for shear flows by a double-expansion technique," *Physics of Fluids A* **4**, p. 1510.
Zeman, O., 1994. "A note on the spectra and decay of rotating homogeneous turbulence," *Physics of Fluids* **6**, p. 3221.
Zhou, Y., Vahala, G., and Thangam, S., 1994. "Development of a turbulence model based on recursion renormalization group theory," *Physical Review E* **49**, p. 5195.
Zhou, Y., 1995. "A phenomenological treatment of rotating turbulence," *Physics of Fluids* **7**, p. 2092.

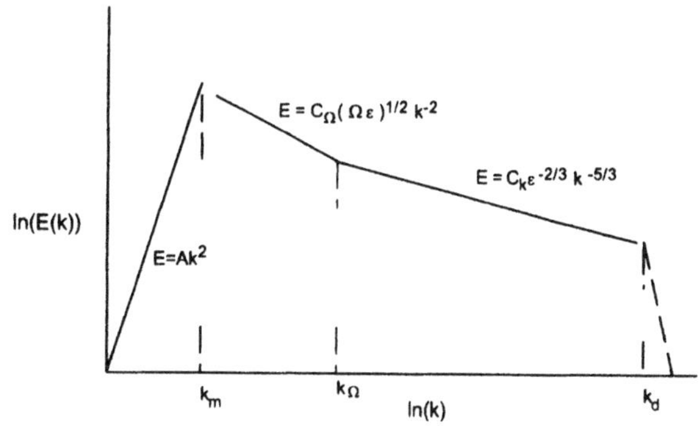

Figure 1. Turbulence energy spectrum subjected to rotation: a simplified sketch

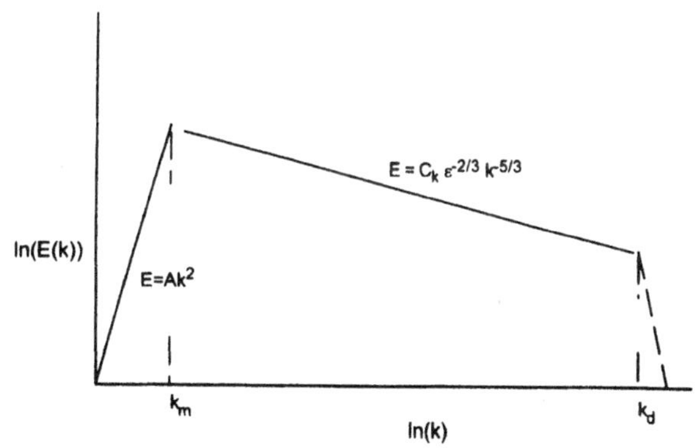

Figure 2. Conventional turbulence energy spectrum

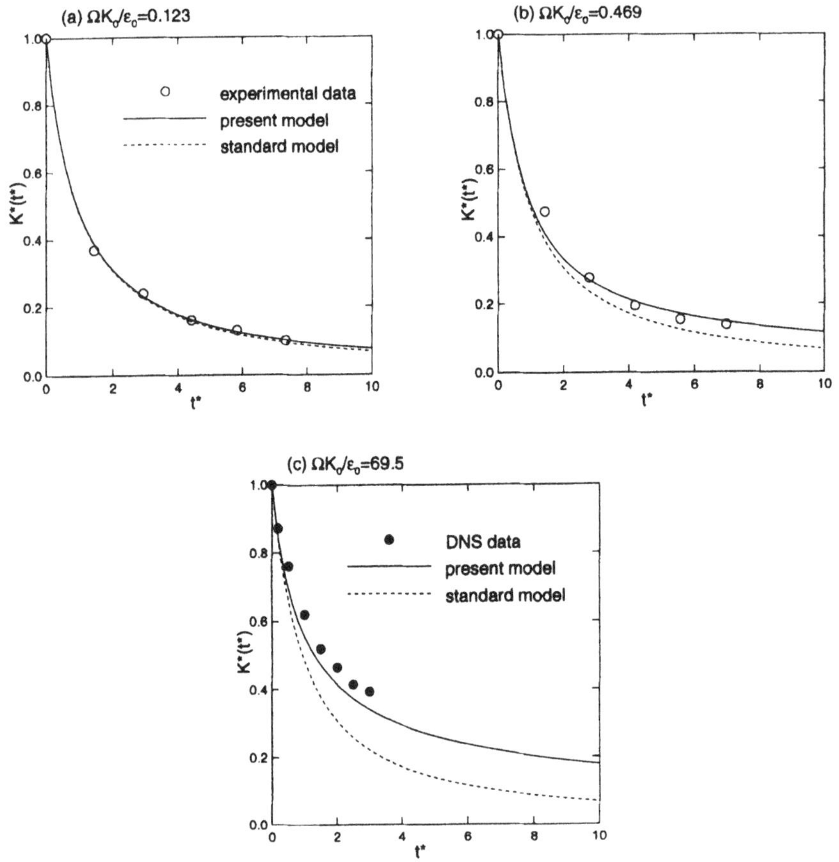

Figure 3. Decay of turbulent kinetic energy in rotating isotropic turbulence: — new modeled dissipation equation; - - - standard modeled dissipation equation; ○ experimental data of Wigeland and Nagib (1978); • direct simulations of Speziale, Mansour, and Rogallo (1987). $(a)\Omega K_0/\epsilon_0 = 0.123$, $(b)\Omega K_0/\epsilon_0 = 0.469$, $(c)\Omega K_0/\epsilon = 69.5$.

MODELING AND ANALYSIS OF TURBULENT FLOWS IN THE PRESENCE OF ROTATION AND CURVATURE

X. WANG AND S. THANGAM
Stevens Institute of Technology
Castle Point on the Hudson, Hoboken, New Jersey

Y. ZHOU
IBM Research Division, T.J. Watson Research Center
Yorktown Heights, New York and
Institute for Computer Applications in Science and Engineering
NASA Langley Research Center, Hampton, Virginia

AND

A. ZIGH
Parsons and Brinckerhoff, Inc., New York and
Stevens Institute of Technology
Castle Point on the Hudson, Hoboken, New Jersey

Abstract. A generalized eddy viscosity model that is formulated using the rotation modified energy spectrum is applied for the prediction of rotating flows. The model directly incorporates rotation and mean shear effects and is not limited by the use of local equilibrium hypothesis. The model is shown to reproduce the dominant effects of rotation on turbulence in rotating homogeneous shear flows and turbulent channel flows subject to spanwise rotation. The general applicability of the model and its implications are also addressed.

1. Introduction

Rotating turbulent flows are quite common in engineering practice as well as in a wide range of practical applications of relevance to geophysical and astrophysical systems. The flow is also of fundamental interest from a turbulence modeling point of view since the influence of rotation on turbulence is quite complex and not completely understood. The formulation of turbu-

M. D. Salas et al. (eds.), Modeling Complex Turbulent Flows, 361–372.

lence models to accurately describe the essential features of rotating flows requires that the energy transfer mechanisms are effectively incorporated in the governing transport equations.

In the case of homogeneous isotropic turbulence subjected to solid body rotation, experimental studies (Wigeland and Nagib, 1978) show that the primary effect of rotation on grid turbulence is to decrease the decay rate of the turbulent kinetic energy. Theoretical interpretations based on spectral analysis of Jacquin *et al.* (1990) indicates that the effect of rotation typically takes place at the level of the third-order velocity correlations in wave-space and rotation effect results in an inhibition of the build up of triple correlations. This is shown to lead to a reduction in energy transfer from the large scales to the dissipative eddies. As discussed in our companion paper (Thangam *et al.*, 1997) the effects of rotation on isotropic turbulence observed by experiments cannot be reproduced by most of the existing models as well as by several of the alternatives proposed earlier (Hanjalić *et al.*, 1980; Bardina *et al.*, 1985). The incorporation of rotation effects through a rotation modified energy spectrum on the other hand is shown to accurately reproduce the decay of isotropic turbulence for large range of rotation rates (Thangam *et al.*, 1997). In this context, it should be noted that several of the recent efforts have in fact lead to the development of a class of algebraic stress models that are obtained from a hierarchy of second-order Reynolds stress closure schemes. These models show considerable promise and include rotation and curvature based on phenomenological considerations (Launder, 1997; Speziale, 1997) or through the inclusion of nonlocal effects (Girimaji, 1995, 1996).

The case of simultaneous action of rotation and shear is of particular importance because of the wide range of applications in engineering systems (for example, flows in blade passages of radical flow pump and compressor impellers, flow past helicopter rotor, etc.). In this context, effects of rotation on turbulent channel flow has been investigated experimentally by Johnston *et al.* (1972), where a high aspect ratio channel is used to isolate the effect of rotation from secondary flows. The effects of rotation about a spanwise axis can have a stabilizing or destabilizing effect on turbulent boundary layers, depending on the direction of rotation compared to the shear. The boundary layer near the wall is destabilized when the wall is rotating into the flow and it is then referred to as the pressure side. A wall rotating out of the flow is referred to as the suction side and for this case, the boundary is stabilized. From the point of turbulence model development, second order closures are able to account for rotation effects since rotation appears explicitly in the production and Coriolis terms of transport equations of individual Reynolds stresses. The major difficulty with this type of models involves the proper representation of the rotation effects in the pressure strain correlation — a crucial requirement for the success of second-order closure schemes. However, in the frame work of two-equation models, most existing approaches fail to account for the effects of system rotation.

In our companion paper (Thangam *et al.*, 1997), a generalized eddy viscosity model for rotating turbulent flow by the use of a rotation modified energy spectrum has been developed. In the present work, this model is combined with a consistent anisotropic turbulent stress representation to develop a new generation of eddy viscosity model that can be applied for rotating turbulent flows. The model is combined with the transport equation for dissipation (ϵ) and the model coefficients are appropriately modified to account for rotation and anisotropy of turbulent stresses. The model is then validated for several applications involving rotating turbulent flows.

2. Formulation of the Physical Problem

The continuity and momentum equation for the mean flow in a non-inertial reference frame that rotates with angular velocity Ω_i relative to an inertial frame can be written as:

$$\frac{\partial U_i}{\partial x_i} = 0, \tag{1}$$

$$\frac{\partial U_i}{\partial t} + U_j \frac{\partial U_i}{\partial x_j} = -\frac{\partial \tau_{ij}}{\partial x_j} - \frac{\partial \overline{p}}{\partial x} - 2e_{mki}\Omega_m U_k + \nu \nabla^2 U_i, \tag{2}$$

where $\tau_{ij} = \overline{u_i u_j}$ is the Reynolds stress tensor and $\overline{p}$ is the modified mean pressure.

The Reynolds stresses are calculated by a nonlinear two-equation model with an eddy viscosity:

$$\begin{aligned} \tau_{ij} &= \frac{2}{3}K\delta_{ij} - \nu_T S_{ij} + C_d^* \frac{K^3}{\epsilon^2}(S_{ik}S_{kj} - \frac{1}{3}S_{mn}S_{mn}\delta_{ij}) \\ &+C_e^* \frac{K^3}{\epsilon^2}(S_{ik}W_{kj} + S_{jk}W_{ki}) \\ &+C_f^* \frac{K^3}{\epsilon^2}(W_{ik}W_{kj} - \frac{1}{3}W_{mn}W_{mn}\delta_{ij}) + \tau_{ij}^{><}, \end{aligned} \tag{3}$$

where $S_{ij} = \frac{1}{2}(\frac{\partial U_i}{\partial x_j} + \frac{\partial U_j}{\partial x_i})$, $W_{ij} = \frac{1}{2}(\frac{\partial U_i}{\partial x_j} - \frac{\partial U_j}{\partial x_i}) + 2.25e_{mji}\Omega_m$, and $\tau_{ij}^{><}$ represents the third order terms (Zhou *et al.*, 1995). The corresponding eddy viscosity is given by (Thangam *et al.*, 1997):

$$\nu_T = \frac{C_6 \frac{K^2}{\epsilon}}{\sqrt{1 + 0.2W_{ij}^* W_{ij}^*}}, \tag{4}$$

where $W^*_{ij} = \frac{K}{\epsilon}W_{ij}$. The corresponding transport equations for K and ϵ are:

$$\frac{\partial K}{\partial t} + U_i\frac{\partial K}{\partial x_i} = -\tau_{ij}\frac{\partial U_i}{\partial x_j} - \epsilon - \frac{\partial}{\partial x_i}(\frac{\nu_T}{\sigma_K}\frac{\partial K}{\partial x_i}) + \nu\nabla^2 K, \tag{5}$$

$$\frac{\partial \epsilon}{\partial t} + U_i\frac{\partial \epsilon}{\partial x_i} = -C^*_{\epsilon 1}\frac{\epsilon}{K}\tau_{ij}\frac{\partial U_i}{\partial x_j} - C^*_{\epsilon 2}\frac{\epsilon^2}{K} + \frac{\partial}{\partial x_i}(\frac{\nu_T}{\sigma_\epsilon}\frac{\partial \epsilon}{\partial x_i}) + \nu\nabla^2 \epsilon. \tag{6}$$

For isotropic eddy viscosity based on the standard model the rotation modified model coefficients are: $C^*_{\epsilon 1} = 1.44, C^*_{\epsilon 2} = \frac{1.83C_3 + \frac{8}{3}\frac{K\Omega}{\epsilon}}{C_3 + \frac{K\Omega}{\epsilon}}, C_\mu = 0.09, \sigma_K = 1, \sigma_\epsilon = 1.3$, $C_6 = 0.13$. A similar approach for anisotropic eddy viscosity in the RNG based $K-\epsilon$ system yields the following model coefficients: $C^*_{\epsilon 1} = 1.42 - \frac{\eta(1-\eta/\eta_0)}{1+\beta\eta^2}(1 + 0.07\frac{P}{\epsilon})$, with $\eta_0 = 5.6, \beta = 0.14, P = -\tau_{ij}S_{ij}$, and $C^*_{\epsilon 2} = \frac{1.68C_3 + \frac{8}{3}\frac{K\Omega}{\epsilon}}{C_3 + \frac{K\Omega}{\epsilon}}, C_\mu = 0.085$, $\alpha_K = 1/\sigma_K = 1.39, \alpha_\epsilon = 1/\sigma_\epsilon = 1.39$, $C_6 = 0.1$. Here, $C^*_{\epsilon 1} = 1.42 - \frac{\eta(1-\eta/\eta_0)}{1+\beta\eta^2}(1 + 0.07\frac{P}{\epsilon})$ was made a function of $\frac{P}{\epsilon}$ and η to account for the anisotropic production terms in the ϵ equation, the term is crucial in the large shear rate case (Yakhot *et al.*, 1992; Thangam *et al.*, 1998).

It should be noted that the quadratic and cubic terms in the expression for the turbulent stresses which arise from the recursion RNG model (Zhou *et al.*, 1995) need to be modified in a consistent manner. This is beyond the scope of this study and will be addressed in a future work by the authors. However, it is suffice to note here that the normal Reynolds stresses ($\overline{u^2}, \overline{v^2}$ and $\overline{w^2}$) are not utilized in the applications considered herein and the eddy viscosity is not influenced by the higher order terms.

Another point that should be noted is that the present model is not applicable directly to near-wall turbulence which is characterized by strong inhomogeneity and large anisotropy. Though the proposed eddy viscosity and the modeled form of transport equations are effective for analyzing turbulence in the presence of strong shear, the formulation does not incorporate the strong inhomogeneity and relaxational effects. Therefore, following the normal practice in such flows wall functions or near-wall models are used for implementing the computations at the boundaries.

3. Method of Solution

Computations were performed for both rotating shear flows and for fully developed turbulent flow in a rotating channel. These computations generally involve the solution of a system of coupled second order nonlinear ordinary differential equations which are first reduced to a system of first-order nonlinear equations and solved though a semi-implicit method.

In addition, for the case of rotating channel flow a discussion of the integration approach in the vicinity of the wall and the implementation of the boundary conditions is in order. The first approach for rotating channel flow computations involve the development a suitable wall function. In this approach, it is assumed that at a point (say, at a distance y_c) just outside the viscous sublayer, the velocity follows the law of the wall and the turbulence is in local equilibrium ($\mathcal{P} = \epsilon$). With further assumptions that uniform shear stress prevails in the viscous sublayer and dissipation ϵ is proportional to $K^{3/2}/y$, the velocity U_c, the turbulent kinetic energy K_c and the turbulent dissipation rate ϵ_c at point y_c are related to the friction velocity U_τ by:

$$U_c = U_\tau \frac{1}{k} ln(E \frac{U_\tau y_c}{\nu}),$$

$$K_c = \frac{U_\tau^2}{\sqrt{C_\mu}}, \ and \ \epsilon_c = \frac{U_\tau^3}{k y_c}.$$

where $E = 9.8$, $k = 0.42$, and the quantity of U_τ can be obtained using mass conservation in the calculation.

Computations typically start with an assumed initial friction velocity on both walls $U_{\tau u}$ and $U_{\tau s}$ and the values of position of the first start point ($y^+ = 30$). Initial values for y_{c1}, the turbulent kinetic energy K_{c1} and the dissipation rate ϵ_{c1} as well as the values of y_{c2}, K_{c2}, and ϵ_{c2} at the last node are calculated using the wall function and the model transport equations. These are used as boundary conditions and computations for the values of $U(y)$, $K(y)$ and $\epsilon(y)$ are performed in the interior of the channel. The friction velocities $U_{\tau u}$ and $U_{\tau s}$ are next updated using the computed results for velocity and turbulence quantities in the channel. Care is taken to ensure that the overall mass-balance constraint ($\int U(y)dy = \frac{QD}{A}$, here Q is the flow rate, A and D are respectively the cross-sectional area and width of the channel) is satisfied. The calculations are repeated until a fully converged solution is obtained. For the wall function approach, computations were performed with 100 cells across the channel.

The second approach involves direct integration to the channel wall. For this case, computations were performed with an extended version of a non-rotating channel flow solver. In this code, unsteady terms are retained in the transport equations to facilitate solution of the two-point boundary value problem. The algorithm used is based on an implicit Crank-Nicolson finite-difference scheme where during each time step, the transport equations are solved iteratively until the solution converges.

The extensions required for rotating channel flow computations include modifications to the turbulence model equations and the momentum equation with a near-wall model. To stabilize the computations, calculations are performed at progressively increasing rotation rates. At each rotation rate, the average wall friction factor, U_τ (defined such that $U_\tau^2 = \frac{1}{2}(U_{\tau u}^2 + U_{\tau s}^2)$) was calculated iteratively to maintain the same Reynolds number (and hence the same flow rate) as the nonrotating case. For computations involving direct integration to the wall 150 cells across channel were utilized with the first grid point at $y_{min}^+ = 0.1$. Grid refinement tests were done to ensure a grid-independent solution.

In general, for the computations presented herein, the direct wall integration approach provides better results than the wall function approach for the prediction of turbulent kinetic energy and turbulent stresses. However, no significant differences are observed for mean flow quantities as well as for integral estimates such as the wall friction factor. Typically, the direct wall integration approach (with damping function) requires about twice the computational time and has difficulty converging at very high rotation rates. For these cases, the wall function approach is preferable until a better near-wall model for direct integration is developed.

4. Results and Discussion

4.1. ROTATING HOMOGENEOUS SHEAR FLOWS

The present eddy viscosity model was first applied for homogeneous shear flows with and without rotation. Initially isotropic turbulence ($b_{ij} = 0, K = K_0, \epsilon = \epsilon_0$) is subjected to a constant uniform shear in a rotating frame with a constant angular velocity. The mean velocity gradient relative to the rotating frame is $\frac{\partial U_i}{\partial x_j} = S\delta_{i1}\delta_{j2}$ and the rotation rate of the frame is $\Omega_i = (0, 0, \Omega)$. The simple flow, which combines shear and rotation, was used for model development and assessment. The solution of the flow $(K/K_0, \epsilon/\epsilon_0)$ only depends on the initial conditions of ϵ_0/SK_0 and dimensionless parameter Ω/S (Shima, 1993).

Rotating homogeneous shear flow calculations are compared to the LES results of Bardina *et al.*, (1983) in the absence of DNS and experimental data. Figure 1 shows the time evolution of turbulent kinetic energy K of homogeneous shear flows for different rotation rates of $\Omega/S = 0, 0.5, -0.5, 0.25$. The prediction of pure shear flow by the new model is in excellent agreement with the LES data. The model correctly shows the stabilizing and destabilizing effects of rotation for $\Omega/S = 0.5, -0.5$. The $\Omega/S = 0.25$ case is the most energetic, and existing models (including second-order closure schemes) underpredict the growth rate (Speziale, 1991). Though the rotation-corrected standard two-equation model overpredicts the initial slope of turbulent kinetic energy it is able to capture the trend involving substantially large growth rate. If the relaxational effects are included, the rotation modified model equation would be expected to provide a better overall prediction for this energetic case as well.

4.2. ROTATING CHANNEL FLOWS

The rotation corrected eddy viscosity model was utilized for the prediction of the fully developed turbulent channel flows that are subjected to spanwise rotation. The computational predictions were validated against the experimental findings of Johnston *et al.* (1972) and the numerical (LES) results of Kim (1983). Computations were also performed with the same algorithm using two of the algebraic stress models that show promise (Girimaji, 1995, 1996; Gatski and Speziale, 1993) for additional validation. While these results are not shown for the sake of clarity, the proposed model performs at least as well as the algebraic stress models that were tested.

Figure 2 shows the model predictions for the velocity profile in the rotating channel (at $Ro = 0.21, Re = 11500$) along with the experimental data of Johnston *et al.* (1972). Compared to the zero rotation case, the velocity profile under rotation shows a higher slope in the near wall region on the unstable side of channel. As can be seen, the present model captures the asymmetric velocity distribution that is characteristic of rotating channel flows. In addition, consistent with experimental findings, the maximum velocity location is seen to shift towards the stable side of the channel.

The comparison of predictions of wall shear stress are displayed in Figure 3. In this figure the variation of the rotating friction velocity U_τ is shown as a ratio made dimensionless with the corresponding value for the nonrotating case as function of the rotation number, *Ro*. The friction coefficient increases with rotation number in the unstable side as a consequence of instability in the shear layer. On the stable side, the friction coefficient decreases with rotation. The present model is seen to reproduce the correct evolution of the friction factors with the rotation rates.

The variation of turbulent kinetic energy profiles at a rotation rate corresponding to the LES computations of Kim (1983) are presented in Figure 4. As can be seen, consistent with the LES data, the proposed model is seen to predict the stabilizing and destabilizing effects of the rotation due to the enhancement of the dissipation rate on the stable side of the channel and its reduction on the unstable side.

It should be emphasized herein that models that do not include the effects of rotation and/or the anisotropy of Reynolds stresses will not be able to predict the mean flow and turbulence quantities in such wall-bounded flows where both strong rotation and shear occur simultaneously in the presence of large inhomogeneity.

5. Conclusions

An eddy viscosity model that is formulated using the rotation corrected energy spectrum is incorporated into the transport equations and applied to the prediction of rotating flows. The wall effects are represented using both modified wall functions and the direct representation of the near wall effects. The model is validated for rotating homogeneous shear flows and turbulent channel flows subject to spanwise rotation. The model is also shown to be capable of accurately predicting the dominant characteristics of rotation and mean shear.

References

Bardina, J., Ferziger, J.H., and Rogallo, R.S., 1985. "Effects of rotation on isotropic turbulence: Computation and modelling," *Journal of Fluid Mechanics* **154**, p. 32.

Gatski, T.B. and Speziale, C.G., 1993. "On explicit algebraic stress models for complex turbulent flows," *Journal of Fluid Mechanics* **254**, p. 59.

Girimaji, S.S., 1995. "Fully explicit and self-consistent algebraic Reynolds stress model," ICASE Report 95-82.

Girimaji, S.S., 1996. "Improved algebraic Reynolds stress models for engineering flows," *Engineering Turbulence Modeling and Experiments*, eds. W. Rodi and G. Bergeles, Elsevier Science B.V., p. 121.

Hanjalić, K., Launder, B.E., and Schiestel, R., 1980. "Multiple time-scale concepts in turbulent transport modeling," *Turbulent Shear Flows* **2**, p. 36.

Jacquin, L., Leuchter, O., and Cambon, C., 1994. "Homogeneous turbulence in the presence of rotation," *Journal of Fluid Mechanics* **220**, p. 1.

Johnston, J.P., Halleen, R.M., and Lezius, D.K., 1972. "Effects of spanwise rotation on the structure of two-dimensional fully developed turbulent channel flow," *Journal of Fluid Mechanics* **56**, p. 533.

Kim, J., 1983. *Proceedings of the 4th Turbulent Shear Flows Symposium*, 6.14, Karlsruhe, Germany.

Launder, B.E., 1997. "The modeling of turbulent flows with significant curvature or rotation effects," *ICASE/LaRC/AFOSR Symposium on Modeling Complex Turbulent Flows*, Hampton, Virginia.

Shima, N., 1993. "Prediction of turbulent boundary layers with a second-momentum closure: Part 2 – effects of streamline curvature and spanwise rotation," *ASME Journal of Fluids Engineering* **115**, p. 64.

Speziale, C.G., Gatski, T.B., and Mac Giolla Mhuiris, N., 1990. "A critical comparison of turbulence models for homogeneous shear flows in a rotating frame," *Physics of Fluids A* **2**, p. 1678.

Speziale, C.G., 1991. "Analytical methods for the development of Reynolds stress closures in turbulence," *Annual Review of Fluid Mechanics* **23**, p. 107.

Speziale, C.G., 1997. "Modeling non-equilibrium turbulent flows," *ICASE/LaRC/AFOSR Symposium on Modeling Complex Turbulent Flows*, Hampton, Virginia.

Speziale, C.G. and Gatski, T.B., 1997. "Analysis and modelling of anisotropies in the dissipation rate of turbulence," *Journal of Fluid Mechanics* **344**, p. 155.

Thangam, S., Zhou, Y., Zigh, A., and Wang, X., 1998. "Modeling of complex turbulent flows in the presence of rotation and curvature," ICASE Report (in preparation).

Thangam, S., Wang, X., and Zhou, Y., 1997. "Development of a turbulence model for flows with curvature and rotation," *ICASE/LaRC/AFOSR Symposium on Modeling Complex Turbulent Flows*, Hampton, Virginia.

Wigland, R.A. and Nagib, H.M., 1978. "Grid generated turbulence with and without rotation about the streamwise direction," *IIT Fluids & Heat Transfer Report R78-1*.

Yakhot, Y., Orszag, S.A., Thangam, S., Gatski, T.B., and Speziale, C.G., 1992. "Development of turbulence models for shear flows by a double-expansion technique," *Physics of Fluids A* **4**, p. 1510.

Zhou, Y., Vahala, G., and Thangam, S., 1994. "Development of a turbulence model based on recursion renormalization group theory," *Physical Review E* **49**, p. 5195.

Zhou, Y., 1995. "A phenomenological treatment of rotating turbulence," *Physics of Fluids* **7**, p. 2092.

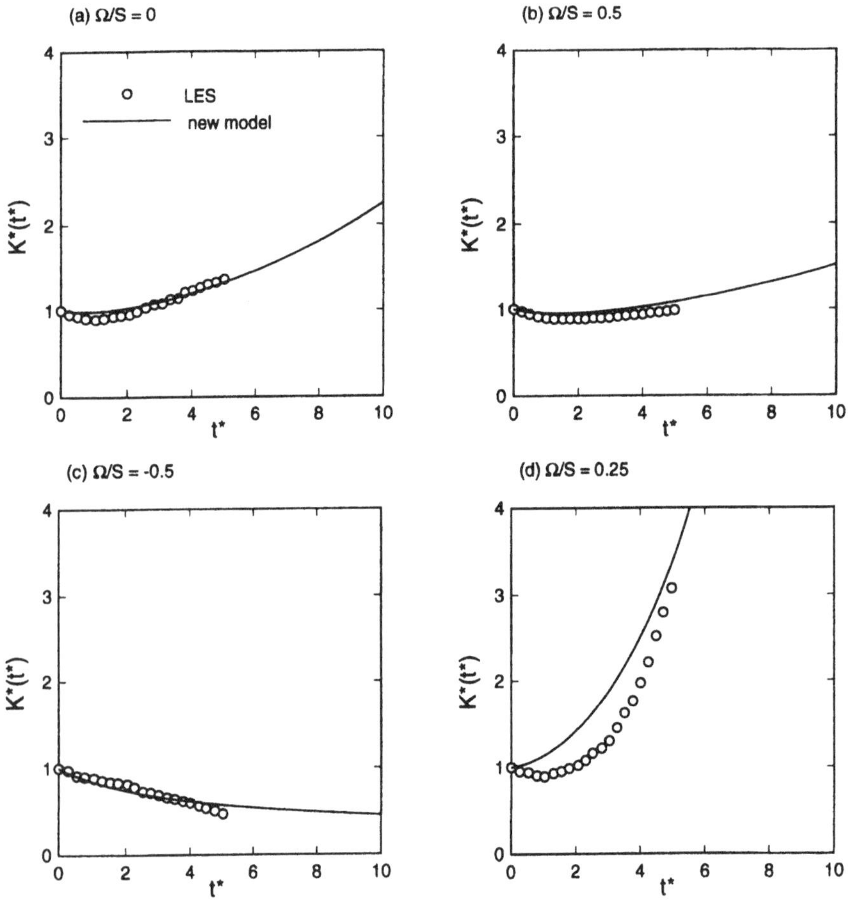

Figure 1. Time evolution of the turbulent kinetic energy in rotating homogeneous shear flow: — new model predictions; ∘ large-eddy simulations of Bardina *et al.* (1983). $(a)\Omega/S = 0.5$, $(b)\Omega/S = -0.5$, $(c)\Omega/S = 0.25$ for $\epsilon_0/SK_0 = 0.296$.

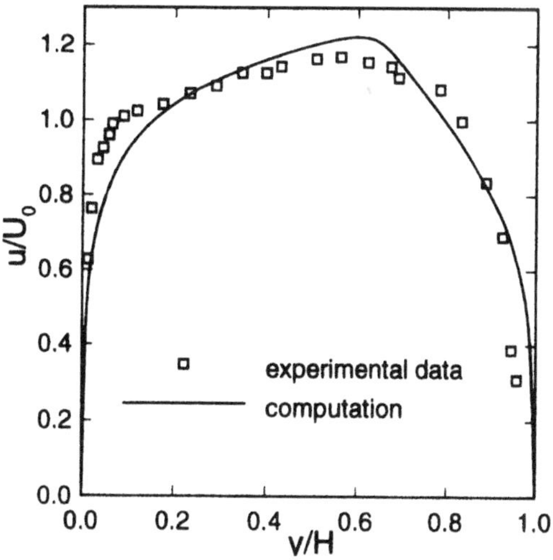

Figure 2. Mean velocity profiles in a rotating channel ($Re = 11500, Ro = 0.21$): $\circ$, experiments by Johnston *et al.* (1972); — present computations.

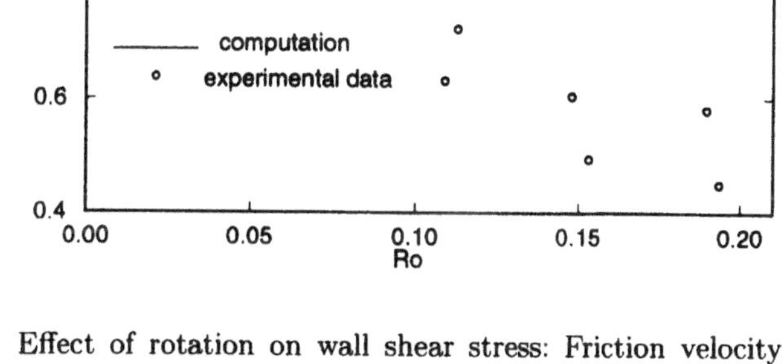

Figure 3. Effect of rotation on wall shear stress: Friction velocity on the two sides of channel, $\circ$ experimental data of Johnston *et al.* (1972) and Moore (1967) with $Re = 11400 \sim 36000$. — present computations.

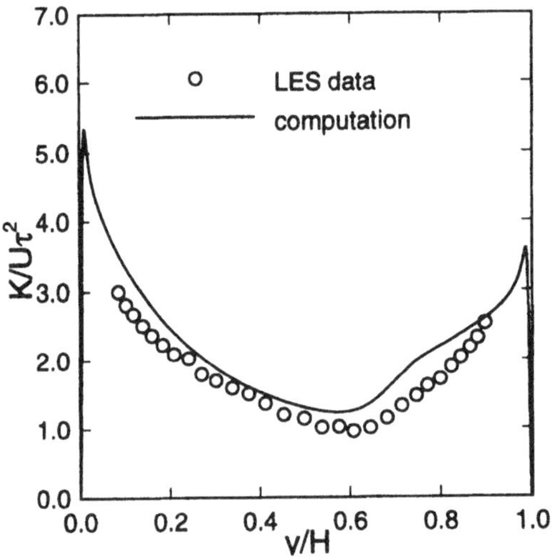

Figure 4. Distribution of turbulent kinetic energy K across a rotating channel, $Re_\tau = 640, Ro_\tau = 1.47$ $(Re = 24600, Ro = 0.076)$. ○, Large-eddy simulation by Kim (1983),— present computations.

LIST OF ATTENDEES

Ihab Adam
Department of Aerospace Engineering
Old Dominion University
Norfolk, VA 23529
(757) 683-5373
ihab@aero.odu.edu

Rayhaneh Akhavan
Department of Mechanical Engineering
University of Michigan
2250 G.G. Brown
Ann Arbor, MI 48109-2125
(313) 763-1048
raa@umich.edu

Eyal Arian
ICASE
Mail Stop 403
NASA Langley Research Center
Hampton, VA 23681-2199
(757) 864-2208
arian@icase.edu

S. Balachandar
Department of Theoretical & Applied Mechanics
University of Illinois
216 Talbot Lab
Urbana, IL 61801
(217) 244-4371
balals@h.tam.uiuc.edu

P. Balakumar
Department of Aerospace Engineering
Old Dominion University
Norfolk, VA 23529
(757) 683-4796
balak_p@aero.odu.edu

Robert Bartels
Mail Stop 340
NASA Langley Research Center
Hampton, VA 23681-2199
(757) 864-2813
r.e.bartels@larc.nasa.gov

*Speaker
†Panel Member

Oktay Baysal
Department of Aerospace Engineering
Old Dominion University
Norfolk, VA 23529-0247
(757) 865-4835
baysal@aero.odu.edu

Rodney Bowersox
Department of Aerospace Engineering & Mechanics
University of Alabama
205 Hardaway Hall
Box 870280
Tuscaloosa, AL 35487-0280
(205) 348-7300
rbowerso@coe.eng.ua.edu

Peter Bernard*
Department of Mechanical Engineering
University of Maryland
College Park, MD 20742
(301) 405-5272
bernard@eng.umd.edu

Peter Bradshaw*
Department of Mechanical Engineering
Stanford University
Stanford, CA 94305
(415) 725-0704
bradshaw@vk.stanford.edu

Arild Bertelrud
AS&M
Mail Stop 170
NASA Langley Research Center
Hampton, VA 23681-2199
(757) 864-5559
arild@rebelle.larc.nasa.gov

Dennis Bushnell
Mail Stop 110
NASA Langley Research Center
Hampton, VA 23681-2199
(757) 864-8987
d.m.bushnell@larc.nasa.gov

Frederick Blottner
Sandia National Laboratory
Department 9115
Mail Stop 0825
Albuquerque, NM 87185-0825
(505) 844-8549
fgblott@sandia.gov

Qian Chen
Department of Mechanical Engineering
Columbia University
500 West 120th St., Mail Code 4703
New York, NY 10027
(212) 854-7306
jc4@columbia.edu

Timothy Clark*
Los Alamos National Laboratory
Mail Stop B216
Los Alamos, NM 87545
(505) 665-4858
ttc@lanl.gov

Thomas J. Coakley
Mail Stop 229-1
NASA Ames Research Center
Moffett Field, CA 94035
(415) 604-6451
coakley@coak.arc.nasa.gov

William Compton
Mail Stop 286
NASA Langley Research Center
Hampton, VA 23681-2199
(757) 864-3048
w.b.compton@larc.nasa.gov

Karen Deere
Mail Stop 499
NASA Langley Research Center
Hampton, VA 23681-2199
(757) 864-8986
k.a.deere@larc.nasa.gov

Feng Ding
Department of Marine, Earth &
Atmospheric Sciences
North Carolina State University
Raleigh, NC 27695-8208
(919) 515-1437
fding@eos.ncsu.edu

Thomas Doligalski†
Army Research Office
Fluid Dynamics Branch
P.O. Box 12211
Research Triangle Park, NC 27709
(919) 549-4251
tom@aro-emh1.army.mil

Philip Drummond
Mail Stop 197
NASA Langley Research Center
Hampton, VA 23681-2199
(757) 864-2298
j.p.drummond@larc.nasa.gov

Gregory Eyink
Department of Mathematics
University of Arizona
Building No. 89
Tucson, AZ 85721
(520) 621-4468
eyink@math.arizona.edu

Abdelkader Frendi
NYMA, Inc.
Mail Stop 352
NASA Langley Research Center
Hampton, VA 23681-2199
(757) 864-1050
a.frendi@larc.nasa.gov

Laurent Gicquel
Department of Mechanical & Aerospace Engineering
State University of New York - Buffalo
Box 604400
Buffalo, NY 14260-4400
(716) 645-2593
gicquel@eng.buffalo.edu

Sharath Girimaji*
ICASE
Mail Stop 403
NASA Langley Research Center
Hampton, VA 23681-2199
(757) 864-2322
girimaji@icase.edu

Mark Glauser†
AFOSR
110 Duncan Avenue
Bolling Air Force Base
Washington, DC 20332-0001
(202) 767-4936
glauser@sun.soe.clarkson.edu

Peter Gnoffo
Mail Stop 408A
NASA Langley Research Center
Hampton, VA 23681-2199
(757) 864-4380
gnoffo@ab13.larc.nasa.gov

Frank A. Haidinger
Daimler-Benz Aerospace AG
P.O. Box 80 11 68
81663 Muenchen
GERMANY
011 49 89 607 24340
frank.haidinger@space.otn.dasa.de

Raimo J. Hakkinen
Department of Mechanical Engineering
Washington University in St. Louis
One Brookings Drive
St. Louis, MO 63130-4899
(314) 935-4084
rjh@mecf.wustl.edu

Jongil Han
Department of Marine, Earth & Atmospheric Sciences
North Carolina State University
Raleigh, NC 27695-8208
(919) 515-1437
han@brunt.meas.ncsu.edu

Kemo Hanjalic*
Faculty of Applied Physics
Delft University of Technology
Lorentzweg 1 Delft 2628CJ
THE NETHERLANDS
011 31 152781735
hanjalic@wt.tm.tudelft.nl

Ehtesham Hayder
CRPC - MS 41
Rice University
6100 Main Street
Houston, TX 77005
(713) 737-5868
hayder@cs.rice.edu

Jerry Hefner†
Mail Stop 462
NASA Langley Research Center
Hampton, VA 23681-2199
(757) 864-3640
j.n.hefner@larc.nasa.gov

George Huang*
Department of Mechanical Engineering
University of Kentucky
Lexington, KY 40506-0052
(606) 257-9313
ghuang@engr.uky.edu

Craig Hunter
Mail Stop 499
NASA Langley Research Center
Hampton, VA 23681-2199
(757) 864-3020
c.a.hunter@larc.nasa.gov

S. Jeyasingham
Department of Aerospace Engineering
Old Dominion University
Norfolk, VA 23529
(757) 683-5295
sxj@aero.odu.edu

Dennis A. Johnson
Mail Stop 229-1
NASA Ames Research Center
Moffett Field, CA 94035
(415) 604-5399
djohnson@cyclone.arc.nasa.gov

Thibauld Jongen
Swiss Federal Institute of Technology
INHEF-EPFL
Lausanne 1015
SWITZERLAND
011 41 21 693 3563
jongen@dgm.epfl.ch

Chun-Teh Kao
Department of Marine, Earth & Atmospheric Sciences
North Carolina State University
Raleigh, NC 27695-8208
(919) 515-1448
kao@brunt.meas.ncsu.edu

Stavros Kassinos
Department of Mechanical Engineering
Stanford University
Stanford, CA 94305
(415) 723-0546
kassinos@eddy.stanford.edu

Christopher Kennedy
Sandia National Laboratories
7011 East Avenue
Mail Stop 9051
Livermore, CA 94551-0969
(510) 294-3813
cakenne@ca.sandia.gov

S.E. Kim*
Fluent, Inc.
10 Cavendish Court
Lebanon, NH 03766
(603) 643-2600
sek@fluent.com

Doyle Knight*
Department of Mechanical & Aerospace Engineering
Rutgers University
P.O. Box 909
Piscataway, NJ 08855-0909
(908) 445-4464
knight@jove.rutgers.edu

Jens Knoell
Department of Mechanical & Aerospace Engineering
State University of New York - Buffalo
321 Jarvis Hall
Buffalo, NY 14260
(716) 645-2593
knoell@acsu.buffalo.edu

Thomas Korjack
Department of Computer & Information Sciences
University of Maryland
College Park, MD 20742
(410) 278-8674
tak@arl.mil

Budugur Lakshminarayana*
Department of Mechanical Engineering
Pennsylvania State University
153 Hammond Building
University Park, PA 16802
(814) 865-5551
bllaer@engr.psu.edu

Brian Launder*
Department of Mechanical Engineering
UMIST
P.O. Box 88
Manchester M60 1QD
UNITED KINGDOM
011 44 16120037 1
brian.launder@umist.ac.uk

Sanjiva K. Lele*
Department of Aeronautics & Astronautics
Stanford University
Durand Building
Stanford, CA 94305-4035
(415) 723-7721
lele@leland.stanford.edu

Geoffrey Lilley
Department of Aeronautics & Astronautics
University of Southampton
Southampton, S017
UNITED KINGDOM
011 44 1 703769109
gml@uk.ac.soton.aero

Shenzhi Liu
DSO National Laboratory
20 Science Park Drive
118230
SINGAPORE
011 65 6607600
lshenzhi@dso.org.sg

Josip Loncaric
ICASE
Mail Stop 403
NASA Langley Research Center
Hampton, VA 23681-2199
(757) 864-2192
josip@icase.edu

James Luckring
Mail Stop 280
NASA Langley Research Center
Hampton, VA 23681-2199
(757) 864-2869
j.m.luckring@larc.nasa.gov

John L. Lumley*
Department of Mechanical & Aerospace Engineering
Cornell University
256 Upson Hall
Ithaca, NY 14853-7501
(607) 255-4050
jll4@cornell.edu

Li-Shi Luo
ICASE
Mail Stop 403
NASA Langley Research Center
Hampton, VA 23681-2199
(757) 864-8006
luo@icase.edu

Douglas MacMartin
United Technologies Research Center
411 Silver Lane
Mail Stop 129-77
East Hartford, CT 06108
(860) 610-2164
macmardg@utrc.utc.com

Mujeeb Malik
High Technology Corporation
28 Research Drive
Hampton, VA 23666
(757) 865-0818
malik@htc-tech.com

Frank Marconi
Advanced Systems & Technology Organization
Northrop Grumman Corporation
Building 14
Bethpage, NY 11714-3580
(516) 575-2228
marconi@gateway.grumman.com

Khalid Mouaouya
Department of Mechanical Engineering
Columbia University
Room 134F
500 West 120th Street
New York, NY 10027
(212) 854-7306

F.T.M. Nieuwstadt
J.M. Burgers Centre for Fluid Mechanics
Delft University of Technology
Rotterdamseweg 145 AL Delft 2628
THE NETHERLANDS
011 31 15 2781005
f.t.m.nieuwstadt@wbmt.tudelft.nl

Matthew Overholt
Department of Mechanical & Aerospace Engineering
Cornell University
Ithaca, NY 14853-7801
(603) 668-9519
overholt@mae.cornell.edu

S. Paul Pao
Mail Stop 499
NASA Langley Research Center
Hampton, VA 23681-2199
(757) 864-3044
s.p.pao@larc.nasa.gov

Stephen Pope*
Department of Mechanical & Aerospace Engineering
Cornell University
Upson Hall
Ithaca, NY 14853
(607) 255-4314
pope@mae.cornell.edu

Fred Proctor
Mail Stop 156A
NASA Langley Research Center
Hampton, VA 23681-2199
(757) 864-6697
f.h.proctor@larc.nasa.gov

Serge Prudhomme
TICAM
The University of Texas at Austin
Taylor Hall 2.400
Austin, TX 78712
(512) 471-1721
serge@ticam.utexas.edu

Patrick Purtell*†
Office of Naval Research
800 N. Quincy Street
Arlington, VA 22217
(703) 696-4308
purtelp@onr.navy.mil

Pradeep Raj
Lockheed Martin Aeronautical Systems
D/73-07, Z/0685
86 S. Cobb Drive
Marietta, GA 30063-0685
(770) 494-3801
raj@mar.lmco.com

R. Ganesh Rajagopalan
Iowa State University
404 Town Engineering Building
Ames, IA 50014
(515) 294-6796
rajagopa@iastate.edu

William C. Reynolds*
Department of Mechanical Engineering
Stanford University
Stanford, CA 94305
(415) 723-3840
wcr@thermo.stanford.edu

J. Raymond Ristorcelli, Jr.*
IGPP, MS C305
Los Alamos National Laboratory
Los Alamos, NM 87545
(505) 667-0920
rayr@kokopelli.lanl.gov

Cord Rossow
DLR - Institute for Design Aerodynamics
Lilienthalplatz 7
D-38022 Braunschweig
GERMANY
011 49531 295 2410
cord.rossow@dlr.de

Robert Rubinstein*
ICASE
Mail Stop 403
NASA Langley Research Center
Hampton, VA 23681-2199
(757) 864-7058
bobr@icase.edu

Christopher Rumsey
Mail Stop 128
NASA Langley Research Center
Hampton, VA 23681-2199
(757) 864-2165
c.l.rumsey@larc.nasa.gov

Leonidas Sakell†
AFOSR/NM
110 Duncan Avenue
Suite B115
Bolling Air Force Base, DC
20332-8050
(202) 767-4935
len.sakell@afosr.af.mil

Manuel Salas†
ICASE
Mail Stop 403
NASA Langley Research Center
Hampton, VA 23681-2199
(757) 864-2174
salas@icase.edu

John Seiner
Mail Stop 165
NASA Langley Research Center
Hampton, VA 23681-2199
(757) 864-6276
j.m.seiner@larc.nasa.gov

Shaohua Shen
Department of Marine, Earth &
Atmospheric Sciences
North Carolina State University
Raleigh, NC 27695-8208
(919) 515-1437
sshen@eady.meas.ncsu.edu

Roger L. Simpson*
Department of Aerospace & Ocean
Engineering
Virginia Polytechnic Institute &
State University
218 Randolph Hall
Blacksburg, VA 24061-0203
(540) 231-5989
simpson@aoe.vt.edu

Bart Singer
Mail Stop 128
NASA Langley Research Center
Hampton, VA 23681-2199
(757) 864-2154
b.a.singer@larc.nasa.gov

Brian R. Smith
Lockheed Martin Tactical Aircraft Systems
P.O. Box 748
MZ 9333
Fort Worth, TX 76101
(817) 935-1127
brian.r.smith@lmtas.lmco.com

Sonya Smith
Department of Mechanical Engineering
Howard University
2300 6th Street, N.W.
Washington, DC 20059
(202) 806-4837
sts6f@vortex.eng.howard.edu

Alexander Smits*
Department of Mechanical & Aerospace Engineering
Princeton University
E-Quad
Princeton, NJ 08544
(609) 258-5117
asmits@princeton.edu

Charles G. Speziale*
Department of Aerospace & Mechanical Engineering
Boston University
110 Cummington Street
Boston, MA 02215
(617) 353-3568
speziale@enga.bu.edu

David G. Stephens
Mail Stop 462
NASA Langley Research Center
Hampton, VA 23681-2199
(757) 864-3640
d.g.stephens@larc.nasa.gov

Chao-Ho Sung
Code 5030
David Taylor Model Basin
West Bethesda, MD 20817-5700
(301) 227-1865
sung@sigma.dt.navy.mil

Ken Tedjojuwono
Mail Stop 493
NASA Langley Research Center
Hampton, VA 23681-2199
(757) 864-4595
k.k.tedjojuwono@larc.nasa.gov

Siva Thangam*
Department of Mechanical Engineering
Stevens Institute of Technology
Castle Point on the Hudson
Hoboken, NJ 07030
(201) 216-5558
sthangam@stevens-tech.edu

Sheldon Tieszen
Sandia National Laboratories
P.O. Box 5800
Albuquerque, NM 87185-0836
(505) 844-6526
srtiesz@sandia.gov

C. Randall Truman
Department of Mechanical Engineering
University of New Mexico
Albuquerque, NM 87131
(505) 277-6296
truman@unm.edu

Stefan Wallin
FFA
Box 11021, SE-16111
Bromma
SWEDEN
011 46 86341318
wns@ffa.se

Bono Wasistho
Department of Applied Mathematics
University Twente
P.O. Box 217, 7500 AE Enschede
THE NETHERLANDS
011 31 53 4893418
b.wasistho@math.utwente.nl

David Weatherly
Department of Mechanical Engineering
University of Kentucky
521 CRMS Building
Lexington, KY 40506-0108
(606) 257-2368
davidw@engr.uky.edu

Jeffery White
Mail Stop 197
NASA Langley Research Center
Hampton, VA 23681-2199
(757) 864-7773
jawhite@hyp00.larc.nasa.gov

Ye Zhou*
IBM Research Division
T.J. Watson Research Center
P.O. Box 218
Yorktown Heights, NY 10598
(914) 945-2313
yzhou@watson.ibm.com

ICASE/LaRC Interdisciplinary Series in Science and Engineering

1. J. Buckmaster, T.L. Jackson and A. Kumar (eds.): *Combustion in High-Speed Flows.* 1994 ISBN 0-7923-2086-X
2. M.Y. Hussaini, T.B. Gatski and T.L. Jackson (eds.): *Transition, Turbulence and Combustion.* Volume I: Transition. 1994 ISBN 0-7923-3084-6; set 0-7923-3086-2
3. M.Y. Hussaini, T.B. Gatski and T.L. Jackson (eds.): *Transition, Turbulence and Combustion.* Volume II: Turbulence and Combustion. 1994 ISBN 0-7923-3085-4; set 0-7923-3086-2
4. D.E. Keyes, A. Sameh and V. Venkatakrishnan (eds): *Parallel Numerical Algorithms.* 1997 ISBN 0-7923-4282-8
5. T.G. Campbell, R.A. Nicolaides and M.D. Salas (eds.): *Computational Electromagnetics and Its Applications.* 1997 ISBN 0-7923-4733-1
6. V. Venkatakrishnan, M.D. Salas and S.R. Chakravarthy (eds.): *Barriers and Challenges in Computational Fluid Dynamics.* 1998 ISBN 0-7923-4855-9
7. M.D. Salas, J.N. Hefner and L. Sakell (eds.): *Modeling Complex Turbulent Flows.* 1999 ISBN 0-7923-5590-3

KLUWER ACADEMIC PUBLISHERS – DORDRECHT / BOSTON / LONDON

SPRINGER NATURE

GPSR Compliance

The European Union's (EU) General Product Safety Regulation (GPSR) is a set of rules that requires consumer products to be safe and our obligations to ensure this.

If you have any concerns about our products, you can contact us on ProductSafety@springernature.com

In case Publisher is established outside the EU, the EU authorized representative is:

Springer Nature Customer Service Center GmbH
Europaplatz 3
69115 Heidelberg, Germany

FSC
www.fsc.org
MIX
Papier | Fördert gute Waldnutzung
FSC® C083411

Kolibri 360 GmbH
Ferdinand-Jühlke-Straße 7
99095 Erfurt, Deutschland
produktsicherheit@kolibri360.de